T0230468

Volunteered
Geographic
Information

Dirk Burghardt • Elena Demidova • Daniel A. Keim
Editors

Volunteered Geographic Information

Interpretation, Visualization
and Social Context

 Springer

Editors
Dirk Burghardt
Cartographic Communication
TU Dresden
Dresden, Germany

Daniel A. Keim
Data Analysis and Visualization
University of Konstanz
Konstanz, Germany

Elena Demidova
Data Science and Intelligent Systems,
Computer Science Institute
University of Bonn
Bonn, Germany

Lamarr Institute for Machine Learning and
Artificial Intelligence
Bonn, Germany

ISBN 978-3-031-35376-5   ISBN 978-3-031-35374-1   (eBook)
https://doi.org/10.1007/978-3-031-35374-1

The publication of this book was supported by the Deutsche Forschungsgemeinschaft (DFG, German Research Foundation) within Priority Research Program 1894 Volunteered Geographic Information: Interpretation, Visualization and Social Computing.

This Springer imprint is published by the registered company Springer Nature Switzerland AG
The registered company address is: Gewerbestrasse 11, 6330 Cham, Switzerland

Paper in this product is recyclable.

Preface

Volunteered geographic information (VGI) has emerged as a novel form of user-generated content in recent years. VGI involves the active generation of geodata, for example, in citizen science projects, and the passive data collection via location-enabled mobile devices. In addition, an increasing number of sensors perceive our environment with ever-greater detail and dynamics. The resulting VGI data can support various applications addressing critical societal challenges, e.g., environment and disaster management, health, transport, and citizen participation.

The interpretation and visualization of VGI are challenging due to the considerable heterogeneity of the underlying multi-source data and the social context. In particular, the utilization of VGI is influenced by the specific characteristics of the underlying data, such as real-time availability, event-driven generation, and subjectivity, all with an implicit or explicit spatial reference. The DFG-funded priority program "VGI: Interpretation, Visualization, and Social Computing"[1] (2016–2022) aims to address these challenges. The results of the research work in this program form the basis for this book publication.

The book includes three parts, "Representation and Analysis of VGI," "Geovisualization and User Interactions Related to VGI," and "Active Participation, Social Context, and Privacy Awareness," representing the principal research pillars within the priority program. The intersection of these three domains offers new research potential. Of particular interest is the link between social behavior and technology during the collective collection, processing, and usage of VGI. This includes, e.g., the consideration of different acquisition and usage contexts, evaluating subjective information, and ensuring privacy-aware data processing.

A total of 30 projects were financed as part of the twice 3-year funding period. Research groups from the fields of geoinformation science, computer science, psychology, and mechanical and safety engineering were involved in the interdisciplinary collaboration. Results of the collaborative research within the priority program have been made available through the VGI repository, the sustainable

---

[1] https://www.vgiscience.org/.

public platform with access to benchmark data, and VGI-related software tools, documentation, and publications. A focus was placed on cross-project collaboration. The flagship was the Young Research Groups, who used their complementary methodological knowledge to identify and answer cross-project research questions in self-organized projects. The results were published in 25 dissertations and are also included in the publications within this book.

Dresden, Germany   Dirk Burghardt
Bonn, Germany   Elena Demidova
Konstanz, Germany   Daniel A. Keim

Contents

Part I
Representation and Analysis of VGI

VGI is a critical data source for various real-world applications, including mobility services in smart cities, remote sensing, tracing animal behavior, environmental protection, and disaster management. Popular examples of VGI data sources include crowdsourced maps such as OpenStreetMap (OSM) and GPS trajectory data. However, VGI is highly challenging to represent, access, and analyze.

The challenges associated with VGI representation, access, and analysis can be attributed to the 4 Vs of big data, such as volume, velocity, variety, and veracity. The volume and velocity of community-created geographic data are rapidly expanding due to the growth of volunteer communities, increased availability of open, editable community-created data sources, and easy access to real-time tracking devices. For example, the number of OSM contributors has grown from approximately 5.6 million in August 2019 to 9.6 million in December 2022.[1] Increased heterogeneity of data representations and a range of quality issues accompany this growth.

Community-created geographic data comes in various formats with heterogeneous annotations, often lacking clear semantics. Whereas standardized formats and representations for geographic information and semantic annotation exist, the openness of VGI sources leads to increased heterogeneity. Furthermore, the representation and access requirements of different downstream applications vary. In particular, data-driven machine learning applications, such as traffic inference and accident prediction, depend on the availability of large-scale, high-quality data, and machine-readable annotations to build predictive models effectively and efficiently.

In this part, we discuss recent approaches to enhance the representation and analysis of VGI. In particular, this includes semantic representation of VGI data in knowledge graphs, machine-learning approaches to VGI mining, completion, and enrichment, as well as to the improvement of data quality and fitness for purpose. Furthermore, we discuss new approaches for more efficient analytics of VGI images.

---

[1] https://planet.openstreetmap.org/statistics/data_stats.html.

Chapter 1
WorldKG: World-Scale Completion of Geographic Information

Alishiba Dsouza, Nicolas Tempelmeier, Simon Gottschalk, Ran Yu, and Elena Demidova

Abstract Knowledge graphs provide standardized machine-readable representations of real-world entities and their relations. However, the coverage of geographic entities in popular general-purpose knowledge graphs, such as Wikidata and DBpedia, is limited. An essential source of the openly available information regarding geographic entities is OpenStreetMap (OSM). In contrast to knowledge graphs, OSM lacks a clear semantic representation of the rich geographic information it contains. The generation of semantic representations of OSM entities and their interlinking with knowledge graphs are inherently challenging due to OSM's large, heterogeneous, ambiguous, and flat schema and annotation sparsity. This chapter discusses recent knowledge graph completion methods for geographic data, comprising entity linking and schema inference for geographic entities, to provide semantic geographic information in knowledge graphs. Furthermore, we present the WorldKG knowledge graph, lifting OSM entities into a semantic representation.

Keywords Geographic knowledge graphs · WorldKG

A. Dsouza (✉) · R. Yu
Data Science & Intelligent Systems Group (DSIS), University of Bonn, Bonn, Germany
e-mail: dsouza@cs.uni-bonn.de; ran.yu@uni-bonn.de

N. Tempelmeier · S. Gottschalk
L3S Research Center, University of Hannover, Hannover, Germany
e-mail: tempelmeier@L3S.de; gottschalk@L3S.de

E. Demidova
Data Science & Intelligent Systems Group (DSIS), University of Bonn, Bonn, Germany

Lamarr Institute for Machine Learning and Artificial Intelligence, Bonn, Germany
e-mail: elena.demidova@cs.uni-bonn.de www.lamarr-institute.org

© The Author(s) 2024
D. Burghardt et al. (eds.), *Volunteered Geographic Information*,
https://doi.org/10.1007/978-3-031-35374-1_1

1.1 Introduction

Geographic information is of crucial importance for a variety of real-world applications, including accident prediction (Dadwal et al. 2021), detection of topological dependencies in road networks (Tempelmeier et al. 2021a), and positioning charging stations (von Wahl et al. 2022). Such applications can substantially profit from standardized machine-readable representations of geographic entities, including monuments, roads, and charging stations. Particularly, such semantic representations should comprise detailed descriptions of geographic entities, including their types, properties, context, relations, and interlinking across sources.

OpenStreetMap (OSM)[1] is a critical source of volunteered and openly available geographic information. OSM provides rich but highly heterogeneous data regarding geographic entities, including fine-grained coordinates of real-world locations and user-defined tags comprising entity types, properties, and relations. At the time of writing, OSM contains over 6.8 billion entities from 188 countries.[2] However, the adoption of OSM data in real-world applications is limited, mainly due to the large, heterogeneous, ambiguous, and flat schema adopted for the OSM tags.

Knowledge graphs (Hogan et al. 2021)—graph-based representations of real-world entities and their relations—provide detailed machine-readable descriptions of real-world entities through ontologies and facilitate interlinking across sources. Information representation in knowledge graphs is based on W3C standards such as the Resource Description Framework (RDF)[3] and established ontologies. This representation facilitates structured semantic access via standardized query languages, such as SPARQL.[4] Although popular general-purpose knowledge graphs such as Wikidata and DBpedia (Auer et al. 2007) contain a number of geographic entities, only a tiny fraction of them include precise location information. Furthermore, whereas some community-defined links between OSM entities and knowledge graphs exist at the instance level, these links are sparse and cover only selected entity types. For example, as of September 2022, only 0.52% of OSM nodes provided links to the Wikidata knowledge graph. In this setting, knowledge graph completion, such as interlinking knowledge graphs and geographic information sources at the entity and schema levels, is inherently challenging due to the representation heterogeneity in OSM and the sparsity of geographic information in popular knowledge graphs.

Table 1.1 illustrates a geographic entity, Cairo, the capital of Egypt, and its representations in OSM and the Wikidata knowledge graph,[5] OSM provides

---

[1] OpenStreetMap, OSM, and the OpenStreetMap magnifying glass logo are trademarks of the OpenStreetMap Foundation and are used with their permission. We are not endorsed by or affiliated with the OpenStreetMap Foundation.

[2] OSMstats: https://osmstats.neis-one.org.

[3] Resource Description Framework: https://www.w3.org/RDF/.

[4] SPARQL 1.1 Query Language: https://www.w3.org/TR/sparql11-query/.

[5] wd and wtd are the prefixes of http://www.wikidata.org/entity/ and http://www.wikidata.org/prop/direct/, respectively.

Table 1.1 Representation of Cairo in OpenStreetMap and Wikidata

Key	Value	Subject	Predicate	Object
name	Cairo	wd:Q85	rdfs:label (*label*)	Cairo
place	city	wd:Q85	wdt:P31 (*instance of*)	wd:Q515 (*city*)
population	9120350	wd:Q85	wdt:P1082 (*population*)	9513330
capital	yes	wd:Q85	wdt:P1376 (*capital of*)	wd:Q79 (*Egypt*)
(a) OpenStreetMap tags.		(b) Wikidata triples. wd:Q85 identifies Cairo.		

information as heterogeneous key-value pairs called "tags." In this example, OSM encodes the entity type information as ⟨place, city⟩, whereas the precise semantics of the tags often remain unclear. In contrast, entities in Wikidata are represented via well-defined statements, also known as RDF triples. A triple has the form ⟨Subject, Predicate, Object⟩ and enables the representation of entity types, properties, and relations. In Wikidata, an entity type is expressed using the instance of property, denoted as wdt:P31.[6] In this example, this property connects a unique entity identifier wd:Q85, representing Cairo, to the entity type city, denoted as wd:Q515.[7]

In this chapter, we discuss recent methods aiming to bridge the gap between OSM and knowledge graphs through semantically enriching geospatial information in OSM and making this information available in WORLDKG—a novel geographic knowledge graph. In particular, we develop methods for knowledge graph completion, to establish links between OSM and knowledge graphs at the entity and schema levels. *Geographic entity linking* discussed in this chapter aims at interlinking the representations of entities in OSM and Wikidata (Cairo in this example). *Geographic class alignment* aims to link the OSM tags that provide entity type information to the corresponding knowledge graph classes. In this example, the OSM tag ⟨place, city⟩ should be aligned to the Wikidata class wd:Q515 (city).

Existing schema alignment and entity linking methods are not directly applicable to geographic data sources such as OSM due to structural differences between OSM and knowledge graphs (Otero-Cerdeira et al. 2015). Generic schema matching methods typically rely on name and schema structure similarities (Madhavan et al. 2001). Other approaches, such as LIMES (Ngomo and Auer 2011), have strict heuristics and consider fixed schemas. Geographic entity linking approaches such as LinkedGeoData (Auer et al. 2009) rely on manually aligned schemas and create links using type information, spatial distance, and name similarity. These approaches often fail due to representation differences, toponym ambiguities, OSM schema flatness, and geographic coordinate variations across sources. Thus, new approaches are required to lift OSM's flat and heterogeneous geographic information into a precise, machine-readable semantic representation.

---

[6] Definition of the instance of Wikidata property: https://www.wikidata.org/wiki/Property:P31.

[7] Definition of the city (Q515) in Wikidata: https://www.wikidata.org/wiki/Q515.

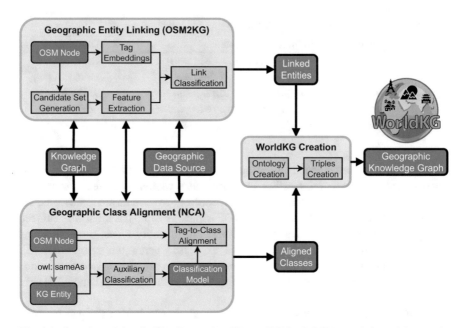

Fig. 1.1 Overview of the pipeline for creating WORLDKG from OSM, consisting of three main steps: (i) geographic entity linking with OSM2KG, (ii) geographic class alignment with NCA, and (iii) WORLDKG geographic knowledge graph creation

In the remainder of this chapter, we first formally define the problem of geographic entity linking and class alignment in Sect. 1.2. Then, we discuss approaches we recently proposed to interlink OSM and knowledge graphs at the entity and schema levels. These approaches are illustrated in Fig. 1.1. In Sect. 1.3, we present OSM2KG—a geographic entity linking approach (Tempelmeier and Demidova 2021) depicted in the upper part of Fig. 1.1. Following that, in Sect. 1.4, we discuss NCA—a neural approach for geographic class alignment between OSM and knowledge graphs that utilizes entity links (Dsouza et al. 2021a). NCA is illustrated in the lower part of Fig. 1.1. Then, in Sect. 1.5, we describe WORLDKG— a novel geographic knowledge graph (Dsouza et al. 2021b) that adopts OSM2KG and NCA to provide semantic representations of OSM entities. Finally, in Sect. 1.6, we discuss open research directions and provide a conclusion.

1.2 Problem Definition

Linking geographic data sources and knowledge graphs at the entity and the schema level can help create a comprehensive source of geographic information, i.e., a geographic knowledge graph. First, we define geographic data sources and knowledge graphs based on the definitions by Tempelmeier and Demidova (2021).

A geographic data source represents geographic entities. Each geographic entity is annotated with an identifier, a location, and a set of key-value pairs called tags. More formally:

Definition 1.1 A *geographic data source* $G = (N, T)$ consists of a set of geographic entities N and a set of tags T. Each tag $t \in T$ is represented as a key-value pair $t = \langle k, v \rangle$. Each node $n \in N$ represents a real-world geographic entity with a geolocation and a set of tags $T_n \subset T$.

A typical example of a geographic data source is OpenStreetMap. Examples of the tags assigned to the node representing Cairo are illustrated in Table 1.1. A geolocation can be represented as a coordinate pair (i.e., latitude and longitude) or a sequence of coordinate pairs (e.g., forming a polygon).

A knowledge graph is a semantic information source containing data regarding real-world entities, their classes, and their properties. Typical examples of popular general-purpose knowledge graphs are Wikidata and DBpedia, which cover an extensive set of real-world entities and their relations in various application domains. Table 1.1 illustrates selected properties of the Wikidata entity representing Cairo.

Definition 1.2 A *knowledge graph* $KG = (E, C, P, L, L_{geo}, F)$ consists of a set of entities E, a set of classes $C \subset E$, a set of properties P, a set of literals L, a set of geolocations L_{geo}, and a set of relations $F \subseteq E \times P \times (E \cup L \cup L_{geo})$.

An entity in a knowledge graph can represent a real-world entity or a class of entities. Literals represent values including strings, numbers, and dates. An entity $e \in E$ can be related to one or multiple geolocations representing different geometries, such as a point or a polygon.

Geographic entity linking aims to align entities from a geographic data source and a knowledge graph representing the same real-world entity.

Definition 1.3 Given a geographic data source $G = (N, T)$ and a knowledge graph $KG = (E, C, P, L, L_{geo}, F)$, the problem of *geographic entity linking* is to identify a node $n \in N$ and an entity $e \in E$ representing the same real-world object.

In the example illustrated in Table 1.1, Cairo from OSM and Cairo from Wikidata represent the same real-world entity. In RDF, this link is typically denoted using the owl:sameAs property.

Schema alignment refers to the interlinking of equivalent schema elements across sources. In the context of this work, we focus on geographic class alignment, which refers to the interlinking of tags of a geographic data source and classes of a knowledge graph representing the same semantic concept. The key or the value of the tag alone is not sufficient to describe the class. For example, tag $\langle$natural, peak$\rangle$ describes the concept *Mountain*. Considering only the key or the value here may not align to the correct class *Mountain* of the knowledge graphs. Hence, we align the tag, i.e., key=value, to the knowledge graph classes.

Definition 1.4 Given a geographic data source $G = (N, T)$ and a knowledge graph $\mathcal{KG} = (E, C, P, L, L_{geo}, F)$, the problem of *geographic class alignment* is to identify a tag $t \in T$ and a class $c \in C$ representing the same semantic concept.

For example, the OSM tag ⟨place, city⟩ and the Wikidata class wd:Q515 illustrated in Table 1.1 represent the same semantic concept.

In the context of this work, a *geographic knowledge graph* refers to a knowledge graph that provides geolocation information for a substantial fraction of the entities it contains. This work aims to create a geographic knowledge graph through geographic entity linking and class alignment.

1.3 Geographic Entity Linking with OSM2KG

Geographic entity linking refers to the task of interlinking geographic entities representing the same real-world entity across data sources (Definition 1.3). Typically, entity linking approaches utilize semantic and syntactic similarity of the different entity representations. In OSM, geographic entity linking is particularly challenging due to the large scale and the ambiguities of location names. For example, "Berlin" can denote the name of the capital of Germany and a restaurant name. Also, the names of geographic entities, such as "Church Road," are often non-distinctive.

1.3.1 Related Work

State-of-the-art entity linking approaches such as LIMES (Ngomo and Auer 2011) and WOMBAT (Sherif et al. 2017) assume that entities are represented through the same number of properties with a 1:1 property mapping. In the case of linking OSM nodes with the knowledge graph entities, these assumptions do not hold. Entity linking methods such as DBpedia Spotlight (Daiber et al. 2013) detect links between textual data and knowledge graphs. LinkedGeoData (Stadler et al. 2012) performs geographic entity linking by creating links between OSM and knowledge graphs such as DBpedia and GeoNames. However, the linking with LinkedGeoData relies on manually aligned schema, syntactic similarity, and spatial distance. To overcome the shortcomings of current approaches, we proposed OSM2KG (Tempelmeier and Demidova 2021), a machine learning algorithm for geographic entity linking based on representation learning of OSM tags.

1.3.2 The OSM2KG Approach

The overall geographic entity linking process of OSM2KG is illustrated in the upper part of Fig. 1.1. First, OSM2KG adopts geographic blocking to reduce the number of potential candidate entities given an OSM node (*candidate set generation*). Then, OSM2KG creates latent representations of OSM tags (*tag embeddings*) and extracts features of the candidate entities (*feature extraction*). Finally, OSM2KG predicts if a node-entity pair represents the same real-world entity (*link classification*). In the following, we present these steps in more detail.

Candidate Set Generation Given a node $n \in N$ of the geographic data source $G = (N, T)$, the goal of the candidate set generation step is to identify potentially matching entities in the knowledge graph $\mathcal{KG}$. The geographic coordinates in OSM and knowledge graphs represent the points of community consensus, rather than an objective metric (Auer et al. 2009). Consequently, the coordinates of geographic entities represented in these sources can deviate. The candidate generation step is based on the intuition that entities and nodes representing the same real-world entities should be located in geographic proximity. Thus, for a given input node n, OSM2KG creates the candidate set by considering all entities within an experimentally determined distance threshold. A spatial index such as R-Tree (Guttman 1984) can be utilized to enable efficient geographic blocking.

Tag Embeddings The set of tags T_n assigned to an OSM node n plays an essential role in detecting the correct matching candidate. As OSM tags are highly heterogeneous, OSM2KG aims at learning their unsupervised latent representations. OSM2KG utilizes a skip-gram-based neural network model (Mikolov et al. 2013). This representation learns the co-occurrences of OSM tags to capture the semantic similarity between OSM nodes. The embedding model is trained in an unsupervised manner based on the tag similarity of geographic entities, meaning geographic entities with similar tags are represented in a closer space. The resulting embeddings can be used to estimate the semantic similarity of the OSM nodes. Geographic coordinates are not considered in this step, such that the embedding reflects semantic similarity independent of the geolocation.

Feature Extraction For each candidate entity from $\mathcal{KG}$ in the candidate set, we extract additional features, namely, the entity type and its popularity, as reflected by the number of incoming edges in the knowledge graph.

In addition to the tag embeddings and the entity features, the Jaro-Winkler distance (Winkler 1999) is calculated between the names of the OSM node and the candidate to measure their name similarity. Furthermore, the logistic distance proposed by Stadler et al. (2012) is used to compute the geographic distance between the OSM node and the candidate entity.

Link Classification Finally, a random forest classification model is utilized to classify whether the input node in G represents the same real-world entity as the candidate entity in $\mathcal{KG}$. To train the model, as positive examples, OSM2KG takes node-entity pairs from the existing links between $\mathcal{G}$ and $\mathcal{KG}$.

1.3.3 Evaluation Results of the OSM2KG Approach

We evaluated OSM2KG (Tempelmeier and Demidova 2021) regarding its inter-linking performance. In particular, we considered the interlinking of OSM entities with Wikidata and DBpedia knowledge graphs in Germany, France, and Italy. This evaluation demonstrated a substantial F1-score improvement achieved by OSM2KG compared to eight different baseline approaches. OSM2KG performed best on all Wikidata datasets, achieving an F1-score of 92.05% on average and outperforming the best-performing baseline by 21.82 percentage points. OSM2KG also achieved the best recall performance and high precision on all datasets.

As a result of OSM2KG, we can infer new links between the OSM nodes and geographic entities in knowledge graphs. Such links can be beneficial for creating and enriching semantic sources, as they can provide complementary information regarding the linked geographic entities. These linked entities can also serve as additional training data to develop supervised methods for geographic schema alignment.

1.4 Geographic Class Alignment with NCA

Geographic class alignment between a geographic data source $\mathcal{G} = (\mathcal{N}, \mathcal{T})$ and a knowledge graph $\mathcal{KG} = (E, C, P, L, L_{geo}, F)$ aims to align the tags and the classes representing the same real-world concepts (Definition 1.4).

The heterogeneous tag-based OSM structure created by volunteers makes it challenging to identify the tags that can be linked to knowledge graph ontologies. For example, the OSM tag $\langle natural, peak \rangle$ corresponds to the "mountain" class in the Wikidata knowledge graph. This match cannot be easily identified using the existing approaches based on syntactic and structural similarity.

1.4.1 Related Work

Ontology alignment methods typically rely on structural and element-level similarity to align schema elements (Otero-Cerdeira et al. 2015). As the OSM schema is flat, approaches that depend on the structural hierarchy (Melnik et al. 2002) do not perform well. Schema alignment methods that depend on the element-level syntactic similarity (Madhavan et al. 2001) do not work well either, due to the essential differences in the syntactic representation of OSM tags and knowledge graph classes. Instance-based alignment approaches (Ngo et al. 2013) rely on the structural similarity of neighboring instances to align schema elements. Machine learning (Doan et al. 2004) and deep learning-based approaches (Bento et al. 2020; Xiang et al. 2015) also rely on the structure. Furthermore, tabular data alignment methods (Cappuzzo et al. 2020) cannot appropriately handle sparse OSM tag annotations. Overall, the lack of a well-defined OSM ontology and the essential differences in the structural as well as syntactic representation of OSM tags and knowledge graph ontologies, along with the sparsity of OSM annotations, hinder the application of state-of-the-art ontology and schema alignment approaches. To overcome these limitations, we proposed NCA, a neural class alignment approach that utilizes existing entity links between geographic data sources and knowledge graphs in a novel shared latent space.

1.4.2   The NCA Approach

At the bottom of Fig. 1.1, we briefly illustrate the building blocks of the class alignment NCA approach. In the first step, NCA aims to create a shared latent space that aligns the feature spaces of G and KG. To this extent, NCA creates an auxiliary neural classification model. This model captures the semantic relations between the OSM tags and the semantic classes in the shared latent space. In the second step, NCA probes the auxiliary model to obtain the tag-to-class alignments between the OSM tags and the knowledge graph classes.

Auxiliary Classification The goal of the first NCA step is to create a shared latent space containing similar latent representations of geographic entities in G and KG. To achieve this aim, NCA creates an auxiliary classification model. This model is trained to classify linked entities from G and KG into the corresponding semantic classes $c \in C$ of KG. During supervised training, the auxiliary classification model adopts known pairs of linked geographic entities from G and KG as a training set. For G, tags and keys having more than 50 occurrences[8] in OSM are selected as features. For KG, top-25 properties of each class are used as features. These features are passed through the fully connected layers to form the shared latent space of the model that aligns the representations of OSM and KG entities. The intuition behind the shared latent space is that linked entities from OSM and KG that belong to the same semantic class will be represented similarly. To create the shared latent space, NCA adopts an adversarial classifier that exploits linked entities of G and KG. This classifier aims to distinguish between G and KG entities in the latent space. NCA aims to make their representations similar by inverting the gradient of the adversarial loss. In this way, as a result of the training, the feature spaces of G and KG are aligned.

---

[8] https://taginfo.openstreetmap.org/tags.

Tag-to-Class Alignment The training of the auxiliary classification model results in a shared latent space. NCA then probes the model with one OSM tag at a time and computes the complete forward pass. NCA selects the results of the classification layer to obtain the tag-to-class alignment. As one OSM tag can be matched to multiple classes in the knowledge graph, NCA selects all classes whose confidences exceed an experimentally determined threshold value.

1.4.3   Evaluation Results of the NCA Approach

We evaluated the NCA approach on OSM as the geographic data source and Wikidata and DBpedia as knowledge graphs (Dsouza et al. 2021a). The NCA performance was compared to six state-of-the-art ontology and tabular data alignment methods. The evaluation was conducted on a dataset with seven countries having the most data available in OSM, namely, Germany, France, Great Britain, Spain, Russia, the USA, and Australia. In terms of tag-to-class alignment, NCA obtained up to 13 and up to 37 percentage point improvement of the F1-score on Wikidata and DBpedia, respectively. On average, we observed 10 (21) percentage point F1-score improvement on Wikidata (DBpedia). As a result, the NCA approach increased the number of OSM entities with semantic class annotations from Wikidata and DBpedia knowledge graphs by over 400%. The resulting tag-to-class annotations are available as part of the WORLDKG knowledge graph presented in the next section.

1.5   The WORLDKG Knowledge Graph

WORLDKG is a geographic knowledge graph that provides semantic information on geographic entities extracted from OSM. While OSM contains rich data regarding such geographic entities, this data is not directly accessible to semantic applications. With WORLDKG, we tackle this problem and provide a geographic knowledge graph. WORLDKG follows the ontology illustrated in Fig. 1.2 and is available online.[9]

1.5.1   Related Work

Geographic knowledge graphs such as LinkedGeoData (Auer et al. 2009) and YAGO2geo (Karalis et al. 2019) either contain only a few geographic classes

---

[9] WORLDKG: https://www.worldkg.org/.

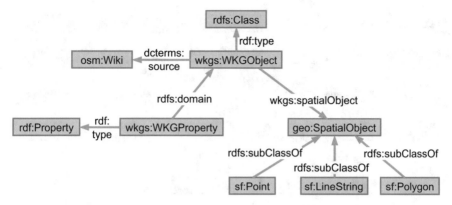

Fig. 1.2 The WORLDKG ontology

or represent data of a restricted geographic area. Specialized geographic knowledge graphs such as the KnowWhereGraph (Janowicz et al. 2022) and EventKG (Gottschalk and Demidova 2019) concentrate on past events and have a limited location coverage. In contrast, WORLDKG is based on OSM and contains over 100 million geographic entities on a world scale typed with over 1,000 semantic classes.

1.5.2 WORLDKG *Creation Approach*

WORLDKG captures geographic entities in OSM and contains links to Wikidata and DBpedia at the entity and class levels. The creation procedure of WORLDKG includes two main tasks depicted in Fig. 1.1, namely, ontology creation and triple creation.

Ontology Creation To infer a class hierarchy from OSM tags, we utilize OSM *map features*[10]—a list of established key-value pairs. We extract classes (keys) and their subclasses (values) from the map features. For example, from the map feature ⟨*place, city*⟩, we infer the class "Place" and its subclass "City." All remaining keys not covered by the map features are considered properties.

We convert the names of the extracted properties and classes according to the OWL naming conventions.[11] We also incorporate the tag-to-class alignment inferred using the NCA approach (Sect. 1.4) into the WORLDKG ontology.

---

[10] https://wiki.openstreetmap.org/wiki/Map_features.

[11] https://www.w3.org/TR/owl-ref/.

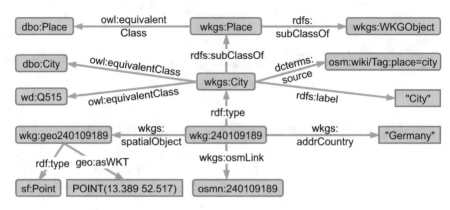

Fig. 1.3 An example of a city and its classes in WORLDKG

The WORLDKG ontology is depicted in Fig. 1.2. Any object in WORLDKG can be connected to a geolocation (geo:SpatialObject—either a point, a line string, or a polygon) via wkgs:spatialObject. Other relations are represented using properties typed as wkgs:WKGProperty. Information regarding the original OSM tags is provided by dcterms:source.

An example of a WORLDKG entity of the class "City" is illustrated in Fig. 1.3. Via the property wkgs:spatialObject, the entity is connected to its geolocation, which provides a coordinate pair. The "City" class is connected to its equivalents in DBpedia and Wikidata.

Triple Creation We add all OSM nodes that have at least one tag and belong to at least one class of the WorldKG ontology to the WORLDKG knowledge graph. To this extent, we create triples that represent the nodes and their properties and adhere to the WorldKG ontology.

1.5.3 WORLDKG *Access, Statistics, Evaluation, and Examples*

Access WORLDKG offers a GeoSPARQL endpoint.[12] This endpoint supports queries in GeoSPARQL[13]—a geographic query language for RDF data—and visualizes geolocations of the query results on a map.

---

[12] WORLDKG GeoSPARQL endpoint: https://www.worldkg.org/sparql.

[13] GeoSPARQL: https://www.ogc.org/standards/geosparql.

Statistics As of September 2022, WORLDKG contains over 800 million triples describing approximately a 100 million entities that belong to over 1,000 distinct classes. The number of unique properties (wgks:WKGProperty) in WORLDKG is over 1,800. As a result of the NCA approach presented in Sect. 1.4, WORLDKG provides links to 40 Wikidata and 21 DBpedia classes.

Evaluation We evaluated the quality of WORLDKG by assessing the type assertions of the geographic entities (Dsouza et al. 2021b). From Wikidata and DBpedia, we randomly selected five classes each, aligned with the WORLDKG ontology. Per each of these classes, we randomly selected 100 example geographic entities in WORLDKG and manually checked if they belong to the assigned knowledge graph class. We observed that WORLDKG achieved over 97% accuracy on average.

Examples Listing 1.1 illustrates the representation of a WORLDKG entity of type wkgs:Restaurant in the Turtle format. Listing 1.2 is an example query that makes use of the GeoSPARQL function bif:st_distance to extract three restaurants closest to the Dresden Central Station.[14] The query results are shown in Table 1.2 and Fig. 1.4. This example illustrates the potential of using WORLDKG in downstream applications such as POI recommendation.

```
wkg:4182951095 a wkgs:Restaurant ;
    rdfs:label "Restaurant Quarre" ;
    wkgs:addrHousenumber "77" ;
    wkgs:addrPostcode "10117" ;
    wkgs:capacity "200" ;
    wkgs:cuisine "regional" ;
    wkgs:delivery "no" ;
    wkgs:openingHours "12:00-15:00,18:00-22:30" ;
    wkgs:osmLink osmn:4182951095 ;
    wkgs:phone "+493022611959" ;
    wkgs:takeaway "no" ;
    wkgs:website "http://restaurant-quarre.de" ;
    wkgs:spatialObject wkg:geo4182951095 .

wkg:geo4182951095 a sf:Point;
    geo:asWKT "POINT(13.3798808 52.5160887)"^^geo:wktLiteral .
```

Listing 1.1 RDF triples in the Turtle format for an example geographic entity of type wkgs:Restaurant in WORLDKG

---

[14] The geographic location of the Dresden Central Station is taken from OSM.

```
SELECT ?restaurant
       (bif:st_distance(?cWKT, ?fWKT) AS ?Distance)
WHERE {
   ?poi rdfs:label "Dresden Hbf".
   ?poi rdf:type wkgs:RailwayStation .
   ?poi wkgs:spatialObject [
      geo:asWKT ?cWKT
   ] .
   ?closeObject rdf:type wkgs:Restaurant .
   ?closeObject rdfs:label ?restaurant .
   ?closeObject wkgs:spatialObject ?fGeom .
   ?fGeom geo:asWKT ?fWKT .
}
ORDER BY ASC(bif:st_distance(?cWKT, ?fWKT))
LIMIT 3
```

Listing 1.2 Example GeoSPARQL query to retrieve three restaurants closest to the Dresden Central Station (Dresden Hbf)

Table 1.2 Result of the example GeoSPARQL query in Listing 1.2

Restaurant	Distance (in kilometers)
Marché	0.02
dean&david	0.08
Dschingis Khan	0.13

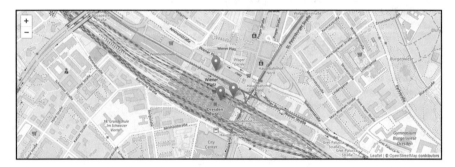

Fig. 1.4 Visualization of three restaurants closest to the Dresden Central Station returned for the query in Listing 1.2

1.6 Discussion and Open Research Directions

In this chapter, we presented WORLDKG—a geographic knowledge graph that we developed to provide a semantic representation of geographic entities in OSM. Furthermore, we described OSM2KG and NCA, novel methods for geographic entity linking and class alignment. These methods enable interlinking geographic entities in OpenStreetMap with other semantic sources of geographic information at the entity and schema levels. Our proposed approaches outperformed state-of-the-art methods when applied to OSM and popular general-purpose knowledge graphs, Wikidata and DBpedia. We made WORLDKG publicly available.

WORLDKG is a comprehensive source of semantic geographic information in its current form; it also opens many directions for future research.

A critical aspect of the knowledge graph creation from volunteered geographic information is data quality. As OSM data builds a basis for the knowledge graph creation, data quality issues of OSM can be propagated into WORLDKG. In WORLDKG, we rely on the existing links between OSM nodes and knowledge graphs as a quality signal. Moreover, to enhance the quality of OSM data, we developed OVID (Tempelmeier and Demidova 2022)—a novel method to detect vandalism in OpenStreetMap automatically. Quality aspects of OSM are also considered in Chap. 2. In future work, we would like to investigate further methods to enhance data quality in OSM and WORLDKG. WORLDKG can also potentially be used for visual reporting solutions discussed in Chap. 7.

To make OSM data more easily accessible to machine learning algorithms, we developed GeoVectors—a reusable openly available dataset of OSM embeddings (Tempelmeier et al. 2021b). GeoVectors approach extends the OSM node embedding algorithms presented in Sect. 1.3 and encodes semantic and geographic similarity of OSM nodes. In future work, we would like to leverage WORLDKG and GeoVectors to provide semantic geographic information for machine learning applications.

Acknowledgments This research was partially funded by the Deutsche Forschungsgemeinschaft (DFG, German Research Foundation)—WORLDKG, 424985896.

References

Auer S, Bizer C, Kobilarov G, Lehmann J, Cyganiak R, Ives Z (2007) DBpedia: a nucleus for a web of open data. In: Proceedings of the 6th International Semantic Web Conference, Asian Semantic Web Conference, ISWC 2007 + ASWC 2007, volume 4825 of Lecture Notes in Computer Science. Springer, Berlin, pp 722–735. https://doi.org/10.1007/978-3-540-76298-0_52

Auer S, Lehmann J, Hellmann S (2009) LinkedGeoData: adding a spatial dimension to the web of data. In: Proceedings of the 8Th International Semantic Web Conference, ISWC 2009, volume 5823 of Lecture Notes in Computer Science. Springer, Berlin, pp 731–746. https://doi.org/10.1007/978-3-642-04930-9_46

Bento A, Zouaq A, Gagnon M (2020) Ontology matching using convolutional neural networks. In: Proceedings of the 12th Language Resources and Evaluation Conference, LREC 2020. European Language Resources Association, pp 5648–5653. https://aclanthology.org/2020.lrec-1.693/

Cappuzzo R, Papotti P, Thirumuruganathan S (2020) Creating embeddings of heterogeneous relational datasets for data integration tasks. In: Proceedings of the 2020 International Conference on Management of Data, SIGMOD 2020. ACM, New York, pp 1335–1349. https://doi.org/10.1145/3318464.3389742

Dadwal R, Funke T, Demidova E (2021) An adaptive clustering approach for accident prediction. In: Proceeding of the 24th IEEE International Intelligent Transportation Systems Conference, ITSC 2021. IEEE, pp 1405–1411. https://doi.org/10.1109/ITSC48978.2021.9564564

Daiber J, Jakob M, Hokamp C, Mendes PN (2013) Improving efficiency and accuracy in multilingual entity extraction. In: Proceeding of the International Conference on Semantic Systems, ISEM '13. ACM, New York, pp 121–124. https://doi.org/10.1145/2506182.2506198

Doan A, Madhavan J, Domingos PM, Halevy AY (2004) Ontology matching: a machine learning approach. In: Handbook on Ontologies, International Handbooks on Information Systems. Springer, Berlin, pp 385–404. https://doi.org/10.1007/978-3-540-24750-0_19

Dsouza A, Tempelmeier N, Demidova E (2021a) Towards neural schema alignment for open-streetmap and knowledge graphs. In: Proceeding of the 20th International Semantic Web Conference, ISWC 2021, volume 12922 of Lecture notes in computer science. Springer, Berlin, pp 56–73. https://doi.org/10.1007/978-3-030-88361-4_4

Dsouza A, Tempelmeier N, Yu R, Gottschalk S, Demidova E (2012b) WorldKG: a world-scale geographic knowledge graph. In: Proceeding of the 30th ACM International Conference on Information and Knowledge Management, CIKM '21. ACM, New York, pp 4475–4484. https://doi.org/10.1145/3459637.3482023

Gottschalk S, Demidova E (2019) EventKG—the Hub of event knowledge on the web- and biographical timeline generation. Semantic Web 10(6):1039–1070. https://doi.org/10.3233/SW-190355

Guttman A (1984) R-trees: a dynamic index structure for spatial searching. In: Proceedings of the Annual Meeting, SIGMOD 1984. ACM Press, New York, pp 47–57. https://doi.org/10.1145/602259.602266

Hogan A, Blomqvist E, Cochez M, d'Amato C, de Melo G, Gutiérrez C, Kirrane S, Gayo JEL, Navigli R, Neumaier S, Ngomo AN, Polleres A, Rashid SM, Rula A, Schmelzeisen L, Sequeda JF, Staab S, Zimmermann A (2021) Knowledge graphs. ACM Comput Surv 54(4):71:1–71:37. https://doi.org/10.1145/3447772

Janowicz K, Hitzler P, Li W, Rehberger D, Schildhauer M, Zhu R, Shimizu C, Fisher CK, Cai L, Mai G, Zalewski J, Zhou L, Stephen S, Estrecha SG, Mecum BD, Lopez-Carr A, Schroeder A, Smith D, Wright DJ, Wang S, Tian Y, Liu Z, Shi M, D'Onofrio A, Gu Z, Currier K (2022) Know, Know Where, KnowWhereGraph: a densely connected, cross-domain knowledge graph and geo-enrichment service stack for applications in environmental intelligence. AI Mag 43(1):30–39. https://doi.org/10.1609/aimag.v43i1.19120

Karalis N, Mandilaras GM, Koubarakis M (2019) Extending the YAGO2 knowledge graph with precise geospatial knowledge. In: Proceedings of the 18th International Semantic Web Conference, ISWC 2019, volume 11779 of Lecture notes in computer science. Springer, Berlin, pp 181–197. https://doi.org/10.1007/978-3-030-30796-7_12

Madhavan J, Bernstein PA, Rahm E (2001) Generic schema matching with cupid. In: Proceedings of the 27th International Conference on Very Large Data Bases, VLDB 2001. Morgan Kaufmann, pp 49–58. https://doi.org/10.5555/645927.672191

Melnik S, Garcia-Molina H, Rahm E (2002) Similarity flooding: a versatile graph matching algorithm and its application to schema matching. In: Proceedings of the 18th International Conference on Data Engineering, 2002. IEEE Computer Society, pp 117–128. https://doi.org/10.1109/ICDE.2002.994702

Mikolov T, Sutskever I, Chen K, Corrado GS, Dean J (2013) Distributed representations of words and phrases and their compositionality. In: Proceedings of the 27th Annual Conference on Neural Information Processing Systems 2013, pp 3111–3119. https://doi.org/10.5555/2999792.2999959

Ngo D, Bellahsene Z, Todorov K (2013) Opening the black box of ontology matching. In: Proceedings of the ESWC 2013, volume 7882 of Lecture Notes in Computer Science. Springer, Berlin, pp 16–30. https://doi.org/10.1007/978-3-642-38288-8_2

Ngomo AN, Auer S (2011) LIMES—a time-efficient approach for large-scale link discovery on the web of data. In: Proceedings of the 22nd International Joint Conference on Artificial Intelligence, IJCAI 2011. IJCAI/AAAI, pp 2312–2317. https://doi.org/10.5591/978-1-57735-516-8/IJCAI11-385

Otero-Cerdeira L, Rodríguez-Martínez FJ, Gómez-Rodríguez A (2015) Ontology matching: a literature review. Expert Syst Appl 42(2):949–971. https://doi.org/10.1016/j.eswa.2014.08.032

Sherif MA, Ngomo AN, Lehmann J (2017) Wombat—a generalization approach for automatic link discovery. In: Proceedings of the Semantic Web—14Th International Conference, ESWC 2017, volume 10249 of Lecture Notes in Computer Science, pp 103–119. https://doi.org/10.1007/978-3-319-58068-5_7

Stadler C, Lehmann J, Höffner K, Auer S (2012) LinkedGeoData: a core for a web of spatial open data. Semantic Web 3(4):333–354. https://doi.org/10.3233/SW-2011-0052

Tempelmeier N, Demidova E (2021) Linking OpenStreetMap with knowledge graphs—link discovery for schema-agnostic volunteered geographic information. Fut Gener Comput Syst 116:349–364. https://doi.org/10.1016/j.future.2020.11.003

Tempelmeier N, Demidova E (2022) Attention-based vandalism detection in OpenStreetMap. In: Proceeding of the ACM Web Conference 2022, WWW 2022. ACM, New York, pp 643–651. https://doi.org/10.1145/3485447.3512224

Tempelmeier N, Feuerhake U, Wage O, Demidova E (2021a) Mining topological dependencies of recurrent congestion in road networks. ISPRS Int J Geo-Inform 10(4):248. https://doi.org/10.3390/ijgi10040248

Tempelmeier N, Gottschalk S, Demidova E (2021b) GeoVectors: a linked open corpus of OpenStreetMap Embeddings on world scale. In: Proceedings of the 30th ACM International Conference on Information And Knowledge Management, CIKM 2021. ACM, pp 4604–4612. https://doi.org/10.1145/3459637.3482004

von Wahl L, Tempelmeier N, Sao A, Demidova E (2022) Reinforcement learning-based placement of charging stations in urban road networks. In: Proceedings of the 28th ACM SIGKDD Conference on Knowledge Discovery and Data Mining, KDD 2022. ACM, New York, pp 3992–4000. https://doi.org/10.1145/3534678.3539154

Winkler WE (1999) The state of record linkage and current research problems. In: Statistical Research Division, US Census Bureau

Xiang C, Jiang T, Chang B, Sui Z (2015) ERSOM: a structural ontology matching approach using automatically learned entity representation. In: Proceedings of the 2015 Conference on Empirical Methods in Natural Language Processing, EMNLP 2015. The Association for Computational Linguistics, pp 2419–2429. https://doi.org/10.18653/v1/d15-1289

Chapter 2
Analyzing and Improving the Quality and Fitness for Purpose of OpenStreetMap as Labels in Remote Sensing Applications

Moritz Schott, Adina Zell, Sven Lautenbach, Gencer Sumbul, Michael Schultz, Alexander Zipf, and Begüm Demir

Abstract OpenStreetMap (OSM) is a well-known example of volunteered geographic information. It has evolved to one of the most used geographic databases. As data quality of OSM is heterogeneous both in space and across different thematic domains, data quality assessment is of high importance for potential users of OSM data. As use cases differ with respect to their requirements, it is not data quality per se that is of interest for the user but fitness for purpose. We investigate the fitness for purpose of OSM to derive land-use and land-cover labels for remote sensing-based classification models. Therefore, we evaluated OSM land-use and land-cover information by two approaches: (1) assessment of OSM fitness for purpose for samples in relation to intrinsic data quality indicators at the scale of individual OSM objects and (2) assessment of OSM-derived multi-labels at the scale of remote sensing patches (1.22×1.22 km) in combination with deep learning approaches. The first approach was applied to 1000 randomly selected relevant OSM objects. The quality score for each OSM object in the samples was combined with a large set of intrinsic quality indicators (such as the experience of the mapper, the number of mappers in a region, and the number of edits made to the object) and auxiliary information about the location of the OSM object (such as the continent or the ecozone). Intrinsic indicators were derived by a newly developed tool

M. Schott (✉) · A. Zipf
Institute of Geography, GIScience, Heidelberg University, Heidelberg, Germany
e-mail: moritz.schott@uni-heidelberg.de; zipf@uni-heidelberg.de

A. Zell · G. Sumbul · B. Demir
Faculty of Electrical Engineering and Computer Science, Remote Sensing Image Analysis Group, Technische Universität Berlin, Berlin, Germany
e-mail: adina.zell@campus.tu-berlin.de; gencer.suembuel@tu-berlin.de; demir@tu-berlin.de

S. Lautenbach
HeiGIT gGmbH at Heidelberg University, Heidelberg, Germany
e-mail: sven.lautenbach@heigit.org

M. Schultz
Institute of Geography, University of Tübingen, Tübingen, Germany
e-mail: michael.schultz@uni-tuebingen.de

© The Author(s) 2024
D. Burghardt et al. (eds.), *Volunteered Geographic Information*,
https://doi.org/10.1007/978-3-031-35374-1_2

based on the OSHDB (OpenStreetMap History DataBase). Afterward, supervised and unsupervised shallow learning approaches were used to identify relationships between the indicators and the quality score. Overall, investigated OSM land-use objects were of high quality: both geometry and attribute information were mostly accurate. However, areas without any land-use information in OSM existed even in well-mapped areas such as Germany. The regression analysis at the level of the individual OSM objects revealed associations between intrinsic indicators, but also a strong variability. Even if more experienced mappers tend to produce higher quality and objects which underwent multiple edits tend to be of higher quality, an inexperienced mapper might map a perfect land-use polygon. This result indicates that it is hard to predict data quality of individual land-use objects purely on intrinsic data quality indicators. The second approach employed a label-noise robust deep learning method on remote sensing data with OSM labels. As the quality of the OSM labels was manually assessed beforehand, it was possible to control the amount of noise in the dataset during the experiment. The addition of artificial noise allowed for an even more fine-grained analysis on the effect of noise on prediction quality. The noise-tolerant deep learning method was capable to identify correct multi-labels even for situations with significant levels of noise added. The method was also used to identify areas where input labels were likely wrong. Thereby, it is possible to provide feedback to the OSM community as areas of concern can be flagged.

Keywords Volunteered geographic information · Data quality · Data analysis · Remote sensing · OpenStreetMap · Machine learning

2.1   Introduction

OpenStreetMap (OSM) has evolved to one of the most used geographic databases and is a prototype for volunteered geographic information (VGI). It is a major knowledge source for researchers, professionals, and the general public to answer geographically related questions. As a free and open community project, the OSM database can not only be edited but also used by any person with very limited restrictions such as internet access or usage citation. This open nature of the project enabled the establishment of a vibrant community that curates and maintains the projects' data and infrastructure, but also a growing ecosystem of tools that use or analyze the data (OpenStreetMap Contributors 2022a,b).

Recently, OSM has become a popular source of labeled data for the remote sensing community. However, spatial heterogeneous data quality provides challenges for the training of machine learning models. Frequently, OSM land-use and land-cover (LULC) data has thereby been taken at face value without critical reflection. And, while the quality and fitness for purpose of OSM data have been proven in many cases (e.g., Jokar Arsanjani et al. 2015; Fonte et al. 2015), these analyses have also unveiled quality variations, e.g., between rural and urban regions. The quality of OSM can thus be assumed to be generally high, but remains unknown for a specific use case. It is therefore of importance to develop both tools that are capable of quantifying data quality of LULC information in OSM and approaches that are capable of dealing with the noise potentially present in OSM.

The IDEAL-VGI project investigated the fitness for purpose of OSM to derive LULC labels for remote sensing-based classification models by two approaches: (1) assessment of OSM fitness for purpose for samples in relation to intrinsic data quality indicators at the scale of individual OSM objects and (2) assessment of OSM-derived multi-labels at the scale of remote sensing patches (1.22 × 1.22 km) in combination with deep learning methods.

2.2 Intrinsic Data Quality Analysis for OSM LULC Objects

One of the most prominent analysis topics in OSM-related research is data quality that has been covered in theory (see, e.g., Barron et al. 2014; Senaratne et al. 2017) as well as in many practical studies (e.g., Jokar Arsanjani et al. 2015; Brückner et al. 2021). The topic of data quality is of concern for many studies working with volunteered geographic information—Chap. 1, for example, deals with data quality in OSM and Wikidata. Senaratne et al. (2017) characterize analyses into extrinsic metrics, where OSM is compared to another dataset, and intrinsic indicators, where metrics are calculated from the data itself. Semi-intrinsic (or semi-extrinsic) metrics use auxiliary information to assess the quality of OSM— population density can, for example, be used to assess the completeness of buildings in OSM, as population density and number of buildings are related. The quality gold standard has frequently been defined for extrinsic metrics through an external dataset of higher or known quality and standards. However, external datasets of high quality—including high up-to-dateness—are frequently not available. Therefore, intrinsic data quality indicators have frequently been used (Barron et al. 2014). These try to capture data quality aspects based on the history of OSM data itself, such as the number of edits to an object. Although OSM objects can be viewed individually, they are always embedded in a larger context of surrounding OSM objects, communities of contributors, and other classification systems, such as biomes or socioeconomic factors. Comparing contributions and communities for selected cities, (Neis et al. 2013), e.g., found a positive correlation between contributor density and gross national product per capita and showed that commu- nity sizes vary between Europe and other regions. In 2021, Schott et al. (2021) described "digital" and "physical locations" in which an OSM object is located. These "locations" consist of, intrinsic, OSM-specific measures such as density and diversity of elements, but also include—semi-intrinsic—aspects of economic status, culture, and population density to describe the surrounding of an object. Such information provides potentially relevant information to help characterize and predict data quality of OSM objects.

LULC information in OSM is a challenging topic. On the one hand, this information provides the background for all other data rendered on the central map.

It can highly benefit from local input, survey mapping, and (live) updates. On the other hand, this information has a difficult position within the OSM ecosystem. While the routing and the building of communities are prominent, LULC is not so frequently mentioned in the ecosystems' communication platforms. LULC information can also be quite cumbersome or even difficult to map, e.g., due to natural ambiguity. The growing tagging scheme provides a collection of sometimes ambiguous or overlapping tag definitions that are not fully compatible with any official LULC legend definition (Fonte et al. 2016). Furthermore, the data is highly shaped by national preferences and imports.

2.2.1   OSM Element Vectorization: Intrinsic and Semi-intrinsic Data Quality Indicators

The OSM element vectorization tool (OEV, Schott et al. 2022) has been developed to ease access to intrinsic and semi-intrinsic indicators, with a specific focus on LULC feature classes. The tool[1] provides access to currently 31 indicators at the level of single OSM objects (c.f. Table 2.1), which cover aspects concerning the element itself, surrounding objects, and the editors of the object.

The usability of the tool was proven on the use case of LULC polygons. One thousand out of the globally existing 62.9 million LULC elements were randomly sampled on 2022-01-01. Only polygonal objects with at least one of the LULC defining tags were considered. These elements' IDs were then fed to the tool to extract the data and calculate the described metrics from Table 2.1. These metrics were used in a cluster analysis to identify structures in the OSM LULC objects. Furthermore, we tested three hypotheses on the triangular relation between the size of OSM objects, their age, and their location in terms of population density. We hypothesized that a general mapping order exists where the OSM community first concentrates on or arises from urban areas before moving to rural areas. This was tested by the hypotheses 1 (H_1): *There is a positive correlation between the object age and the population density*. Second, we tested the hypotheses that areas with higher population density are more fragmented and therefore exhibit smaller elements, while areas with low population density, such as forest, are often larger objects: (H_2) *there is a negative correlation between the object size and the population density*. Third, we tested the effect between the OSM LULC objects' age and population density, assuming a non-significant correlation. This was based on two opposing assumptions: Large geographical entities may be mapped first, and regions may be first coarsely drafted before adding details. This would lead to old objects being of larger size. Yet, hypotheses 1 and 2 contradict this tendency: according to H_1 and H_2, younger objects would be in areas with less population density and therefore tend to be larger. All three hypotheses were tested separately

---

[1] https://oev.geog.uni-heidelberg.de/.

Table 2.1 Intrinsic and semi-intrinsic indicators calculated by the OEV tool. The indicators are grouped in the categories of semantic, geometric, surrounding, OSM surrounding, temporal, mapper, mapping process, and linting tools

Indicator	Description	Logical link to LULC quality
osm_type	Type of the OSM object (node, way, relation)	Relations in OSM are more complex and need higher skill to be mapped but also are more error-prone
primtag	Primary tag of the object	Some tags may be more difficult to map, ambiguous, less established, or less documented than others
corine_class	Aggregation of primary tags at CORINE level 2	
invalidity	Gravity of geometric invalidity	Invalid geometries describe bad data
Detail	How detailed is the object drawn?	
Complexity	How complex is the geometry?	The more detailed/less course an object is drawn, the higher was the effort by the mapper and a higher quality can be assumed
obj_size	Size of the object	The smaller the element, the better (up to a certain point)
continent	The country or continent an element is located on	Regions have different communities, physical geographic contexts, etc. that influence quality
pop5k	Population density in the objects' surrounding	Findings are that rural areas often have lower quality; how many potential local mappers are available and what features are plausible given the population density?
hdi	Economic situation of a region	The economic status influences mapping (leisure time, equipment, etc.)
biome	The "physical appearance" of the objects' surrounding	The biome influences the probability/validity of certain objects
rs_variance	How homogeneous is the object in remote sensing imagery?	Homogeneous reflectance -> homogeneous area -> object well drawn (depending on object type)

(continued)

Table 2.1 (continued)

Indicator	Description	Logical link to LULC quality
overlap	How much does the element overlap with elements in the surrounding?	Less overlap -> better geometry
shared_border	What part of the object's border is shared with surrounding objects?	If the object is well placed, other objects might be closely attached to it. As LULC is optimally space filling, a well-mapped area would be represented by connected LULC objects
surrounding_mapped	How well is the area mapped in general?	Well-mapped regions might indicate good quality
surrounding_diversity	How many different primary tags relevant for LULC are present in the surrounding?	Diversity of mapped LULC objects might indicate high quality—if real-world LULC is characterized by some diversity. Often some LULC classes are mapped first—if surrounding diversity is high, this could indicate that the mapping has progressed beyond this point
surrounding_obj_dens	How many objects are there in general in the area?	Well-mapped areas have many objects (in relation to population density, etc.)
how_many_eyes	How many contributors were involved in this area?	Does this area even deserve the "crowdsourced" tag?
oqt_mapping_saturation	Estimated area mapping completeness?	Saturated areas may have better quality as more effort by mappers can be put into improvement instead of the creation of missing objects

outofdateness	How outdated is the object?	Recently changed objects are potentially more up to date
object_age	How old is the object?	Old objects are either outdated and were mapped with superseded mapping practices or had much time to mature with many small improvements
changes_per_year	How frequently has the object been changed/updated?	Objects that are regularly changed are well maintained
user_mean_exp	How experienced were the users that changed this object?	More experienced users may have a higher probability to make changes that improve quality
user_remoteness	How far away were other edits done by the mapper(s) of the object?	Local users are more familiar with the region and should create better data
user_diversity	How specialized is the user?	Diverse users may be general experts, but specialized users may be experts on this topic
data_source	What sources were given for the changes?	Certain sources may be more reliable
cs_editor	Editor/editing software used	Some editors might have bugs or are not suited for certain editing
cs_median_area	Was the object edited by local edits or large-scale edits?	Changesets that span large areas hint toward bad quality, imports, or unspecific changes
validator_issued_density	Density of issues reported by linting tools	Less errors indicate a well-maintained region and good quality
n_osm_notes	How many OSM notes are there?	Many notes either show bad data or an active community
osm_cha_flags	What OSMCha flags were added to the changesets?	Many flags are related to suspicious changesets or edits

using Kendall's τ (Hollander and Wolfe 1973), with the p-values adjusted for multiple testing (Benjamini and Hochberg 1995).

Results of this first exploratory analysis provided interesting insights into the complex and multifaceted structure of OSM land-use objects and its relation to OSM mapping communities. The main hypotheses regarding the mapping order (H_1) could not be confirmed. In fact, the estimated correlation was slightly negative, meaning that for the used sample, objects in urban areas were younger than in rural areas. Yet, this does not imply that this mapping order does not exist in certain sub-regions. In addition, the age of an object is a fragile metric that highly depends on the mapping style of local mappers. Mappers frequently decide to delete and redraw elements instead of changing the original object, especially if the object was only a coarse approximation. This "resets" the object age, meaning that urban areas may have a high share of young objects because they are still actively mapped and maintained, even though they started their map appearance relatively early. H_3 was equally confirmed, but only after p-value correction (p-value $= 0.14$). Regional specialities may exist in this aspect and need further investigation.

The negative correlation between the object size and the population density (H_2) was confirmed with a p-value <0.01 though the τ was only -0.096 implying a small effect. At the global scale, many influencing factors may overlap or intervene with each other, hindering the extraction of single detailed effects. For the example at hand, we can assume that there are multiple regional communities or active mappers with individual mapping styles. The mapping detail in urban or rural regions will therefore be linked to these and other factors as well, not only the population density. Population density itself may not be generalizable on a global scale. The same level of fragmentation, meaning object size distribution, may be reached at different population density values, depending, e.g., on the continent.

The cluster analysis revealed interesting aspects, as some clusters could be associated with imports. Especially, a large import of North American lakes could be separated. This element group made up a considerable share of the global data and must therefore be taken into account when analyzing or describing the global dataset.

One thousand LULC objects were manually checked against high-resolution imagery. A combined quality score was assigned based on the thematic and the geometric correctness of the object. A quantile random forest was used to identify relationships between the data quality score and the 31 indicators calculated by the OEV tool.

While the overall quality of the model was intermediate, we were able to identify a series of interesting relationships between the indicators and the quality of the land-use objects based on the visual inspection of partial dependency plots. The most important features in the model (c.f. Fig. 2.1) were the size of the OSM object, contributor characteristics (such as experience and remoteness), rare OSM tags, and regional OSM mapping aspects (e.g., number of OSM objects in the surrounding of the object).

Element size had by far the highest feature importance. However, the effect on data quality of the OSM object had no clear direction. The indicator had to

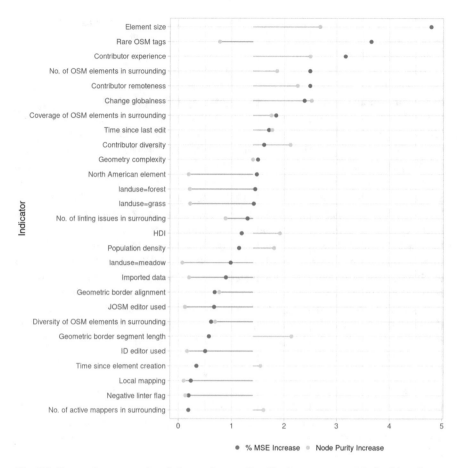

Fig. 2.1 Feature importance in relation to data quality. The importance was derived based on a quantile random forest for 1000 randomly selected OSM objects. The features are sorted by the percentage increase in squared mean error if the feature would be dropped. In addition, node purity is provided as a second feature importance indicator. To ease interpretation, the second indicator is displayed together with its position relative to the median node purity value across all selected features

be interpreted in combination with other indicators such as the primary tag (e.g., *landuse* or *natural*) as some land-use tags were characterized by big objects (e.g., forest) and others by small objects (e.g., urban grass). Objects with rare primary tags indicated, in general, a lower object quality—presumably as these tags are less well established and possibly poorly defined and thereby harder to map consistently. The effect of contributor experience showed a multimodal distribution: contributors with very little experience (newcomers) were associated with objects of medium quality and contributors with medium experience (stable mappers) with high quality of the land-use objects. Interestingly, highly experienced mappers were associated with poor quality of the land-use object. One possible explanation is that these extraordinary active users might represent bots or importers. One has, however, to keep in mind that only a few contributors fell into this category, which led to a low statistical power.

With respect to the number of elements surrounding an OSM LULC object, areas with more elements were—expectedly—better mapped. The larger the share of the surrounding of an OSM object mapped by LULC, the higher the quality of the OSM object. However, shares above 100% expectedly were associated with a lower quality of the OSM object. These regions are characterized by areas mapped using highly overlapping polygons, which typically indicate mapping errors. OSM edits are contributed as changesets. Larger changesets (more elements, often from different regions) were associated with lower quality of the OSM land-use objects. This is in line with the expectation that local, concise, and coherent edits are better. With respect to the primary tag (the LULC class), the model indicated differences in the quality of some classes: while forest objects were of higher quality, grass-dominated LULC classes were of lower quality. This could be explained by the clear distinction of forest from other LULC classes and the diversity of tags used to characterize different grass-dominated LULC classes. LULC objects in North America had a tendency for lower quality, presumably due to the large imports in this region. Besides that, LULC objects mapped in regions with higher Human Development Index or higher population density were associated with better data quality. This presumably reflects the larger OSM community in the Global North, especially in countries such as Austria, Germany, or France, as well as the higher number of potential mappers available in urban areas compared to the countryside. With respect to out-of-dateness, complex interactions were identified; however, generally recently changed objects were associated with higher quality. Except for newcomers, lower user diversity—i.e., users focusing on one aspect of OSM—was associated with higher data quality.

2.3 Label Noise Robust Deep Learning for Remote Sensing Data with OSM Tags

2.3.1 OSM as the Source of Training RS Image Labels in ML

Supervised ML methods have attracted great attention for Earth observation applications on ever-growing RS image archives. Due to their capability to automatically model higher-level RS image semantics in large scale, they are applied to many problems in RS such as multi-label image classification and land-cover map generation. These methods generally require the availability of a high quantity of annotated training RS images. However, the manual collection of RS image annotations by domain experts for a large amount of data can be time-consuming, complex, and costly. Accordingly, the use of volunteered geographic information as crowdsourced data such as OSM to automatically derive annotated training data has been drawing significant attention in RS. As an example, in (Kaiser et al. 2017; Wan et al. 2017; Comandur and Kak 2021), it is shown that the direct use of OSM tags as pixel-level land-use class labels is useful for the automatic map generation of RS images through support vector machines and convolutional neural networks (CNNs). In (Li et al. 2020), OSM tags are utilized as scene-level class labels of training RS images to automatically predict land-use/land-cover classes of RS image scenes through CNNs. In (Audebert et al. 2017), OSM data is also utilized as an auxiliary information source by fusing it with optical data from very-high-resolution satellite imagery through dual-stream CNNs. In (Lin et al. 2022), an active learning strategy is introduced to partially annotate RS images with salient multi-labels based on OSM tags. In this study, an adaptive temperature-associated model is also proposed to apply multi-label RS image classification by utilizing partially annotated training data and automatically assigning missing labels to training images during training.

Thanks to the publicly available OSM database, collection of RS image annotations for a high quantity of training data to be utilized for ML methods can be achieved at lower costs. However, OSM tags can be outdated regarding RS images due to possible changes on the ground; or there can be annotation errors. Accordingly, using OSM tags as the source of training image annotations may increase the chance of including noisy labels in training data of ML methods. As an example, for multi-label image annotations of RS images, two types of noise can exist. Noise can be associated with missing labels or wrong labels. A missing label means that although a land-use/land-cover class exists in an RS image, the corresponding class label is not assigned. A wrong label means that although a class label is assigned to RS image, the corresponding class is not present in the image.

2.3.2 Label Noise Robust ML Methods

When a ML model is trained on noisy training data, there is a risk of overfitting of the model parameters to noisy labels and thus suboptimal inference performance. To this end, a few methods are presented in RS to improve the robustness of ML models toward noisy labels in training data. As an example, in (Zhang et al. 2020a), a noisy label knowledge distillation method is introduced for single-label RS image classification problems to leverage the knowledge learned through a teacher model on images with noisy labels for a student model. In this method, two CNNs are employed as a teacher-student framework, while a clean and trustworthy subset of a training set is assumed to be available for the student CNN. In (Aksoy et al. 2022), a collaborative learning framework is proposed to identify and exclude images with noisy multi-labels during training. To this end, it employs two CNNs operating collaboratively, while they are forced to characterize distinct image representations and to produce similar predictions. In (Burgert et al. 2022), the effects of the abovementioned label noise types in multi-label RS image classification problems are investigated, while different single-label noise robust methods are integrated to multi-label classification problems in RS. In (Dong et al. 2022), for land-cover map generation through semantic segmentation, an online noise correction approach is introduced to detect and correct pixel-level noisy labels via information entropy at the early stage of model training and thus to continue training with corrected labels. Although all these methods are potentially effective, the development of label noise robust ML methods when the OSM tags are utilized as the source of image annotations has not yet been investigated in RS literature.

It is worth mentioning that learning the parameters of ML models under noisy labels in training data has been studied more extensively in computer vision (CV) literature than RS. Recent research directions in CV can be grouped into the development of (1) deep neural network (DNN) architectures, (2) ML loss functions, (3) regularization strategies for training ML models, and (4) training sample selection and label adjustment techniques for single-label image classification problems. The first set of methods are concentrated on the development of DNN architectures designed for training data with noisy labels. For example, a contrastive-additive noise network is introduced in (Yao et al. 2019) to model trustworthiness of noisy training labels. This network consists of a probabilistic latent variable model as a contrastive layer in order to measure the quality of annotations and an additive layer to aggregate the class predictions and noisy labels. The second set of methods is mostly focused on the development of ML loss functions, which embody robust characteristics toward noisy labels. As an example, in (Ridnik et al. 2021), an asymmetric loss function is proposed to dynamically decrease the weights of negative classes in multi-labels. This allows to decrease the effect of images with missing labels on ML model parameter updates during training. The third set of methods are concentrated on regularizing the whole ML model training to prevent overfitting of model parameters to noisy labels. For instance, a regularization term is integrated into the cross-entropy loss function in (Liu et al. 2020) to utilize the class predictions from an early stage of ML model training to prevent the memorization of noisy labels. The fourth set of methods aim to first select images with correct labels or adjust noisy labels and then to learn through samples with correct labels. As an example, a joint training with co-regularization approach is introduced in (Wei et al. 2020) to employ collaborative learning of two CNNs for the selection of correct labels by an agreement strategy.

2.3.3 Proposed Methods

Due to the public availability of OSM, RS images can be automatically associated with multiple land-use/land-cover classes (i.e., multi-labels) by using OSM tags. This allows to create large training sets for deep learning (DL)-based multi-label RS image classification methods at lower costs. Let $X=\{x_1, \ldots, x_M\}$ be an RS image archive that includes M images, where x_k is the kth image in the archive. We assume that a training set $\mathcal{T} = \{(x_i, y_i)\}_{i=1}^{D}$ is available. Each training image

x_i is associated with a set of class labels $y_i \in \{0, 1\}^S$ based on the corresponding OSM tags, where S is the total number of classes. Let $\phi : \theta, X \mapsto \hat{\mathcal{Y}}$ be any type of convolutional neural network (CNN) that generates the multi-label $\hat{y}_k$ of an image $x_k \in X$. Training ϕ on $\mathcal{T}$, which may include noisy labels due to noisy OSM tags, can lead to learning suboptimal model parameters θ and inaccurate inference performance, as discussed in the previous sections.

To address this issue, we aim to first automatically detect noisy OSM tags based on the CNN ϕ trained on $\mathcal{T}$ and then adjust training labels associated with noisy OSM tags for label noise robust learning of the CNN model parameters θ.

2.3.3.1 Noisy OSM Tag Detection

Region-based RS image representations combining both local information and the related spatial organization of land-use/land-cover classes are important for the accurate detection of noisy OSM tags. However, multi-label RS image predictions $\hat{\mathcal{Y}}$ of the considered CNN ϕ do not provide spatial information regarding the class location.

Accordingly, we employed class activation maps (CAM) introduced in (Zhou et al. 2016) since they are capable of deriving the regions most relevant for a given class with respect to the DL model trained for image classification. Let $\mathcal{F}_i \in \mathbb{R}^{C \times W \times H}$ be a set of feature maps for an image x_i obtained from the last convolutional layer of the CNN backbone where C, H, and W represent the number of channels, height, and width of the feature maps, respectively. CAMs associated with x_i can be obtained by applying a 1x1 convolutional layer, which takes the feature maps $\mathcal{F}_i$ of x_i as input and produces a set of feature maps $\mathcal{A}_i \in \mathbb{R}^{S \times W \times H}$. The sth feature map $\mathcal{A}_i^s \in \mathbb{R}^{W \times H}$ is the localization map associated with a class s, which can be obtained as follows:

$$\mathcal{A}_i^s = \sum_{c=1}^{C} w_s^c \mathcal{F}_i^c \qquad (2.1)$$

where w_s^c is the weight of importance for the cth feature map $\mathcal{F}_i^c$ regarding the sth class. The obtained CAMs are forwarded through a global average pooling (GAP) layer to obtain multi-label class predictions. However, multi-label classification models from which CAMs can be derived are trained only to identify the presence of a given class within the image. Thus, CAMs tend to focus only on the most discriminative features within the image, leading to the incomplete coverage of the target class within the image (Zhang et al. 2020b).

Self-enhancement maps (SEMs) introduced in (Zhang et al. 2020b) address this issue and improve the localization maps derived from CAMs by including the similarity of feature maps in the localization map calculation. This is achieved by first defining seed coordinates, which are the image regions with the largest activation values on CAMs for a given class. Then, a similarity map is created for each seed point based on the cosine similarity between the seed point feature vector and all other feature vectors. For a given class s of an image x_i, the final class activation map $\mathcal{E}_i^s$ is obtained by taking the maximum value at each pixel across the similarity maps. After obtaining all the class activation maps, we generate a class prototype P^s for the class s by following (Lee et al. 2018). This is achieved by averaging all the feature maps, which are extracted from $\mathcal{T}$ and associated with the class s. Class prototypes allow us to obtain more accurate class predictions based on spatial information regarding the location of image classes and the corresponding region-based image representations. Accordingly, to define whether the class s is present in the image x_i, the extracted features of the image for the class $\mathcal{F}_i^s$ are compared with the corresponding class prototype based on their cosine similarity as follows:

$$\hat{y}_i^s = \begin{cases} 1, & \cos(P^s, \mathcal{F}_i^s) > 0.5 \\ 0, & \text{otherwise} \end{cases} \qquad (2.2)$$

To detect if x_i is associated with noisy labels, we compare the class predictions $\hat{y}_i$ with the associated OSM tags. If the CNN model predicts a class which is not in the list of class labels derived from OSM tags, it is assumed to be a missing class. This missing class can be localized through SEMs. If the CNN model does not predict a class, but it is in the list of class labels derived from OSM tags, it is assumed to be a wrong class label. It is noted that automatically defining noisy OSM tags and the localization of missing classes via SEMs allow providing feedback to the OSM community. Such feedback together with further investigations in the OSM community can lead to correcting noisy OSM tags by human mappers.

2.3.3.2 Label Noise Robust Multi-label RS Image Classification

It is worth noting that the abovementioned method for the detection of noisy OSM tags relies on the model parameters θ of the considered CNN, which is obtained through training on $\mathcal{T}$. If a small trustworthy subset $C \in \mathcal{T}$ of the training set is available, this method can also be used to automatically find training images associated with noisy labels.

To this end, we divide the whole learning procedure into two stages. In the first stage, θ is learned by training ϕ only on C. After this stage is finalized, we first automatically divide the rest of the training set $\mathcal{T} \setminus C$ into training images with noisy labels $\mathcal{N}$ and training images with correct labels $\mathcal{L}$ (i.e., $\mathcal{N} \cup \mathcal{L} = \mathcal{T} \setminus C$). Then, class labels associated with each image in $\mathcal{N}$ are automatically corrected based on (2.2) leading to training images with corrected labels $\mathcal{N}^*$. This leads to automatically correcting noisy labels in $\mathcal{T} \setminus C$ derived from OSM tags. Then, the training set of the second stage $\mathcal{T}^*$ is formed by combining $\mathcal{L}$, C, and $\mathcal{N}^*$.

In the second stage, all the model parameters of ϕ are fine-tuned on $\mathcal{T}^*$. Thanks to the first stage, noisy labels included in the training set of this stage are significantly reduced. This allows to overcome overfitting on noisy labels of the whole training set in the second stage. Due to this two-stage learning of the model parameters, abundant training RS images annotated with OSM tags can be facilitated for label noise robust learning of multi-label RS image classification through CNNs.

2.3.4   Results and Discussion

In this subsection, we first describe the considered dataset and the experimental setup and then provide our analysis of the experimental results.

2.3.4.1   Dataset Description and Experimental Setup

To conduct experiments, we selected a Sentinel-2 tile acquired over South-West Germany including parts of France on 2021-06-13. This region spans from the Palatinate Forest in the west to the Odenwald in the east and includes large forested areas as well as areas dominated by agriculture or by built-up areas. This tile was divided into 81001.22 × 1.22-km-sized image patches. Each image patch is annotated with multi-labels based on the presence or absence of four major land-use classes that are defined with OSM tags (c.f. Table 2.2). While assigning the labels, small OSM objects were filtered (see Table 2.2 for thresholds used). The resulting labels of 910 image patches were manually validated against Sentinel-2 imagery.

Table 2.2   OSM land-use classes used for the multi-label image classification

Class	Description and filter	OSM tags
Water bodies	Continuous 0.2 ha of non-intermittent surface water; smaller ponds and all pools were not considered	landuse=reservoir, natural=water, waterway=dock, waterway=riverbank
Forests	Continuous 0.5 ha closed tree cover; smaller tree groups were not considered	landuse=forest, natural=wood
Agricultural areas	Continuous 0.5 ha meadow; arable land or vineyards, non-agricultural areas (parks, etc.), and smaller isolated elements were not considered assuming they are non-agricultural gardens or similar	landuse=farmland, landuse=meadow, landuse=vineyard
Built-up areas	Continuous 0.5 ha containing mostly impermeable features (buildings, roads, etc.); single isolated buildings are not considered, and large permeable objects like parks or sports grounds are not part of the built-up area	landuse=civic_admin, landuse=commercial, landuse=depot, landuse=education, landuse=farmyard, landuse=garages, landuse=industrial, landuse=residential, landuse=retail

Table 2.3 Multi-label image classification results in terms of mean average precision (mAP) obtained by the direct use of OSM tags (OSM) with different values of synthetic label noise rate (SLNR) of test set and DeepLabV3+ with different values of SLNR of training set

Method	SLNR	mAP (%)
OSM	0%	94.2
	10%	89.4
	20%	83.1
	30%	77.3
	40%	72.9
DeepLabV3+	0%	99.2
	20%	96.8
	40%	70.9
	60%	66.6
	80%	60.4

The OSM quality and completeness in the region were high (c.f. Table 2.3). OSM-based multi-label assignments had a mean average precision of 94.2%. The patches were clustered with respect to the correct assignment of the multi-labels: Clusters of correct OSM data were often due to monotonous landscapes, e.g., the Palatinate Forest. Clusters of flawed data were often due to missing data, e.g., in the region around Kaiserslautern. To perform experiments, 200 manually labeled patches were used as the test set, while the rest of the image patches were utilized as the training set.

In the experiments, we utilized the DeepLabv3+ (Chen et al. 2018) CNN architecture as the DL model. It is worth noting that DeepLabv3+ is originally designed for semantic segmentation problems. We replaced its semantic segmentation head with a fully connected layer followed by a GAP layer that forms the multi-label classification head with four output classes. We trained DeepLabv3+ for 20 epochs with Adam optimizer and the initial learning rate of 0.001. For the proposed label noise robust learning method, the same number of training epochs is used for each of the first and second stages. Experimental results are provided in terms of micro mean average precision (mAP) scores and noise detection accuracies.

We conducted experiments to (i) compare the considered DL model with the direct use of OSM tags for multi-label RS image classification, (ii) analyze the effectiveness of the proposed label noise detection method, and (iii) assess the effectiveness of the proposed label noise robust learning method. For the proposed label noise robust learning method, which requires the availability of a small trustworthy subset of the training set, we included the manually labeled image patches to the training set. However, for the comparison between the DL model and the OSM tags, only non-verified training data was utilized. To assess the robustness of the CNN model toward label noise and to detect noisy samples, we injected synthetic label noise to the training and test sets at different percentages (which were 20%, 40%, 60%, and 80% for the training set and 10%, 20%, 30%, and 40% for the test set) by following (Burgert et al. 2022).

2.3.4.2   Comparison Between Direct Use of OSM Tags and DL-Based Multi-label Image Classification

In this subsection, we assess the effectiveness of the considered DL model (DeepLabV3+) compared to the direct use of OSM tags for multi-label RS image classification. To this end, the model parameters of the DL model were learned on abundant non-verified training data without considering label noise robust learning. Table 2.3 shows the corresponding results in terms of mAP values, when the different values of synthetic label noise rate (SLNR) were applied to the training set of DeepLabV3+ and the OSM tags. One can observe from the table that when SLNR equals to 0% for training and test sets, DeepLabV3+ achieves 5% higher mAP values compared to directly using OSM tags. Even when 20% label noise is synthetically added to the training set of the CNN model, it is still capable of achieving higher results compared to OSM when SLNR value equals to 0%. It is worth mentioning that directly using OSM of such low quality leads to missing or wrong classes. It can be seen from the table that when synthetic noise is added to OSM tags, its multi-label image classification performance is significantly reduced. As an example, when SLNR value is increased to 20% from 0%, multi-label image classification performance of OSM is reduced by more than 10%. These results show the effectiveness of using OSM as a training source of CNN models compared to directly using OSM tags for multi-label image classification. This is relevant because preliminary OSM data analyses may not be able to confidently identify such malicious areas of bad quality.

It is worth noting that further increasing the SLNR value of the training set of DeepLabV3+ significantly reduces multi-label image classification performance. This is due to the fact that when a training set of a DL model includes a higher rate of noisy labels, the model parameters are overfitted on noisy labels that lead to suboptimal learning of multi-label image classification. Figure 2.2 shows the self-enhancement maps (SEMs) of an RS image obtained on DeepLabV3+ trained under different values of SLNR. One can see from the figure that as the SLNR value of the training set increases, the capability of CNN model to characterize the semantic content of the image reduces due to noisy labels.

2.3.4.3   Label Noise Detection

In this subsection, we assess the effectiveness of the proposed label noise detection method when different rates of synthetic label noise are applied to the test set. We also analyze the effect of the level of label noise (which is present in the training data) on our method. Table 2.4 shows the corresponding label noise detection accuracies obtained on DeepLabV3+ trained with abundant non-verified training data at different SLNR values and a small data, which is verified in terms of label noise. One can observe from the table that when SLNR equals to 0%, our label noise detection method, which is applied to DeepLabV3+ and trained on abundant data, achieves the highest label noise detection accuracies. For example, when synthetic

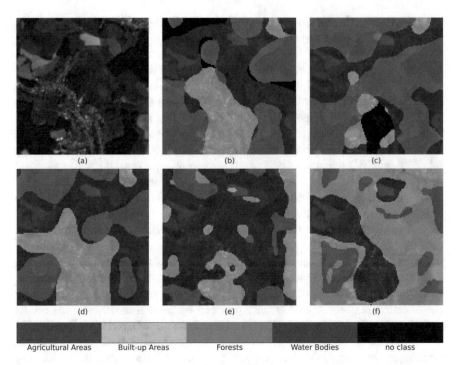

Agricultural Areas   Built-up Areas   Forests   Water Bodies   no class

Fig. 2.2 (**a**) An example of RS image and its self-enhancement maps obtained on DeepLabV3+ trained under synthetic label noise rates (**b**) 0%, (**c**) 10%, (**d**) 20%, (**e**) 30%, and (**f**) 40%

Table 2.4 Label noise detection results in terms of accuracy (%) obtained by the proposed label noise detection method applied to the DeepLabV3+ trained with abundant non-verified data at different values of synthetic label noise rate (SLNR) and small verified data

Training Set	SLNR (Test Set)				
	0%	10%	20%	30%	40%
Abundant non-verified data (SLNR = 0%)	80.0	88.5	87.0	89.0	94.0
Abundant non-verified data (SLNR = 20%)	63.0	72.5	79.0	82.5	86.5
Abundant non-verified data (SLNR = 40%)	0.5	30.0	50.5	64.5	71.5
Abundant non-verified data (SLNR = 60%)	0.0	31.0	55.5	65.5	74.0
Abundant non-verified data (SLNR = 80%)	0.0	31.5	55.5	65.0	74.0
Small verified data	45.0	58.5	67.0	78.5	84.0

label noise is applied to the test set with 40%, our label noise detection method is capable of achieving 94% accuracy. As the SLNR value increases on abundant training data, label noise detection accuracy of our method decreases. This is in line with our conclusion from the previous subsection about the effect of label noise level in training data. In greater details, when SLNR value reaches to 40% for abundant training data, noise detection accuracy of our method decreases by more than 50% compared to SLNR = 0%. When a small verified training data with the size of 10% of the whole training set is used for DeepLabV3+, our noise detection method achieves higher accuracies compared to abundant training data with SLNR $\geq$ 40%. These results show that our label noise detection method is capable of effectively detecting noisy labels without requiring a small verified data when the amount of label noise in the training data is small. However, if the level of label noise in training data is greater than a certain extent, our method requires the availability of a small trustworthy subset of the training set for accurate label noise detection.

2.3.4.4   Label Noise Robust Multi-label Image Classification

In this subsection, we compare the proposed label noise robust learning method with the standard learning procedure, in which label noise of a training set is not considered during training. Table 2.5 shows the corresponding multi-label RS image classification scores when synthetic label noise is injected to the training set at different values of SLNR. It can be seen from the table that when the label noise level in the training set is small (SLNR $\leq$ 20%), standard learning of CNN model parameters achieves higher mAP values compared to label noise robust learning. As an example, when there is no synthetic label noise added to the training set, standard learning leads to more than 3% higher mAP score compared to label noise robust learning. However, as the SLNR value of the training set is higher than a particular value (20%), the considered CNN model with label noise robust learning provides higher multi-label RS image classification accuracies compared to standard learning. For example, when SLNR equals to 80% for the training set, label noise robust learning leads to almost 27% higher mAP value compared to standard learning. These results show that our learning method provides more robust learning of the model parameters for the considered CNN model toward label noise in the training set. Due to the two-stage learning procedure in our method, a

Table 2.5 Multi-label image classification results in terms of mean average precision (mAP (%)) obtained by standard learning and our label noise robust learning for different values of SLNR

SLNR (Training Set)	Standard Learning	Label Noise Robust Learning
0%	99.2	95.6
20%	96.8	89.9
40%	70.9	91.0
60%	66.6	88.3
80%	60.4	87.0

small trustworthy subset of the training set is effectively utilized in its first stage to automatically define noisy labels in the whole training set, which are accurately corrected. Then, employing corrected labels for fine-tuning CNN model parameters on the whole training set leads to leveraging abundant training data without being significantly affected by the label noise.

2.4 Conclusion and Outlook

While OSM provides ample opportunities for use as labels in machine learning-based remote sensing applications, it is necessary to be aware of the challenges the dataset provides. Intrinsic and semi-intrinsic data quality indicators provide insights into the complexity of the OSM mapping process. Meaningful relationships between the indicators and data quality for a test set were derived. The complexity of the interactions did, however, not allow for a reliable prediction of data quality at the level of individual OSM objects. This might change if bigger sample sizes are used. And, while object-level quality prediction requires further research, the developed quality indicators referencing the data region can already support regional quality predictions which are successfully in use in production today.

The proposed deep learning method showed its potential to perform label noise robust multi-label image classification if at least a small set of high-quality labels is available. This shows the potential of the method (i) to overcome the challenges of OSM land-use labels in remote sensing applications and (ii) to provide quality-related feedback for the OSM community. As the OSM community is skeptical toward imports, especially based on automatic labeling, areas flagged as potentially problematic will when presumable be investigated by human mappers and potentially corrected in OSM. Furthermore, these areas can further be analyzed in combination with the intrinsic data quality indicators developed during the project. Approaches described in Chap. 7 might become helpful for this communication. The remote sensing community, on the other hand, can profit from this work through the automated creation of regionalized high-quality image classification models.

Acknowledgments This research was supported by the German Research Foundation DFG within Priority Research Program 1894 *Volunteered Geographic Information: Interpretation, Visualization, and Social Computing* (VGIscience). In this chapter, results from the project *Information Discovery from Big Earth Observation Data Archives by Learning from Volunteered Geographic Information* (IDEAL-VGI, Project number 424966858) are reviewed.

References

Aksoy AK, Ravanbakhsh M, Demir B (2022) Multi-label noise robust collaborative learning for remote sensing image classification. IEEE Trans Neural Netw Learn Syst 1–14. https://doi.org/10.1109/TNNLS.2022.3209992

Audebert N, Le Saux B, Lefèvre S (2017) Joint learning from earth observation and openstreetmap data to get faster better semantic maps. In: IEEE Conference on Computer Vision and Pattern Recognition Workshops, pp 1552–1560. https://doi.org/10.1109/CVPRW.2017.199

Barron C, Neis P, Zipf A (2014) A comprehensive framework for intrinsic openstreetmap quality analysis. Trans GIS 18(6):877–895. https://doi.org/10.1111/TGIS.12073

Benjamini Y, Hochberg Y (1995) Controlling the false discovery rate: a practical and powerful approach to multiple testing. J R Stat Soc B (Methodological) 57(1):289–300. https://doi.org/10.1111/j.2517-6161.1995.tb02031.x

Brückner J, Schott M, Zipf A, Lautenbach S (2021) Assessing shop completeness in openstreetmap for two federal states in Germany. AGILE: GIScience Series 2:20. https://doi.org/10.5194/agile-giss-2-20-2021

Burgert T, Ravanbakhsh M, Demir B (2022) On the effects of different types of label noise in multi-label remote sensing image classification. IEEE Trans Geosci Remote Sensing 60:1–13. https://doi.org/10.1109/TGRS.2022.3226371

Chen L-C, Zhu Y, Papandreou G, Schroff F, Adam H (2018) Encoder-decoder with atrous separable convolution for semantic image segmentation. In: European Conference on Computer Vision. https://doi.org/10.1007/978-3-030-01234-2_49

Comandur B, Kak AC (2021) Semantic labeling of large-area geographic regions using multiview and multidate satellite images and noisy OSM training labels. IEEE J Sel Topics Appl Earth Obs Remote Sensing 14:4573–4594. https://doi.org/10.1109/JSTARS.2021.3066944

Dong R, Fang W, Fu H, Gan L, Wang J, Gong P (2022) High-resolution land cover mapping through learning with noise correction. IEEE Trans Geosci Remote Sensing 60:1–13. https://doi.org/10.1109/TGRS.2021.3068280

Fonte C, Minghini M, Antoniou V, See L, Patriarca J, Brovelli M, Milcinski G (2016) An automated methodology for converting OSM data into a land use/cover map. In: 6th International Conference on Cartography & GIS, 13–17 June 2016, Albena, Bulgaria

Fonte CC, Bastin L, See L, Foody G, Lupia F (2015) Usability of VGI for validation of land cover maps. Int J Geograph Inform Sci 29(7):1269–1291. ISSN 1365-8816. https://doi.org/10.1080/13658816.2015.1018266

Hollander M, Wolfe DA (1973) Nonparametric statistical methods. Wiley series in probability and mathematical statistics: applied probability and statistics. Wiley, New York. ISBN 0-471-40635-X and 978-0-471-40635-8. https://doi.org/10.1002/9781119196037

Jokar Arsanjani J, Mooney P, Zipf A, Schauss A (2015) Quality assessment of the contributed land use information from openstreetmap versus authoritative datasets. In: Jokar Arsanjani J, Zipf A, Mooney P, Helbich M (eds) OpenStreetMap in GIScience: Experiences, Research, and Applications. Springer, Cham, pp 37–58. https://doi.org/10.1007/978-3-319-14280-73

Kaiser P, Wegner JD, Lucchi A, Jaggi M, Hofmann T, Schindler K (2017) Learning aerial image segmentation from online maps. IEEE Trans Geosci Remote Sensing 55(11):6054–6068. https://doi.org/10.1109/TGRS.2017.2719738

Lee K-H, He X, Zhang L, Yang L (2018) Cleannet: Transfer learning for scalable image classifier training with label noise. In: IEEE/CVF Conference on Computer Vision and Pattern Recognition, pp 5447–5456. https://doi.org/10.1109/CVPR.2018.00571

Li H, Dou X, Tao C, Wu Z, Chen J, Peng J, Deng M, Zhao L (2020) RSI-CB: a large-scale remote sensing image classification benchmark using crowdsourced data. Sensors 20(6). https://doi.org/10.3390/s20061594

Lin J, Yu T, Wang ZJ (2022) Rethinking crowdsourcing annotation: partial annotation with salient labels for multilabel aerial image classification. IEEE Trans Geosci Remote Sensing 60:1–12. https://doi.org/10.1109/TGRS.2022.3191735

Liu S, Niles-Weed J, Razavian N, Fernandez-Granda C (2020) Early-learning regularization prevents memorization of noisy labels. In: International Conference on Neural Information Processing Systems. https://doi.org/10.48550/arXiv.2007.00151

Neis P, Zielstra D, Zipf A (2013) Comparison of volunteered geographic information data contributions and community development for selected world regions. Fut Int 5(2):282–300. https://doi.org/10.3390/fi5020282

OpenStreetMap Contributors (2022a) List of OSM-based services. https://wiki.openstreetmap.org/wiki/List_of_OSM-based_services

OpenStreetMap Contributors (2022b) Category:OSM processing. https://wiki.openstreetmap.org/wiki/Category:OSM_processing

Ridnik T, Ben-Baruch E, Zamir N, Noy A, Friedman I, Protter M, Zelnik-Manor L (2021) Asymmetric loss for multi-label classification. In: IEEE/CVF International Conference on Computer Vision, pp 82–91. https://doi.org/10.1109/ICCV48922.2021.00015

Schott M, Grinberger AY, Lautenbach S, Zipf A (2021) The impact of community happenings in OpenStreetMap—establishing a framework for online community member activity analyses. ISPRS Int J Geo-Inform 10(3):164. https://doi.org/10.3390/ijgi10030164

Schott M, Lautenbach S, Größchen L, Zipf A (2022) Openstreetmap element vectorisation—a tool for high resolution data insights and its usability in the land-use and land-cover domain. Int Arch Photogramm Remote Sensing Spatial Inform Sci 48:4. https://doi.org/10.5194/isprs-archives-XLVIII-4-W1-2022-395-2022

Senaratne H, Mobasheri A, Ali AL, Capineri C, Haklay MM (2017) A review of volunteered geographic information quality assessment methods. Int J Geograph Inform Sci 31(1):139–167. https://doi.org/10.1080/13658816.2016.1189556

Wan T, Lu H, Lu Q, Luo N (2017) Classification of high-resolution remote-sensing image using openstreetmap information. IEEE Geosci Remote Sensing Lett 14 (12):2305–2309. https://doi.org/10.1109/LGRS.2017.2762466

Wei H, Feng L, Chen X, An B (2020) Combating noisy labels by agreement: a joint training method with co-regularization. In: IEEE/CVF Conference on Computer Vision and Pattern Recognition, pp 13723–13732. https://doi.org/10.1109/CVPR42600.2020.01374

Yao J, Wang J, Tsang IW, Zhang Y, Sun J, Zhang C, Zhang R (2019) Deep learning from noisy image labels with quality embedding. IEEE Trans Image Process 28(4):1909–1922. https://doi.org/10.1109/TIP.2018.2877939

Zhang R, Chen Z, Zhang S, Song F, Zhang G, Zhou Q, Lei T (2020a) Remote sensing image scene classification with noisy label distillation. Remote Sensing 12(15). https://doi.org/10.3390/rs12152376

Zhang X, Wei Y, Yang Y, Wu F (2020b) Rethinking localization map: towards accurate object perception with self-enhancement maps. arXiv preprint arXiv:2006.05220. https://doi.org/10.48550/arXiv.2006.05220

Zhou B, Khosla A, Lapedriza A, Oliva A, Torralba A (2016) Learning deep features for discriminative localization. In: IEEE Conference on Computer Vision and Pattern Recognition, pp 2921–2929. https://doi.org/10.1109/CVPR.2016.319

Chapter 3
Efficient Mining of Volunteered Trajectory Datasets

Axel Forsch, Stefan Funke, Jan-Henrik Haunert, and Sabine Storandt

Abstract With the ubiquity of mobile devices that are capable of tracking positions (be it via GPS or Wi-Fi/mobile network localization), there is a continuous stream of location data being generated every second. These location measurements are typically not considered individually but rather as sequences, each of which reflects the movement of one person or vehicle, which we call trajectory. This chapter presents new algorithmic approaches to process and visualize trajectories both in the network-constrained and the unconstrained case.

Keywords Trajectories · Data mining · Indexing · Driving preferences · Map matching · Anonymization · Isochrones · Processing pipeline

3.1 Introduction

An abundance of volunteered trajectory data was made openly available in the last decades, fueled by the development of cheap sensor technology and widespread access to tracking devices. The OpenStreetMap project alone collected and published some 2.43 million GPS trajectories from around the world in the past 17 years. Such datasets enable a wealth of applications, as, e.g., movement pattern extraction, map generation, social routing, or traffic flow analysis and prediction.

Dealing with huge trajectory datasets poses many challenges, though. This is especially true for volunteered trajectory datasets which are often heterogeneous in terms of geolocation accuracy, duration of movement, or underlying transportation mode. A raw trajectory is typically represented as a sequence of time-stamped geolocations, potentially enriched with semantic information. To enable trajectory-based applications, the raw trajectories need to be processed appropriately. In the

A. Forsch (✉) · S. Funke · J.-H. Haunert · S. Storandt
Universität Bonn, Bonn, Germany
e-mail: forsch@igg.uni-bonn.de; funke@fmi.uni-stuttgart.de; haunert@igg.uni-bonn.de; storandt@informatik.uni-konstanz.de

43
D. Burghardt et al. (eds.), *Volunteered Geographic Information*,
https://doi.org/10.1007/978-3-031-35374-1_3

following, we describe the core steps of a general processing pipeline. Of course, this pipeline may be refined or adapted depending on the target application.

1. **Anonymization** Trajectories are personal data and can reveal sensitive information about the recording user. As such, the user's privacy needs to be protected.
2. **Preprocessing** Given the raw data, trajectories are usually first filtered and cleaned. For trajectories that stem from movement in an underlying network (e.g., car trajectories), the preprocessing typically includes map matching, which aims at identifying the path in the network that most likely led to the given movement sequence. Map matching helps to reduce noise that stems, e.g., from sensor inaccuracies and enables more efficient storage.
3. **Storing and Indexing** For huge trajectory sets, scalable storage systems and indexing methods are crucial. Storing often involves (lossy or lossless) compression of the trajectories. Furthermore, several query types, such as spatiotemporal range queries or nearest neighbor queries, should be supported by such a system to allow for effective mining later on. Therefore, indexing structures that allow for efficient retrieval of such query result sets need to be built.
4. **Mining** Trajectory mining tasks include, among others, clustering, classification, mode detection, and pattern extraction. Mining tasks often benefit from the availability of a diverse set of trajectory data, covering, for example, large spatial regions or time intervals. Most applications depend on successfully performing one or more mining tasks.
5. **Visualization** The results of the processing steps are often hard to interpret for non-expert users. To improve the accessibility of the results, appropriate visualizations are needed.

This chapter is a review of the achievements made inside the projects "Dynamic and Customizable Exploitation of Trajectory Data" and "Inferring personalized multi-criteria routing models from sparse sets of voluntarily contributed trajectories" inside the DFG priority program "Volunteered Geographic Information: Interpretation, Visualization, and Social Computing" (SPP 1894). We will discuss achievements for each step of the pipeline.

First, we describe a method to protect the privacy of the user who provided their trajectories. In specific, an algorithm is presented that anonymizes sensitive locations, such as the user's home location, by truncating the trajectory. For this, a formal attack model is introduced. To maximize the utility of the anonymized data, the algorithm truncates as little information as possible while still guaranteeing that the user's privacy cannot be breached by the defined attacks. This section is based on Brauer et al. (2022).

Then, we present a new map-matching method that is able to deal with so-called semi-restricted trajectories. Those are trajectories in which the movement is only partially bound to an underlying network, for example, stemming from a pedestrian who walked along some hiking path but crossed a meadow in between. The challenge is to identify the parts of the trajectory that happened within the network and to find a concise representation of the unrestricted parts. A method based on the careful tessellation of open spaces and multi-criteria path selection that copes with this challenge is presented. This section is based on Behr et al. (2021).

Afterward, we present the PATHFINDER storage and indexing system. It allows dealing with huge quantities of map-matched trajectories with the help of a novel data structure that augments the underlying network with so-called shortcut edges. Trajectories are then represented as sequences of such shortcut edges, which automatically compresses and indexes them. Integrating spatial and temporal search structures allows for answering space-time range queries within a few microseconds per trajectory in the result set. This section is based on Funke et al. (2019).

Then, we review new algorithms for driving preference mining from trajectory sets. Based on a linear preference model, these algorithms identify the driving preferences from the trajectories and use this information to compress and cluster the trajectories. This section is based on Forsch et al. (2022) and Barth et al. (2021).

Finally, we present a method to visualize the results of preference mining to improve their interpretability. For this, an isochrone visualization is used. The isochrones show which areas are easy to access for a user with a specific routing profile and which areas are more difficult to access. This information is especially useful for infrastructure planning. This section is based on Forsch et al. (2021).

The overarching vision is to build an integrated system that enables uploading, preprocessing, indexing, and mining of trajectory sets in a flexible fashion and which scales to the steadily growing OpenStreetMap trajectory dataset. We conclude the chapter by discussing some open problems on the way to achieving this goal.

3.2 Protection of Sensitive Locations

Trajectories are personal data, and, as such, they come within the ambit of the General Data Protection Regulation (GDPR). Therefore, the user's privacy must always be considered when collecting or analyzing users' trajectories. For this, location privacy-preserving mechanisms (LPPMs) are developed. In this section, we review the LPPM presented in Brauer et al. (2022), which focuses on protecting sensitive locations along the trajectory.

Publishing trajectories anonymously, i.e., without giving the name of the user recording the trajectory, is not sufficient to protect the user's privacy. An adversary can still extract personal information from this data, such as the user's home and workplace. This extracted information can link the published trajectories back to the user's identity by using so-called re-identification attacks.

A widely used concept to prevent re-identification attacks is k-anonymity (Sweeney 2002). A k-anonymized trajectory cannot be distinguished from at least $k - 1$ other trajectories in the same dataset. LPPMs that k-anonymize trajectory datasets (e.g., Abul et al. 2008; Yarovoy et al. 2009; Monreale et al. 2010; Dong and Pi 2018) make use of generalization, suppression, and distortion. While k-anonymized trajectory datasets still retain the characteristics of the trajectories when analyzing the dataset as a whole, the utility of single trajectories in the dataset is greatly diminished. To counteract this, anonymization can be restricted to protecting sensitive locations along the trajectories. Sensitive locations are either user-specific, e.g., the user's home or workplace, or they are universally sensitive, such as hospitals, banks, or casinos. LPPMs that focus on the protection of sensitive locations (e.g., Huo et al. 2012; Dai et al. 2018; Wang and Kankanhalli 2020) are more utility preserving and thus allow for better analysis in cases where no strict k-anonymity is needed. In this section, a truncation-based algorithm for protecting sensitive locations is reviewed that transfers the concept of k-anonymity to sensitive locations.

3.2.1 Privacy Concept

In this section, the problem of protecting the users' privacy is formalized. At first, a formal attacker model is introduced that defines the kind of adversary considered. Then, the privacy model to prevent the attacks defined in the attacker model is presented.

Attacker Model The attacker has access to a set of trajectories $\Theta = \{T_1, \ldots, T_l\}$ that have a common destination $\tilde{s}$. Furthermore, the attacker knows a set of sites S that is guaranteed to contain $\tilde{s}$. In this context, sites can be any collection of points of interest, such as addresses or buildings, and S could contain all buildings for a given region. The attacker's objective is to identify $\tilde{s}$ based on Θ and S. For this, the attacker utilizes several attack functions $f_1, \ldots, f_a$. For each trajectory T, an attack function f yields a set of candidate sites: $f(S, T) \subseteq S$. By applying all attack functions to all trajectories in Θ, the attacker collects a joint set of candidates, which enables him to infer $\tilde{s}$. For example, the attacker could assume that $\tilde{s}$ is the site that occurs most often in the joint set.

Privacy Model Brauer et al. (2022) introduced a privacy concept called k-site-unidentifiability which transfers the concept of k-anonymity to sites. Given a set S of sites and a destination site $\tilde{s} \in S$, k-site-unidentifiability requires that $\tilde{s}$ is indistinguishable from at least $k - 1$ other sites in S. Put differently, if k-site-unidentifiability is satisfied, $\tilde{s}$ is hidden in a set $C(\tilde{s})$ of k sites, which is termed *protection set*.

Recall that the attacker's attack functions return sets of candidate sites. The sites in the protection set $C(\tilde{s})$ are indistinguishable with respect to a given attack function f if either all sites or none of the sites in $C(\tilde{s})$ are returned as part of the candidate set for f. In order to preserve k-site-unidentifiability, this property must be guaranteed for all attack functions. The sites in S are available to the attacker and thus cannot be changed. Therefore, this can only be done by altering the trajectories in Θ. In conclusion, the problem is defined as follows:

Given a set of sites S, a set of trajectories Θ with a common destination $\tilde{s} \in S$, and a set of attack functions F, transform Θ into a set of trajectories Θ' that fulfills k-site-unidentifiability for all attack functions in F, i.e., either all or none of the sites in $C(\tilde{s})$ are part of the attack's result for each trajectory in Θ.

In the following, an algorithm that truncates trajectories such that they fulfill k-site-unidentifiability is presented.

3.2.2   The S-TT Algorithm

In this section, the **S**ite-dependent **T**rajectory **T**runcation algorithm (S-TT) is explained. This algorithm truncates a trajectory T such that k-site-unidentifiability with respect to a set F of given attack functions is guaranteed. The truncated trajectory T' is obtained by iteratively suppressing the endpoint of T until each attack function either contains all sites of $C(\tilde{s})$ in its candidate set or none of them. The S-TT algorithm is simple to implement, yet it guarantees that none of the attack functions can be used to single out $\tilde{s}$ from the other sites in $C(\tilde{s})$. For using the algorithm, two further considerations need to be made. Firstly, the protection set $C(\tilde{s})$ needs to be selected. Secondly, assumptions on the used attack functions in F need to be made. In the following, both of these aspects are discussed.

Obtaining the Protection Sets The choice of the protection set $C(\tilde{s})$ greatly influences the quality of the anonymized data. There are two requirements for the protection sets to guarantee good anonymization results. Firstly, the sites in the protection set should be spatially close to each other. This maximizes the utility of the anonymized data, as the truncated part of the trajectory gets minimized. Secondly, the choice of the protection set should not depend on $\tilde{s}$. Otherwise, an attacker can infer $\tilde{s}$ by reverse engineering based on its protection set. Both requirements can be fulfilled by computing a partition of S into pairwise disjoint subsets, where each subset contains at least k sites. Each of these subsets becomes the protection set for all sites included in it. The partition of the sites is a clustering problem with minimum cluster size k and spatial compactness as the optimization criterion. Possible approaches can be purely geometric-based, e.g., by computing a minimum-weight spanning forest of the sites such that each of its trees spans at least k sites (e.g., Imielińska et al. 1993), or they can take additional information into account, such as the road network (e.g., Haunert et al. 2021). A polygonal representation of the protection sets, called *protection cell*, is obtained by unioning the Voronoi cells of the sites in the protection set.

Geometric Attack Functions The S-TT algorithm truncates a trajectory based on the attack functions in F. In the following, a geometric-based S-TT algorithm is presented by defining attack functions for a geometric inference of the trajectories' destination site $\tilde{s}$. Two important characteristics of a trajectory that can be used to identify its destination site are the proximity to and the direction toward the destination site. In the following, attacks using these characteristics are formalized into attack functions.

A trajectory T most likely ends very close to its destination site. Therefore, it is reasonable to assume that the trajectory's endpoint $\text{end}(T)$ is closer to $\tilde{s}$ than to all other sites in S. Thus, the proximity-based attack function f_p returns the site closest to $\text{end}(T)$ as the candidate site:

$$f_p(S, T) := \arg\min_{s \in S} d(\text{end}(T), s), \tag{3.1}$$

where the function d is the Euclidean distance. Note that this attack function can easily be extended to return the n-nearest sites to $\text{end}(T)$ to introduce some tolerance.

A second aspect that can be used to infer the destination is the direction in which the trajectory is headed. Specifically, the trajectory's last segment e is of interest. However, the segment e, most likely, does not point exactly toward $\tilde{s}$. Therefore, a V-shaped region $R(t)$ anchored in $\text{end}(T)$ that is mirror-symmetric with respect to e and has an infinite radius is considered (Fig. 3.1). The opening angle of the V-shape is fixed and denoted with α.

The direction-based attack function f_d returns the sites that are inside $R(T)$ as its candidate set:

$$f_d(S, T) := s \cap R(T). \tag{3.2}$$

The attacks f_p and f_d are simple, yet common sense suggests that they may be fairly successful. Figure 3.2 displays the geometric S-TT algorithm on an example trajectory. In the starting situation (Fig. 3.2a), the closest site to $\text{end}(T)$ is $\tilde{s}$. Thus, the algorithm suppresses the endpoint of T until the site closest to its endpoint is not part of the protection set $C(\tilde{s})$ of $\tilde{s}$ anymore (Fig. 3.2b). At this point, some, but not all, of the sites in $C(\tilde{s})$ are part of the candidate set for the direction-based attack (Fig. 3.2b), violating k-site-unidentifiability for this attack function. Thus, truncation continues until either all the sites or none of the sites in $C(\tilde{s})$ are part of

Fig. 3.1 Schematic representation of a direction-based attack. The sites (round dots) inside $R(T)$ (colored black) are candidate sites this attack function returns. $R(T)$ is based on the terminating segment e of trajectory T

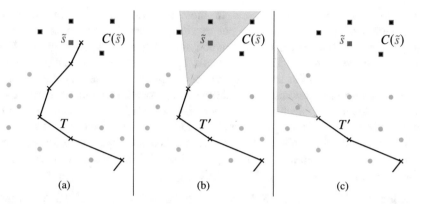

Fig. 3.2 Schematic illustration of the geometric S-TT algorithm. (**a**) The original trajectory T leads to a destination site $\tilde{s}$. This destination has to be hidden among the sites in the protection set $C(\tilde{s})$, depicted as squares. All other sites are shown as gray points. The S-TT algorithm suppresses the trajectory's endpoint until neither (**b**) the site closest to the endpoint is part of the protection set $C(\tilde{s})$ nor (**c**) a V-shaped region that is aligned with the last segment of the truncated trajectory T' contains some, but not all, of the sites in $C(\tilde{s})$

the candidate set of f_d anymore (Fig. 3.2c). Note that at this point, neither f_p nor f_d allow for a distinction between the sites in $C(\tilde{s})$. Thus, k-site-unidentifiability is preserved, and the algorithm is done.

3.2.3   Experimental Results

In this section, experimental results carried out with an implementation of the geometric-based S-TT algorithm are presented. The evaluation focuses on the algorithm's utility and the amount of data lost during anonymization. In this experimental study, it is assumed that the origin and destination of the trajectories were sensitive by default and should be protected under k-site-unidentifiability. For this, the S-TT algorithm is applied to both ends of the trajectories.

Input Data A dataset consisting of 10,927 synthetically generated trajectories in Greater Helsinki, Finland, is used for the evaluation. The trajectories are generated in a three-step process. First, the trajectory endpoints are randomly sampled using a weighted random selection algorithm with the population density as weight. This means that locations in densely populated areas are more likely to be selected as endpoints than locations in sparsely populated areas. In the second step, the endpoints are connected to the road network using a grid-based version of Dijkstra (1959)'s shortest path algorithm. Finally, using generic Dijkstra's algorithm, the shortest path between the two points on the road network is computed and used to connect the two points to a full trajectory.

Selection of the Protection Sets In most scenarios, the sites of origin and destination of the trajectories are not explicitly given. For truncation, the S-TT algorithm does not require the destination site $\tilde{s}$ but only its protection set $C(\tilde{s})$. A naive approach to selecting $C(\tilde{s})$ would be to select the site cluster belonging to the site closest to the trajectory's endpoint. However, trajectory data is prone to positioning uncertainties. Using the naive approach, these uncertainties could lead to selecting a protection set that does not contain $\tilde{s}$, severely diminishing the anonymization quality. A circular buffer of radius b around the endpoint is used to account for these uncertainties. The protection sets of all protection cells intersecting this buffer are unioned into one large protection set. This unioned protection set is then used as the protection set for the trajectory.

Evaluation The geometric S-TT algorithm, as outlined in Sect. 3.2.2, has three major configuration parameters: the minimum size of the protection sets k, the buffer radius b, and the opening angle α for the direction-based attack function. In the following, the effect of these three parameters is analyzed.

Figure 3.3 displays the mean length $\bar{\delta}_{supp}$ of the truncated trajectory parts for different parameterizations. The parameters k and b influence the size of the protection cell. Raising these parameters has a significant impact on the results, with the suppressed length of the trajectories increasing almost at a linear rate (Fig. 3.3a and c). While these results show that small values for k and b preserve more of the data, the choice of these parameters mainly depends on the application and the needed level of anonymity. For example, to eliminate the effect of GNSS accuracies, values for b of approximately 50 meters should be sufficient, while trajectory datasets with a higher inaccuracy need a larger buffer radius.

Regarding the parameterization of the V-shaped region for the direction-based attack, the results indicate that the highest data loss lies around $\alpha = 40°$ to $60°$ (Fig. 3.3b). The reason for this non-monotone behavior is that the direction-based attack function f_d can be satisfied by two conditions: either all or none of the sites in the protection set must be covered by the V-shaped region. In the case that α is

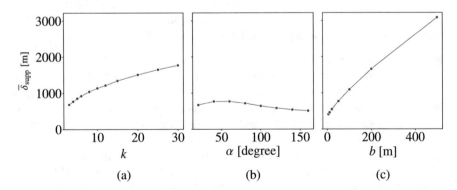

Fig. 3.3 Mean length of suppressed trajectory parts over different parametrization (**a**) of k, $\alpha = 60°$, $b = 50$m; (**b**) of α, $k = 4$, $b = 50$m; and (**c**) of b, $k = 4$, $\alpha = 60°$

very small, and therefore the V-shaped region is very narrow, it is very likely that the V-shape covers none of the sites in the protection set. Analogously, if α is very large, the V-shape is very broad, and all the sites are likely to be covered. Comparing the influence of α to the protection set parameters k and b, α has a significantly smaller impact on $\overline{\delta}_{supp}$.

For each suppressed trajectory point, the attack function that triggered the suppression is stored to evaluate the importance of the two attack functions against each other. Given the parameters $k = 4$, $\alpha = 60°$, and $b = 0$ m, the direction-based attack function did not trigger the suppression of any trajectory points in 24% of the cases. In other words, the algorithm stopped truncating the trajectories at the first trajectory point located outside their protection cell. Likewise, the share of trajectory points suppressed due to the direction-based criterion was 38%. This means that 62% of the suppressed trajectory points were located in the protection cells.

3.3 Map Matching for Semi-restricted Trajectories

Map matching is the process of pinpointing a trajectory to the path in an underlying network that explains the observed measurements best. Map matching is often the first step of trajectory data processing, as there are several benefits when dealing with paths in a known network instead of raw location measurement data:

- Location measurements are usually imprecise. Thus, constraining the trajectory to a path in a network is useful to get a more faithful movement representation.
- Storing raw data is memory-intensive (especially with high sampling densities). On the other hand, paths in a given network can be stored very compactly; see also Sect. 3.4.
- Matching a trajectory to the road network enables data mining techniques that link attributes of the road network to attributes of the trajectories. This is used to, e.g., deduce routing preferences from given trajectories; see also Sect. 3.5.

However, suppose the assumption that a given trajectory was derived from restricted movement in a certain network is incorrect. In that case, map matching might heavily distort the trajectory and erase important semantic characteristics. For example, there could be two trajectories of pedestrians who met in the middle of a market square, arriving from different directions. After map matching, not only the aspect that the two trajectories got very close at one point would be lost, but the visit to the market square would be removed completely (if there are no paths across it in the given network). Not applying map matching might result in misleading results as well, as the parts of the movement that actually happened in a restricted fashion might not be discovered and—as outlined above—having to store and query huge sets of raw trajectories is undesirable.

Hence, the goal is to design an approach that allows for sensible map matching of trajectories that possibly contain on- and off-road sections, also referred to as semi-restricted trajectories. In the following, an algorithm introduced by Behr et al. (2021) is reviewed that deals with this problem.

3.3.1 Methodology

The algorithm is based on a state transition model where each point of a given trajectory is represented by a set of matching candidates—the possible system states, i.e., possible positions of a moving subject. This is similar to hidden Markov model (HMM)-based map-matching algorithms (Haunert and Budig 2012; Koller et al. 2015; Newson and Krumm 2009), which aim to find a sequence of positions maximizing a product of probabilities. In contrast, the presented algorithm minimizes the sum of energy terms of a carefully crafted model.

Input The algorithm works on the spatial components of a given trajectory $T = \langle p_1, \ldots, p_k \rangle$ of k points in $\mathbb{R}^2$, referred to as GPS points in the following.

As base input, we are given a directed, edge-weighted graph $G(V, E)$ that models the underlying transport network. Every directed edge $uv \in E$ corresponds to a directed straight-line segment representing a road segment with an allowed direction of travel. For an edge e, let $w(e)$ be its *weight*.

Additionally, we assume to be given a set of open spaces such as public squares, parks, or parking lots represented as polygonal areas. The given transport network is extended by triangulating all open spaces and adding tessellation edges as arcs into the graph (Fig. 3.4).

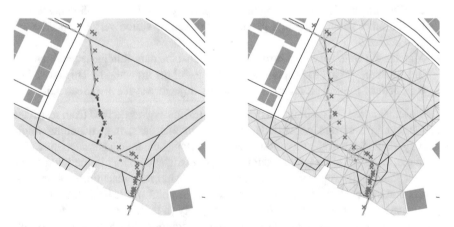

Fig. 3.4 Left: Movement data (blue crosses) on the open space (green) cannot be matched appropriately (dashed orange). An unmatched segment (dashed red) remains. Right: Extended network where open spaces are tessellated to add appropriate off-road candidates. Note that the green polygon has a hole, which remains untessellated since it is not open space

To accurately represent polygons of varying degrees of detail and to provide an appropriate number of off-road candidates, CGAL's meshing algorithm (Boissonnat et al. 2000) is used with an upper bound on the length of the resulting edges and a lower bound on the angles inside triangles.

System States For every p_i, a set of matching *candidates* in the given network is computed by considering a disk D_i of prescribed radius r around p_i and extracting the set of *nodes* $V_i \subseteq V$ within the disk. Furthermore, tessellation nodes (i.e., triangle corners) inside the disk are considered possible matches and included in V_i. Finally, the GPS point itself is added as a candidate to have a fallback. The optimization model ensures that input points are only chosen as output points if all network and tessellation candidates are too far away to be a sensible explanation for this measurement.

Energy Model To ensure that the output path P matches well to the trajectory, an energy function is set up that aggregates a state-based and transition-based energy:

- The *state-based energy* is $\sum_{i=1}^{k} \| p_i - v_i \|^2$, meaning that the energy increases quadratically with the Euclidean distance between a GPS point p_i and the matching candidate v_i selected for it.
- The *transition-based energy* is $\sum_{i=1}^{k-1} w(P_{i,i+1})$, where $P_{a,b}$ is a minimum-weight v_a-v_b-path in G and $w(P)$ is the total weight of a path P (i.e., the sum of the weights of the edges of P).

Note that we have three different sets of edges in the graph: the original edges of the road network E_r, the edges incident to unmatched candidate nodes E_u, and the off-road edges on open spaces E_t. On-road matches are preferred over off-road matches, while unmatched candidates are used as a fallback solution and thus should only be selected if no suitable path over on- and off-road edges can be found. To model this, the energy function is adapted by changing the edge weighting w of the graph. Two weighting terms α_t and α_u are introduced that scale the weight $w(e)$ of each edge e in E_t or E_u, respectively.

The state-based and transition-based energies are subsequently aggregated using a weighted sum, parametrized with a parameter α_c. This yields the overall objective function quantifying the fit between a trajectory $\langle p_1, \ldots, p_k \rangle$ and an output path defined with the selected sequence $\langle v_1, \ldots, v_k \rangle$ of nodes:

$$\text{Minimize} \quad \mathcal{E}(\langle p_1, \ldots, p_k \rangle, \langle v_1, \ldots, v_k \rangle) = \alpha_c \cdot \sum_{i=1}^{k} \| p_i - v_i \|^2 + \sum_{i=1}^{k-1} w(P_{i,i+1})$$

(3.3)

To favor matches in the original road network and keep unmatched candidates as a fallback solution, $1 < \alpha_t < \alpha_u$ should hold. Together with the weighting factor α_c for the state-based energy, the final energy function thus comprises three different weighting factors.

Matching Algorithm An optimal path, minimizing the introduced energy term, is computed using k runs of Dijkstra's algorithm on a graph that results from augmenting G with a few auxiliary nodes and arcs. More precisely, an incremental algorithm that proceeds in k iterations is used. In the ith iteration, it computes for the subtrajectory $\langle p_1, \ldots, p_i \rangle$ of T and each matching candidate $v \in V_i$ the objective value $\mathcal{E}_i^v$ of a solution $\langle v_1, \ldots, v_i \rangle$ that minimizes $\mathcal{E}(\langle p_1, \ldots, p_i \rangle, \langle v_1, \ldots, v_i \rangle)$ under the restriction that $v_i = v$. This computation is done as follows. For $i = 1$ and any node $v \in V_1$, $\mathcal{E}_1^v$ is simply the state-based energy for v, i.e., $\mathcal{E}_1^v = \alpha_c \cdot \| p_1 - v \|^2$. For $i > 1$, a dummy node s_i and a directed edge $s_i u$ for each $u \in V_{i-1}$ whose weight we set as $\mathcal{E}_{i-1}^u$ are introduced. With this, for any node $v \in V_i$, $\mathcal{E}_i^v$ corresponds to the weight of a minimum-weight s_i-v-path in the augmented graph, plus the state-based energy for v. Thus, for s_i and every node in V_i, a minimum-weight path needs to be found. All these paths can be found with one single-source shortest path query with source s_i and, thus, with a single execution of Dijkstra's algorithm.

3.3.2 Experimental Results

The experimental region is the area around Lake Constance, Germany. For this region, data is extracted from OSM to build a transport network feasible for cyclists and pedestrians. In total, a graph with 931,698 nodes and 2,013,590 directed edges is extracted. The open spaces used for tessellation are identified by extracting polygons with special tags. Spaces with unrestricted movement (e.g., parks) and *obstacles* (e.g., buildings) are identified by lists of tags representing these categories, respectively. Following these tags, 6827 polygons representing open spaces are extracted. Including tessellation edges, the final graph for matching consists of 1,148,213 nodes and 3,345,426 directed edges.

The quality of the approach is analyzed using 58 cycling or walking trajectories. Off-road sections were annotated manually to get a ground truth. The length of the trajectories varies from 300 meters to 41.3 kilometers, and overall, they contain 66 annotated off-road segments.

The energy model parameters were set to $a_u = 10$, $a_t = 1.1$, and $a_c = 0.07$. Using these values, the precision and recall for off-road section identification were both close to 1.0. A sensitivity study revealed that similar results are achieved for quite large parameter ranges. Furthermore, movement shapes are far better preserved with the presented approach compared to using map matching only on the transport network. This leads to the conclusion that the combined graph consisting of the transport network plus tessellation nodes and edges is only about 50% larger than the road network, but enables a faithful representation of restricted and unrestricted movement at the same time. This then allows applying storage and indexing methods that demand the movement to happen in an underlying network to be also applicable to semi-restricted trajectories.

3.4 Indexing and Querying of Massive Trajectory Sets

Storing and indexing huge trajectory sets in an efficient manner is the very basis for large-scale analysis and mining tasks. An important goal is to compress the input to deal with millions or even billions of trajectories (which individually might consist of hundreds or even thousands of time-stamped positions). But at the same time, we want to be able to retrieve specific subsets of the data (e.g., all trajectories intersecting a given spatiotemporal range) without the necessity to decompress huge parts of the data.

In this section, we present a novel index structure called PATHFINDER that allows to answer range queries on map-matched trajectory sets on an unprecedented scale. Current approaches for the efficient retrieval of trajectory data make use of different dedicated data structures for the two main tasks, compression and indexing. In contrast, our approach elegantly uses an augmented version of the so-called contraction hierarchy (CH) data structure (Geisberger et al. 2012) for both of these tasks. CH is typically used to accelerate route planning queries, but has also proved successful in other settings like continuous map simplification (Funke et al. 2017). This saves space and makes our algorithms relatively simple without the need of too many auxiliary data structures. Only this slenderness allows for scalability to continent-sized road networks and huge trajectory sets. Other existing indexing schemes only work on small network sizes (e.g., PRESS (Song et al. 2014), network of Singapore; PARINET (Sandu Popa et al. 2011), cities of Stockton and Oldenburg; TED (Yang et al. 2018), cities of Singapore and Beijing) or moderate sizes (e.g., SPNET (Krogh et al. 2016): network of Denmark with 800k vertices). Our approach efficiently deals with networks containing millions of nodes and edges.

3.4.1 Methodology

Throughout this section, we assume to be given a trajectory collection $\mathcal{T}$ that already was map matched to an underlying directed graph $G(V, E)$ with V embedded in $\mathbb{R}^2$. We also assume the edges in the graphs to have non-negative costs $c : E \rightarrow R^+$. Accordingly, each element in $t \in \mathcal{T}$ is a path $\pi = v_0 v_1 \ldots v_k$ in G annotated with timestamps $\tau_0, \tau_1, \ldots, \tau_k$.

The goal is to construct an index for $\mathcal{T}$ which allows to efficiently answer queries of the form $[x_l, x_u] \times [y_l, y_u] \times [\tau_l, \tau_u]$ that aim to identify all trajectories which in the time interval $[\tau_l, \tau_u]$ traverse the rectangular region $[x_l, x_u] \times [y_l, y_u]$. In the literature, this kind of query is often named *window* query (Yang et al. 2018) or *range* query (Krogh et al. 2016), where the formal definitions may differ in detail; also see Zheng (2015).

Contraction Hierarchies Our algorithms heavily rely on the *contraction hierarchy* (CH) (Geisberger et al. 2012) data structure, which was originally developed to speed up shortest path queries. A nice property of CH is that, as a by-product, it also constructs compressed representations of shortest paths.

The CH augments a given weighted graph $G(V, E, c)$ with *shortcuts* and *node levels*. The elementary operation to construct *shortcuts* is the so-called node contraction, which removes a node v and all of its adjacent edges from the graph. To maintain the shortest path distances in the graph, a shortcut $s = (u, w)$ is created between two adjacent nodes u, w of v if the only shortest path from u to w is the path uvw. We define the cost of the shortcut to simply be the sum of the costs of the replaced edges, i.e., $c(s) = c(uv) + c(vw)$. The construction of the CH is the successive contraction of all $v \in V$ in some order; this order defines the *level* $l(v)$ of a node v. The order in which nodes are contracted strongly influences the resulting speed-up for shortest path queries, and hence, many ordering heuristics exist. In our work, we choose the probably most popular heuristic: nodes with low *edge difference*, which is the difference between the number of added shortcut edges and the number of removed edges when contracting a node, are contracted first. We also allow the simultaneous contraction of non-adjacent nodes. As a result, the maximum level of even a continent-sized road network like the one of Europe never exceeds a few hundred in practice. The final CH data structure is defined as $G(V, E^+, c, l)$ where E^+ is the union of E and all shortcuts created.

We also define the nesting depth $nd(e)$ of an edge $e = (v, w)$. If e is an original edge, then $nd(e) = 0$. Otherwise, e is a shortcut replacing edges e_1, e_2, and we define its nesting depth $nd(e) := \max\{nd(e_1), nd(e_2)\} + 1$. Clearly, the nesting depth is upper bounded by the maximum level of a node in the network.

Compression Given a pre-computed CH graph, we construct a CH representation for each trajectory $t \in \mathcal{T}$, that is, we transform the path $\pi = e_0 e_1 \ldots e_{k-1}$ with $e_i \in E$ in the original graph into a path $\pi' = e'_0 e'_1 e'_2 \ldots e'_{k'-1}$ with $e'_i \in E^+$ in the CH graph.

Our algorithm to compute a CH representation is quite simple: We repeatedly check if there is a shortcut bridging two neighboring edges e_i and e_{i+1}. If so, we substitute them with the shortcut. We do this until there are no more such shortcuts. See Fig. 3.5 for an example.

Note that uniqueness of the CH representation can be proven, and therefore, it does not matter in which order neighboring edges are replaced by shortcuts. The running time of that algorithm is linear in the number of edges. Note that by switching to the CH representation, we can achieve a considerable compression rate in case the trajectory is composed of few shortest paths.

Spatial Indexing and Retrieval The general idea is to associate a trajectory with all edges of its *compressed representation* in E^+. Only due to that compression, it becomes feasible to store a huge number of trajectories within the index. Answering a spatial query then boils down to finding all edges of the CH for which a corresponding path in the original graph intersects the query rectangle. Typically, an additional query data structure would be used for that purpose. Yet, we show how to utilize the CH itself as a geometric query data structure. This requires two central definitions:

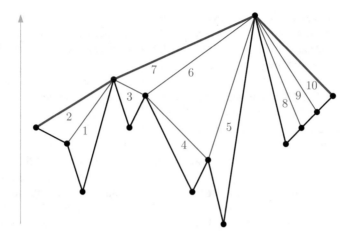

Fig. 3.5 Original path (black, 13 edges) and derivation of its CH representation (bold blue, 3 edges) via repeated shortcut substitution (in order according to the numbers). y-coordinate corresponds to CH level

- With $PB(e)$, we denote the *path box* of an edge e. It is defined as the bounding box for the path that e represents in the original graph G in case $e \in E^+$ is a shortcut or simply the bounding box for the edge e if $e \in E$.
- We define the *downgraph box* $DB(v)$ of a node v as the bounding box of all nodes that are reachable from v on a down-path (only visiting nodes of decreasing CH level), ignoring the orientation of the edges.

Both $PB(e)$ and $DB(v)$ can be computed for all nodes and edges in linear time via a bottom-up traversal of the CH in a preprocessing step and *independent* of the trajectory set to be indexed.

For a spatial-only window query with query rectangle Q, we start traversing the CH level by level in a top-down fashion, first inspecting all nodes which do not have a higher-level neighbor (note that there can be several of them in case the graph is not a single connected component). We can check in constant time for the intersection of the query rectangle and the downgraph box of a node, only continuing with children of nodes with non-empty intersection. We call the set of nodes with non-empty intersection V_I. The set of candidate edges E_Q are then all edges adjacent to a node in V_I.

Having retrieved the candidate edges, we have to filter out edges e for which only the path box $PB(e)$ intersects but not the path represented by e. For this case, we recursively unpack e to decide whether e (and the trajectories associated with e) has to be reported. As soon as one child edge reports a non-empty intersection in the recursion, the search can stop, and e must be reported. We call the set of edges that results from this step E_r. Our final query result is all the trajectories which are referenced at least once by the retrieved edges, i.e., those $t \in \bigcup_{e \in E_r} \mathcal{T}_e$.

Temporal Indexing Timestamps of a trajectory t are annotations to its nodes. In the CH representation of t, we omit nodes via shortcuts, hence losing some temporal information. Yet, PATHFINDER will always answer queries conservatively, i.e., returning a superset of the exact result set.

Like the spatial bounding boxes $PB(e)$, we store time intervals to keep track of the earliest and latest trajectory passing over an edge. Similar to $DB(v)$, we compute minimal time intervals containing all time intervals associated with edges on a down-path from v. This allows us to efficiently answer queries which specify a time interval $[\tau_l, \tau_u]$. Like the spatial bounding boxes, we use these time intervals to prune tree branches when they do not intersect the time interval of the query.

An edge is associated with a set of trajectories, each of which we could check for the time when the respective trajectory traverses the edge. A more efficient approach is to store for all trajectories traversing an edge their time intervals in a so-called interval tree (Berg et al. 2008). This allows us to efficiently retrieve the associated trajectories matching the time interval constraint of the query for a given edge. An interval tree storing ℓ intervals has space complexity $O(\ell)$, can be constructed in $O(\ell \log \ell)$, and can retrieve all intervals intersecting a given query interval in time $O(\log \ell + o)$ where o is the output size.

3.4.2 Experimental Results

As input graph data, we extracted the road and path network of Germany from OpenStreetMap. The network has about 57.4 million nodes and 121.7 million edges. CH preprocessing took only a few minutes and roughly doubled the number of edges.

For evaluation, we considered all trajectories within Germany from the OpenStreetMap collection, only dropping low-quality ones (e.g., due to extreme outliers, non-monotonous timestamps, etc.). As a result, we obtained 350 million GPS measurements which were matched to the Germany network with the map matcher from Seybold (2017) to get a dataset with 372,534 trajectories which we call $\mathcal{T}_{\text{ger,real}}$.

Our experiments reveal that on average, a trajectory from our dataset can be represented by 11 shortest paths in the CH graph. While the original edge representation of $\mathcal{T}_{\text{ger,real}}$ consists of 121.8 million edges (992 MB on disk), the CH representation only requires 13.8 million edges (112 MB on disk). The actual compression for the 372k trajectories took 42 seconds, that is, around 0.1ms per trajectory.

To test window query efficiency, we used rectangles of different sizes and different scales of time intervals. Across all specified queries, PATHFINDER was orders of magnitude faster than its competitors, including the naive linear scan baseline but also an inverted index that also benefits from the computed CH graph. Indeed, queries only take a few microseconds per trajectory in the output set.

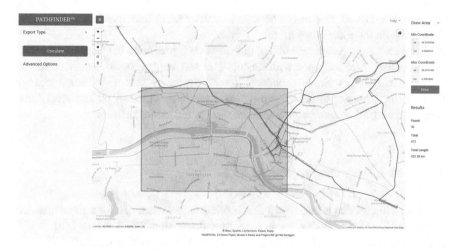

Fig. 3.6   Visualization of trajectory data as developed in Rupp et al. (2022)

Furthermore, the CH-based representation also allows reporting aggregated results easily, which is especially useful for the efficient visualization of large result sets; see, for example, Fig. 3.6, which is a screenshot of our visualization tool presented in Rupp et al. (2022).

3.5   Preference-Based Trajectory Clustering

It is well observable in practice that drivers' preferences are not homogeneous. If we have two alternative paths π_1, π_2 between a given source-target pair, characterized by 3 costs/metrics (travel time, distance, and ascent along the route) each, e.g., $c(\pi_1) = (27\,\text{min}, 12\,\text{km}, 150\,\text{m})$ and $c(\pi_2) = (19\,\text{min}, 18\,\text{km}, 50\,\text{m})$, there are most likely people who prefer π_1 over π_2 and vice versa. The most common model to formalize these preferences assumes a linear dependency on the metrics. This allows for *personalized route planning*, where a routing query consists of not only source and destination but also a weighting of the metrics in the network.

The larger a set of paths is, though, the less likely it is that a single preference/weighting exists which explains all paths, i.e., for which all paths are optimal. One might, for example, think of different driving styles/preferences when commuting versus leisure trips through the countryside. So, a natural question to ask is as follows: what is the minimum number of preferences necessary to explain a set of given paths in a road network with multiple metrics on the edges? This can also be interpreted as a trajectory clustering task, where routes are to be classified according to their purpose. In our example, one might be able to differentiate between commute and leisure. Or in another setting, where routes of different drivers are analyzed, one might be able to cluster them into speeders and cruisers depending on the routes they prefer.

A use case from recent research of our methods is presented in Chap. 11. In a field study, cyclists were equipped with sensors tracking their exposure to environmental stressors. The study's aim is to evaluate to what degree cyclists are willing to change their routing behavior to decrease their exposure to these stressors. The algorithms presented in the following section can be used to aid this evaluation.

3.5.1 A Linear Preference Model

The following section is based on a *linear* preference model, i.e., for a given directed graph $G(V, E)$, for every edge $e \in E$, a d-dimensional cost vector $c(e) \in \mathbb{R}^d$ is given, where $c_1(e), c_2(e), \ldots, c_d(e) \geq 0$ correspond to non-negative quantities like travel time, distance, non-negative ascent, etc., which are to be minimized. A path $\pi = e_1 e_2 \ldots e_k$ in the network then has an associated cost vector $c(\pi) := \sum_{i=1}^{k} c(e_i)$.

A preference to distinguish between different alternative paths is specified by a vector $\alpha \in [0, 1]^d$, $\sum \alpha_i = 1$. For example, $\alpha^T = (0.4, 0.5, 0.1)$ might express that the respective driver does not care much about ascents along the route, but considers travel time and distance similarly important. Alternative paths π_1 and π_2 are compared by evaluating the respective scalar products of the cost vectors of the path and the preference, i.e., $c(\pi_1)^T \cdot \alpha$ and $c(\pi_2)^T \cdot \alpha$. Smaller scalar values in the linear model correspond to a preferred alternative. An st-path π (which is a path with source s and target t) is optimal for a fixed preference α if no other st-path π' exists with $c(\pi')^T \cdot \alpha < c(\pi)^T \cdot \alpha$. Such a path is referred to as α-*optimal*.

From a practical point of view, it is very unintuitive (or rather almost impossible) for a user to actually express their driving preferences as such a vector α, even if they are aware of the units of the cost vectors on the edges. Hence, it would be very desirable to be able to *infer* their preferences from paths they like or which they have traveled before. In Funke et al. (2016), a technique for preference inference is proposed, which essentially instruments linear programming to determine an α for which a given path π is α-optimal or certify that none exists. Given an st-path π in a road network, in principle, their proposed LP has non-negative variables $\alpha_1, \ldots, \alpha_d$ and one constraint for each st-path π' which states that for the α we are after, π' should not be preferred over π". So the LP looks as follows:

$$
\begin{aligned}
\max \quad & \alpha_1 \\
\forall st\text{-paths } \pi' : \quad & \alpha^T(c(\pi) - c(\pi')) \leq 0 \qquad \textit{optimality constraints} \\
& \alpha_i \geq 0 \qquad \textit{non-negativity constraints} \\
& \sum \alpha_i = 1 \qquad \textit{scaling constraint}
\end{aligned}
$$

As such, this linear program is of little use, since typically, there is an exponential number of st-paths, so just writing down the complete LP seems infeasible. Fortunately, due to the equivalence of optimization and separation, it suffices to have an algorithm at hand which—for a given α—decides in polynomial time whether all constraints are fulfilled or, if not, provides a violated constraint (such an algorithm is called a *separation oracle*). In our case, this is very straightforward: we simply compute (using, e.g., Dijkstra's algorithm) the optimum st-path for a given α. If the respective path has better cost (w.r.t. α) than π, we add the respective constraint and resolve the augmented LP for a new α; otherwise, we have found the desired α. Via the ellipsoid method, this approach has polynomial running time; in practice, the dual simplex algorithm has proven to be very efficient.

3.5.2 Driving Preferences and Route Compression

The method to infer the preference vector α from a trajectory presented in Sect. 3.5.1 requires that the given trajectory is optimal for at least one preference value α. In practice, for many trajectories, this is not the case. One prominent example is round trips, which are often recorded by recreational cyclists to share with their community. The corresponding trajectories end at the same location as they started. As such, the optimal route for any preference value with respect to the linear model is to not leave the starting point at all, thus having a cost of zero. Note, however, that this problem of not being optimal for any preference value also occurs for many one-way trips. In this section, we review a method by Forsch et al. (2022) that allows us to infer the routing preferences from these trajectories as well. The approach is loosely based on the minimum description length principle (Rissanen 1978), which states that the less information is needed to describe a dataset, the better it is represented. Applied to trajectories, the input trajectory, represented as a path in the road network, is segmented into as few as possible subpaths, such that each of these subpaths is α-optimal. In the following, a compression-based algorithm for the bicriteria case, i.e., when considering two cost functions, is presented. The algorithm is evaluated on cycling trajectories by inferring the importance of signposted cycling routes. Additionally, the trajectories are clustered using the algorithm's results.

3.5.2.1 Problem Formulation

In the bicriterial case, the trade-off between two cost functions c_0 and c_1 for routing is considered. The linear preference model as described in Sect. 3.5.1 simplifies to $c_\alpha = \alpha_0 \cdot c_0 + \alpha_1 \cdot c_1$, which can be rewritten to:

$$c_\alpha = (1 - \alpha_1) \cdot c_0 + \alpha_1 \cdot c_1.$$

Thus, only a single preference value α_1 is present, which is simply denoted with α in the remainder of this section. For each path, the α interval for which the path is α-optimal is of interest. This interval is called the *optimality range* of the given path. The search for the optimality range is formalized in the following problem:

> Given a graph G with two edge cost functions c_0 and c_1, and a path π, find $\mathcal{I}_{\text{opt}} = \{\alpha \in [0, 1] \mid \pi \text{ is } \alpha\text{-optimal}\}$.

Many paths are not optimal for any α, and as such, their optimality range will be empty. For these paths, a minimal milestone segmentation and the corresponding preference value are searched. A *milestone segmentation* is a decomposition of a path π into the minimum number of subpaths $\{\pi_1, \ldots, \pi_h\}$, such that every subpath π_i with $i \in \{1, \ldots, h\}$ is α-optimal for the given α. Thus, the following optimization problem is solved:

> Given a graph G with two edge cost functions c_0 and c_1 and a path π, find $\alpha \in [0, 1]$ and a milestone segmentation of π with respect to α that is as small as possible. That is, there is no $\tilde{\alpha}$ that yields a milestone segmentation of π with a smaller number of subpaths.

The splitting points between the subpaths are called *milestones*. A path π in G can be fully reconstructed given its starting point and endpoint, α, and the corresponding milestones. By minimizing the number of milestones, we achieve the largest compression of the input data over all α values and thus, according to the minimum description length principle, found the preference value that best describes the input path.

3.5.2.2   Solving Milestone Segmentation

Using a separation oracle as described in Sect. 3.5.1, the optimality range of every (sub)path of π is computed in $O(\text{SPQ} \cdot \log(Mn))$ time, where SPQ is the running time of a shortest path query in G, n is the number of vertices in G, and M is the maximum edge cost among all edges regarding c_0 and c_1. For a more detailed description of the algorithm, we refer to Forsch et al. (2022).

Retrieving the milestone segmentation with the smallest number of subpaths is done by first computing a set that contains a milestone segmentation of π for every α and then selecting the optimal milestone segmentation. Existing works in the field of trajectory segmentation use *start-stop matrices* (Alewijnse et al. 2014; Aronov et al. 2016) to evaluate the segmentation criterion. In our case, α-optimality is the segmentation criterion, and the segmentation is not solved for a single α, but the α value minimizing the number of subpaths over all $\alpha \in [0, 1]$ is searched. Therefore, the whole optimality interval for the corresponding subpath is stored in the start-stop matrix $\mathcal{M}$. This allows us to use the same matrix for all α values. Figure 3.7 shows the (Boolean) start-stop matrix for a single α, retrieved by applying the expression $\alpha \in \mathcal{M}[i, j]$ to all elements of the matrix. For a path π consisting of k vertices, a $(k \times k)$-matrix $\mathcal{M}$ of subintervals of $[0, 1]$ is considered. The entry $\mathcal{M}[i, j]$ in row i and column j corresponds to the optimality range of the subpath of π starting at its

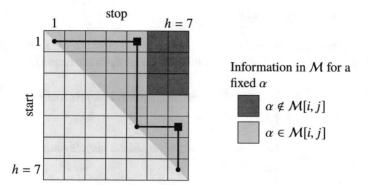

Fig. 3.7 Depiction of a start-stop matrix of a path consisting of $h = 7$ vertices for a fixed $\alpha \in$ [0, 1]. The black line represents a minimum segmentation of the path into two subpaths (v_1, v_5, v_7)

ith vertex and ending at its jth vertex. Since we are only interested in subpaths with the same direction as π, we focus on the upper triangle matrix with $i \leq j$.

Due to the optimal substructure of shortest paths (Cormen et al. 2009), for $i < k < j$, both $\mathcal{M}[i, j] \subseteq \mathcal{M}[i, k]$ and $\mathcal{M}[i, j] \subseteq \mathcal{M}[k, j]$ hold. This results in the structure visible in Fig. 3.7, where no red cell is below or left of a green cell. The start-stop matrix is therefore computed starting in the lower-left corner, and each row is filled starting from its main diagonal. That way, the search space for $\mathcal{M}[i, j]$ is limited to the intersection of the already computed values of $\mathcal{M}[i + 1, j]$ and $\mathcal{M}[i, j - 1]$. This is especially useful if one of these intervals is empty, meaning the current interval is empty as well and computation in this row can be stopped.

As a consequence of the substructure optimality explained above, once the start-stop matrix is set up, it is easy to find a solution to the segmentation problem for a fixed α. According to Buchin et al. (2011), for example, an exact solution to this problem can be found with a greedy approach in $O(h)$ time; see Fig. 3.7.

Since only a finite set of intervals is considered, we know that if a minimal solution exists for an $\alpha \in [0, 1]$, it also exists for one of the values bounding the intervals in $\mathcal{M}$. Consequently, we can discretize our search space without loss of exactness, and each of the $O(h^2)$ optimality ranges yields at most two candidates for the solution. For each of these candidates, a minimum segmentation is computed in $O(h)$ time. Thus, we end up with a total running time of $O(h^2 \cdot (h + \text{SPQ} \log(Mn)))$ where n denotes the number of vertices in the graph and h denotes the number of vertices in the considered path. Thus, the algorithm is efficient and yields an exact solution. The solution consists of the intervals of the balance factor producing the best personalized costs with respect to the input criteria.

3.5.2.3 Experimental Results

The experiments are set up for an examination of the extent to which cyclists stick to official bicycle routes. For this, the following edge costs are defined:

$$c_0(e) = \begin{cases} 0, & \text{if } e \text{ is part of an official bicycle route} \\ d(e), & \text{else.} \end{cases}$$

$$c_1(e) = \begin{cases} d(e), & \text{if } e \text{ is part of an official bicycle route} \\ 0, & \text{else.} \end{cases}$$

Hence, in the cost function, edge costs c_0 and c_1 are used that, apart from an edge's geodesic length d, depend only on whether the edge in question is part of an official bicycle route. Using this definition, a minimal milestone segmentation for $\alpha = 0.5$ suggests that minimizing geodesic length outweighs the interest in official cycle paths. A value of $\alpha < 0.5$ indicates avoidance of official cycle paths, whereas $\alpha > 0.5$ indicates the opposite, i.e., preference.

For the analysis, a total of 1016 cycling trajectories from the user-driven platform GPSies[1] are used. The map-matching algorithm described in Sect. 3.3 (without the extended road network) is used to generate paths in the road network, resulting in 2750 paths.[2]

First, the results of the approach on a single trajectory are presented. Figure 3.8 shows the path resulting from map matching the trajectory, as well as a minimal milestone decomposition for $\alpha = 0.55$. In this example, a minimal milestone segmentation is found for $\alpha \in [0.510, 0.645]$, resulting in the milestones a, b, and c. For $\alpha = 0.5$, i.e., the geometric shortest path, a milestone segmentation takes an additional milestone between a and b. For this specific trajectory, the minimal milestone segmentation is found for α values greater than 0.5. This relates to a preference toward officially signposted cycle routes. The increase in milestones for $\alpha > 0.645$ indicates the upper boundary on the detour the cyclist is willing to take to stick to cycle ways.

Based on the compression algorithm, the set of trajectories is divided into three groups PRO, CON, and INDIF. The group PRO comprises trajectories for which a milestone segmentation of minimal size is only found for $\alpha > 0.5$. Such a result is interpreted that way that the trajectory was planned in favor of official cycle routes. Conversely, it is considered as avoidance of official cycle routes if milestone segmentations of minimal size exist only for $\alpha < 0.5$. The group CON represents these trajectories. Considering trajectories lacking this clarity, no strict categorization into one of the two groups is done. Instead, these trajectories form a third group, INDIF. The results of this classification are displayed in Fig. 3.9. The

---

[1] www.gpsies.com, [no longer available], downloaded May 2018.

[2] Trajectories are split by the algorithm at unmatched segments.

Fig. 3.8 The user's path has an optimal milestone segmentation into four optimal subpaths (s, a, b, c, t) presented by white nodes. Officially signposted paths are highlighted in green. Disregarding minor shiftings of a and/or b, this is a valid optimal segmentation for each $\alpha \in [0.510, 0.645]$. For $\alpha = 0.5$, i.e., the geometric shortest path, a milestone segmentation takes an additional milestone between a and b

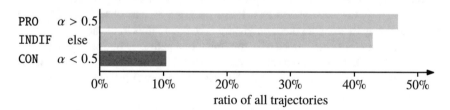

Fig. 3.9 Size of the categories PRO (green, 998 paths), CON (red, 230 paths), and INDIF (gray, 1522 paths)

group PRO being over four times larger than the group CON is a first indicator that cyclists prefer official cycle routes over other roads and paths. The group CON is the smallest group with a share of 8% of all paths. One assumption for this result is that this group mainly consists of road cyclists who prefer using roads over cycle ways. In future research, this could be verified by analyzing a dataset of annotated trajectories.

Overall, more than 50% of the paths are segmented into five α-optimal subpaths or less, resulting in significant compression of the data.

3.5.3 Minimum Geometric Hitting Set

The approach described in Sect. 3.5.1 can easily be extended to decide for a *set* of paths (with different source-target pairs) whether there exists a single preference α for which they are optimal (i.e., which explains this route choice). It does not work, though, if different routes for the same source-target pair are part of the input or simply no single preference can explain all chosen routes. The latter seems quite plausible when considering that one would probably prefer other road types on a leisure trip on the weekend versus the regular commute trip during the week. So, the following optimization problem is quite natural to consider:

> Given a set of trajectories T in a multiweighted graph, determine a set A of preferences of minimal cardinality, such that each $\pi \in T$ is optimal with respect to at least one $\alpha \in A$.

We call this problem *preference-based trajectory clustering* (PTC).

For a concrete problem instance from the real world, one might hope that each preference in the set A then corresponds to a driving style, like speeder or cruiser. Note that while a real-world trajectory often is not optimal for a single α, studies like in Barth et al. (2020) show that it can typically be decomposed into very few optimal subtrajectories if multiple metrics are available.

In Funke et al. (2016), a sweep algorithm is introduced that computes an approximate solution of PTC. It is, however, relatively easy to come up with examples where the result of this sweep algorithm is by a factor of $\Omega(|T|)$ worse than the optimal solution. In Barth et al. (2021), we aim at improving this result by finding practical ways to solve PTC optimally. Our (surprisingly efficient) strategy is to explicitly compute for each trajectory π in T the polyhedron of preferences for which π is optimal and to translate PTC into a geometric hitting set problem.

Fortunately, the formulation as a linear program as described in 3.5.1 already provides a way to compute these polyhedra. The constraints in the above LP exactly characterize the possible values of α for which one path π is optimal. These values are the intersection of half-spaces described by the optimality constraints and the non-negativity constraints of the LP. We call this (convex) intersection *preference polyhedron*.

Using the preference polyhedra, we are armed to rephrase our original problem as a *geometric hitting set (GHS)* problem. In an instance of GHS, we typically have geometric objects (possibly overlapping) in space, and the goal is to find a set of points (a *hitting set*) of minimal cardinality, such that each of the objects contains at least one point of the hitting set. Figure 3.10 shows an example of how preference polyhedra of different optimal paths could look like in the case of three metrics. In terms of GHS, our PTC problem is equivalent to finding a hitting set for the preference polyhedra of minimum cardinality, and the "hitters" correspond to respective preferences. In Fig. 3.10, we have depicted two feasible hitting sets (white squares and black circles) for this instance. Both solutions are minimal in that no hitter can be removed without breaking feasibility. However, the white squares

Fig. 3.10 Example of a geometric hitting set problem as it may occur in the context of PTC. Two feasible hitting sets are shown (white squares and black circles)

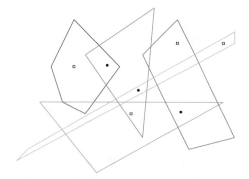

(in contrast to the black circles) do not describe a minimum solution, as one can hit all polyhedra with fewer points.

While the GHS problem allows picking arbitrary points as hitters, it is not hard to see that it suffices to restrict to vertices of the polyhedra and intersection points between the polyhedra boundaries or more precisely vertices in the arrangement of feasibility polyhedra.

The GHS instance is then formed in a straightforward manner by having all the hitting set candidates as ground set and subsets according to containment in respective preference polyhedra. For an exact solution, we can formulate the problem as an integer linear program (ILP). Let $\alpha^{(1)}$, $\alpha^{(2)}$, ..., $\alpha^{(l)}$ be the hitting set candidates and $\mathcal{U} := \{P_1, P_2, \ldots, P_k\}$ be the set of preference polyhedra. We create a variable $X_i \in 0, 1$ indicating whether $\alpha^{(i)}$ is picked as a hitter and use the following ILP formulation:

$$\min \sum_i X_i$$

$$\forall P \in \mathcal{U} : \sum_{\alpha^{(i)} \in P} X_i \geq 1$$

$$\forall i : X_i \in \{0, 1\}$$

While solving ILPs is known to be NP-hard, it is often feasible to solve ILPs derived from real-world problem instances, even of non-homeopathic size.

3.5.3.1 Theoretical and Practical Challenges

In theory, the complexity of a single preference polyhedron could be almost arbitrarily high. In this case, computing hitting sets of such complex polyhedra is extremely expensive. To guarantee bounded complexity, we propose to "sandwich" the preference polyhedron between approximating inner and outer polyhedra of bounded complexity.

Fig. 3.11 Inner (yellow) and
outer approximation (gray) of
the preference polyhedron
(black)

For d metrics, our preference polyhedron lives in $d - 1$ dimensions, so we
uniformly ϵ-sample the unit $(d - 2)$-sphere using $O((1/\epsilon)^{d-2})$ samples. Each of
the samples gives rise to an objective function vector for our linear program; we
solve each such LP instance to optimality. This determines $O((1/\epsilon)^{d-2})$ extreme
points of the polyhedron in equally distributed directions. Obviously, the convex
hull of these extreme points is *contained within* and with decreasing ϵ converges
toward the preference polyhedron. Guarantees for the convergence in terms of ϵ
have been proven before, but are not necessary for our (practical) purposes. We call
the convex hull of these extreme points the *inner approximation* of the preference
polyhedron.

What is interesting in our context is the fact that each extreme point is defined
by $d - 1$ half-spaces. So we can also consider the set of half-spaces that define
the computed extreme points and compute their intersection. Clearly, this half-
space intersection *contains* the preference polyhedron. We call this the *outer
approximation* of the preference polyhedron.

Let us illustrate our approach for a graph with $d = 3$ metrics, so the preference
polyhedron lives in the *two*-dimensional plane; see the black polygon/polyhedron in
Fig. 3.11. Note that we do not have an explicit representation of this polyhedron, but
can only probe it via LP optimization calls. To obtain inner and outer approximation,
we determine the extreme points of this implicitly (via the LP) given polyhedron,
by using objective functions $\max \alpha_1$, $\max \alpha_2$, $\min \alpha_1$, $\min \alpha_2$. We obtain the four
solid red extreme points. Their convex hull (in yellow) constitutes the inner
approximation of the preference polyhedron. Each of the extreme points is defined
by *two* constraints (half-planes supporting the two adjacent edges of the extreme
points of the preference polyhedron). In Fig. 3.11, these are the light green, blue,
dark green, and cyan pairs of constraints. The half-space intersection of these
constraints form the *outer approximation* in gray.

3.5.3.2 Experimental Results

For our experiments, we extracted a weighted graph from OpenStreetMap of the
German state of Baden-Württemberg with the cost types *distance, travel time for
cars*, and *travel time for trucks* containing about 4M nodes and 9M edges. A set of
50 preferences were chosen u.a.r. per instance and created a random source target

Table 3.1 Instance generation and solving for various polyhedra approximations. Car graph with 1000 paths. Time in seconds

Algo.	Polyh. Time	Arr. Time	ILP Sol.	ILP Time
Inner-16	11.3	4.8	45	15.5
Inner-64	26.9	4.9	39	1.8
Exact	6.5	5.2	36	0.7
Outer-64	26.9	5.1	36	0.8
Outer-16	11.3	5.2	36	1.8

optimal for those preferences. Our implementation was evaluated on 24-core Xeon E5-2630 running Ubuntu Linux 20.04.

In Table 3.1, we see the results for a set of 1000 paths. Constructing all the preference polyhedra exactly costs 6.5 seconds and setting up the hitting set instance 5.2 seconds. Solving the hitting set ILP took 0.7 seconds and resulted in a hitting set of size 36. While in theory, the complexity of the preference polyhedra can be almost arbitrarily high, this is not the case in practice. Hence, exact preference polyhedra construction is also more efficient than our proposed approximation approaches (in the table the rows denoted by Inner-XX and Outer-XX), which also only can provide an approximation to the actual hitting set size.

3.6   Visualizing Routing Profiles

In the previous sections, we reviewed algorithms to infer the routing preferences of a user or a group of users from user-generated trajectories. When applying these methods to large trajectory datasets, the trajectories, and thus the users, can be clustered according to their routing behavior. Each cluster has a characteristic routing behavior, expressed as a vector $\alpha \in [0, 1]^d$ of weighting factors of the d different influence factors. On the one hand, these weighting factors are very important for improving route recommendation because they can directly be used to compute optimized future routes. On the other hand, the information on the weighting can be used for planning purposes, e.g., for planning new cycling infrastructure. For this use case, the mathematical representation of the routing preference is not useful, as it is hard to interpret. In this section, we review an alternative representation of routing preferences based on isochrone visualizations geared specifically at a fast and comprehensive understanding of the cyclist's needs. Isochrones are polygons that display the area that is reachable from a given starting point in a specific amount of time and are often used for network analysis, e.g., for public transportation (O'Sullivan et al. 2000) or the movement of electric cars (Baum et al. 2016). For most routing profiles, time is not the only influencing factor. Therefore, the definition of isochrones is slightly altered, and polygons that display areas reachable within a certain amount of *effort* a user is willing to spend instead of a fixed time are used. This effort is dependent on the specific routing preferences.

Even though these polygons do not display times anymore, they will be called isochrones in the following. The visual complexity of the isochrones is reduced by using schematized polygons where the polygon's outline is limited to horizontal, vertical, and diagonal segments. This property is called *octilinearity*. The isochrones created with the presented approach guarantee a formally correct classification of reachable and unreachable parts of the network.

3.6.1 Methodology

The methodology for computing schematic isochrones for specific routing profiles can be split up into two major components: the computation of reachability and the generation of the isochrones.

Computing Reachable Parts of the Road Network The first step for displaying the routing profiles is computing the reachability in the road network, modeled as a directed graph $\mathcal{R}$, for the given profile. Recall the routing model from Sect. 3.5.1, where the personalized cost is the linear combination of the inherent costs in the road network and the user's preference, expressed as α. Thus, given a specific preference, e.g., computed by the methods presented in Sect. 3.5.2 or Sect. 3.5.3, we can compute the personalized costs in our graph. The sum of the personalized costs of the edges along an optimal route from s to t is called the *effort* needed to get from s to t. Shortest path algorithms such as Dijkstra's algorithm (Dijkstra 1959) can be used to compute optimal routes from a given starting location to all other locations in the road network. Such an algorithm is used to compute all nodes in $\mathcal{R}$ that are reachable within a given maximum effort. This information is stored by coloring the reachable nodes blue and the unreachable nodes red. Further, the road network graph $\mathcal{R}$ is planarized to obtain a graph $\overline{\mathcal{R}}$ whose embedding is planar, that is, there are no crossings between two edges. This is done by replacing each edge crossing with an additional node and splitting the corresponding edges at these crossings.

Generating the Isochrones Given the node-colored road network, we proceed by generating the polygon that encloses all reachable nodes, the isochrones. For this, we first extend the node coloring of the graph to an edge coloring by performing the following steps. For each edge e that is incident to both a blue and a red node, we determine the location on e that is still reachable from s; see Fig. 3.12a. At this location, we subdivide e by a blue *dummy node*. A special case occurs when the sum of the remaining distances at two adjacent, reachable nodes u and v is smaller than the length of the edge $e = \{u, v\}$; see Fig. 3.12b. In this case, e is subdivided by two additional dummy nodes, with the middle part being colored red. Further, the coloring and the newly inserted nodes are transferred into the planarized graph $\overline{\mathcal{R}}$, additionally coloring the nodes that have been introduced to planarize $\mathcal{R}$; see Fig. 3.12c. Each such node becomes blue if it subdivides an edge e in $\mathcal{R}$ that is only incident to blue nodes; otherwise, it becomes red. Altogether, we obtain a coloring

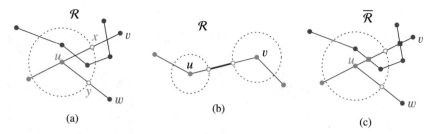

Fig. 3.12 The reachability is modeled by subdividing edges. (**a**) The node x (resp. y) subdivides the edge (u, v) (resp. (u, w)) at the last reachable position. (**b**) Special case: The nodes u and v are reachable from different directions. (**c**) The coloring of $\mathcal{R}$ is transferred to $\overline{\mathcal{R}}$, and the nodes that are introduced for the planarization (squares) are colored

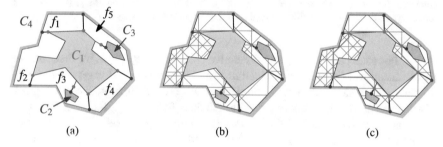

Fig. 3.13 Creating an octilinear polygon enclosing the component C_1 of all reachable nodes and edges. (**a**) The faces $f_1, \ldots, f_5$ surround the component C_1. (**b**) Each face is subdivided by an octilinear grid (Step 1). Furthermore, these grids are connected to one large grid G that is split by the port between f_1 and f_5 (Step 2). (**c**) An octilinear polygon is constructed by computing a bend-minimal path through G (Step 3)

of $\overline{\mathcal{R}}$, defining all nodes either blue or red. Due to the insertion of the dummy nodes, this node coloring induces an edge coloring: edges that are incident to two reachable nodes are also reachable, and all other edges are unreachable.

Given the colored planar graph $\overline{\mathcal{R}}$, we first observe that $\overline{\mathcal{R}}$ has faces that have both red and blue edges. An edge that is incident to both a blue and a red node is called a *gate*. Further, the reachable node of a gate is denoted its *port*. Removing the gates from $\overline{\mathcal{R}}$ decomposes the graph into a set of components $C_1, \ldots, C_\ell$ such that component C_1 is blue and all other components are red. These components are called the *colored components* of $\overline{\mathcal{R}}$. Figure 3.13a shows the gates and the colored components for an example graph.

Given the colored components, we are looking for a single octilinear polygon such that C_1 lies inside and $C_2, \ldots, C_\ell$ lie outside of the polygon. Our method consists of three steps, displayed in Fig. 3.13. In the first step, we create for each pair v_i, v_{i+1} of consecutive ports an octilinear grid G_i contained in f_i. In the second step, we fuse these grids to one large grid G. In the final step, we use G to determine an octilinear polygon by finding an optimal path through G.

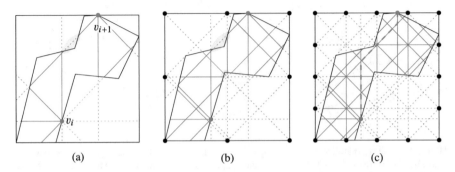

(a)   (b)   (c)

Fig. 3.14 The face f_2 of the example shown in Fig. 3.13 is subdivided by octilinear rays based on vertices on the bounding box. (**a**) Shooting octilinear rays from v_i and v_{i+1} does not yield a connected line arrangement (see yellow highlight). (**b**–**c**) The bounding box of f_1 is successively refined by vertices shooting octilinear rays until they connect v_i and v_{i+1}

Step 1 For each pair v_i, v_{i+1} of consecutive ports with $1 \leq i \leq k$, we first compute an octilinear grid G_i that is contained in f_i and connects v_i and v_{i+1}; see Fig. 3.14. To that end, we shoot from both ports octilinear rays and compute the line segment arrangement $\mathcal{L}$ of these rays restricted to the face f_i; see Fig. 3.14a. If $\mathcal{L}$ forms one component, we use it as grid G_i. Otherwise, we refine $\mathcal{L}$ as follows. We uniformly subdivide the bounding box of f_i by further nodes, from which we shoot additional octilinear rays; see Fig. 3.14b. We insert them into $\mathcal{L}$ restricting them to f_i. We call the number of nodes on the bounding box the *degree of refinement d*. We double the degree of refinement until $\mathcal{L}$ is connected or a threshold $d_{\max}$ is exceeded; see Fig. 3.14c. In the latter case, we also insert the boundary of f_i into $\mathcal{L}$ to guarantee that $\mathcal{L}$ is connected. Later on, when creating the octilinear polygon, we only use the boundary edges of f_i if necessary.

Step 2 In the following, each grid G_i is interpreted as a geometric graph, such that the grid points are the nodes of the graph and the segments connecting the grid points are the edges of the graph. These graphs are unioned into one large graph G. More precisely, G is the graph that contains all nodes and edges of the grids $G_1, \ldots, G_k$. In particular, each port v_i is represented by two nodes x_i and y_i in G such that x_i stems from G_{i-1} and y_i stems from G_i; for $i = 1$, we define $x_1 = v_k$. Two grids G_{i-1} and G_i in G are connected by introducing the directed edge (x_i, y_i) in G for $2 \leq i \leq k$.

Step 3 In the following, let $s := y_1$ and $t := x_1$. A path P from s to t in G is computed such that P has a minimum number of bends, i.e., there is no other path from s to t that has fewer bends. To that end, Dijkstra's algorithm on the linear dual graph of G is used, allowing us to penalize bends in the cost function. In case the choice of P is not unique because there are multiple paths with the same number of bends, the geometric length of the path is used as a tie-breaker, preferring paths that are shorter. As s and t have the same geometric location, the path P forms an octilinear polygon.

In some cases, C_1 contains one or multiple holes, i.e., one or more of the components $C_2, \ldots, C_\ell$ are enclosed by C_1. We deal with this in the following way. For each enclosed component C^*, the road network $\overline{\mathcal{R}}$ is recolored, such that C^* is blue and all other parts of the network are red. We then proceed by computing the enclosing polygon for C^* as outlined for C_1 above and subtracting the resulting polygon from the isochrone of C_1. After this step, the isochrone is guaranteed to contain all parts of the road network that are reachable and none of the parts that are unreachable. We call this the *separation property*.

3.6.2 Application

Recall the classification of cyclists in Sect. 3.5.2. The cyclists are clustered according to their usage of officially signposted cycle routes. Two different classes are introduced, PRO and CON, with preference values of $\alpha_{PRO} > 0.5$ and $\alpha_{CON} < 0.5$. In Fig. 3.15, these two classes are visualized by isochrones with $\alpha = 0.65$ for group PRO and $\alpha = 0.45$ for group CON. In Fig. 3.15a, the isochrones are stretched out very far in the east-west direction along the course of an official cycling route, reflecting the preference of a cyclist in group PRO for officially signposted cycle ways. Moreover, the isochrones are clinched in the north-south direction, revealing a deficit in the road network in this direction for this group of cyclists: while it is possible to get to these areas, the perceived distance will be much longer than in the east-west direction. Figure 3.15b highlights the routing profile for cyclists of the group CON. The isochrones for this routing profile are almost circular, stretching a similar distance in all directions. This is because in most cases, there is a non-signposted road very close to a signposted road, such that avoiding these signposted

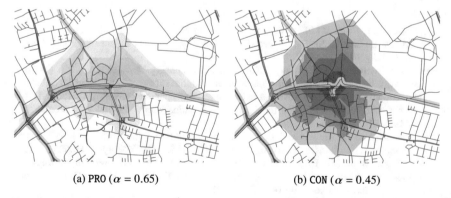

(a) PRO ($\alpha = 0.65$) (b) CON ($\alpha = 0.45$)

Fig. 3.15 Two routing profiles visualized as isochrones. Roads (black) highlighted in blue are part of signposted cycling routes. Displayed are the areas that are reachable for a cyclist in group PRO (**a**) and CON (**b**) within four distinct values of maximum effort (encoded by different brightness values)

roads does not involve large detours and is often as easy as just switching to the road from the existing cycling lane. This routing profile is thus probably much less relevant for planning purposes.

3.7 Conclusion and Future Work

We have presented efficient algorithmic approaches to process and mine huge collections of trajectory datasets and demonstrated their usefulness on volunteered data. The next step is integrating our methods as well as other existing ones into an openly available trajectory processing pipeline, which will be flexibly adapted and augmented to cater for a specific application scenario by the user. This would allow others to easily access and benefit from our developed tools. For example, our anonymization as well as indexing and storing methods could be very useful for the VGI challenges discussed in Part III of this book. In the very next chapter, the focus will be on animal trajectories. Here, anonymization is less of an issue, but efficient mining and visualization tools are crucial. With the rich set of applications for trajectory mining occurring across all VGI domains, we see a clear demand for further development and improvements of algorithms to explore and learn from the information hidden in volunteered trajectory collections in the best possible way.

Acknowledgments This research was supported by the German Research Foundation DFG within Priority Research Program 1894 *Volunteered Geographic Information: Interpretation, Visualization, and Social Computing* (VGIscience). In this chapter, results from the projects VGI-Routing (424960421) and Trajectory 2 (314699172) are reviewed.

References

Abul O, Bonchi F, Nanni M (2008) Never walk alone: Uncertainty for anonymity in moving objects databases. In: 2008 IEEE 24th International Conference on Data Engineering, pp. 376–385. https://doi.org/10.1109/ICDE.2008.4497446

Alewijnse SPA, Buchin K, Buchin M, Kölzsch A, Kruckenberg H, Westenberg MA (2014) A framework for trajectory segmentation by stable criteria. In: Proc. 22nd ACM SIGSPATIAL International Conference on Advances in Geographic Information Systems (SIGSPATIAL '14), Dallas Texas, 2014. Association for Computing Machinery, pp. 351–360. https://doi.org/10.1145/2666310.2666415

Aronov B, Driemel A, Kreveld MV, Löffler M, Staals F (2016) Segmentation of trajectories on nonmonotone criteria. ACM Trans Algorithms 12(2):1–28. https://doi.org/10.1145/2660772

Barth F, Funke S, Jepsen TS, Proissl C (2020) Scalable unsupervised multi-criteria trajectory segmentation and driving preference mining. In: BigSpatial@SIGSPATIAL. ACM, New York, pp 6:1–6:10. https://doi.org/10.1145/3423336.3429348

Barth F, Funke S, Proissl C (2021) Preference-based trajectory clustering—an application of geometric hitting sets. In: ISAAC, volume 212 of LIPIcs. Schloss Dagstuhl - Leibniz-Zentrum für Informatik, pp 15:1–15:14. https://doi.org/10.4230/LIPIcs.ISAAC.2021.15

Baum M, Bläsius T, Gemsa A, Rutter I, Wegner F (2016) Scalable exact visualization of isocontours in road networks via minimum-link paths. In: Sankowski P, Zaroliagis C (eds) 24th Annual European Symposium on Algorithms (ESA 2016), volume 57 of Leibniz International Proceedings in Informatics (LIPIcs), Dagstuhl, Germany, 2016. Schloss Dagstuhl–Leibniz-Zentrum fuer Informatik, pp. 7:1–7:18. https://doi.org/10.4230/LIPIcs.ESA.2016.7

Behr T, van Dijk TC, Forsch A, Haunert J-H, Storandt S (2021) Map matching for semi-restricted trajectories. In: 11th International Conference on Geographic Information Science (GIScience 2021, online). https://doi.org/10.4230/LIPIcs.GIScience.2021.II.12

Berg MD, Cheong O, Kreveld MV, Overmars M (2008) Computational geometry: algorithms and applications, 3rd edn. Springer-Verlag TELOS. ISBN 3540779736, 9783540779735. https://doi.org/10.1007/978-3-540-77974-2

Boissonnat J-D, Devillers O, Teillaud M, Yvinec M (2000) Triangulations in CGAL. In: Proc. 16th Annual Symposium on Computational Geometry (SoCG '00), pp 11–18, 2000. https://doi.org/10.1145/336154.336165

Brauer A, Mäkinen V, Forsch A, Oksanen J, Haunert J-H (2022) My home is my secret: concealing sensitive locations by context-aware trajectory truncation. Int J Geograph Inform Sci 1–29. https://doi.org/10.1080/13658816.2022.2081694

Buchin M, Driemel A, Van Kreveld MJ, Sacristan V (2011) Segmenting trajectories: a framework and algorithms using spatiotemporal criteria. J Spatial Inform Sci 3(3):33–63. https://doi.org/10.5311/JOSIS.2011.3.66

Cormen TH, Leiserson CE, Rivest RL, Stein C (2009) Introduction to algorithms. MIT Press, Cambridge

Dai Y, Shao J, Wei C, Zhang D, Shen HT (2018) Personalized semantic trajectory privacy preservation through trajectory reconstruction. World Wide Web 21(4):875–914. https://doi.org/10.1007/s11280-017-0489-2

Dijkstra EW (1959) A note on two problems in connexion with graphs. Numerische Mathematik 1(1):269–271. https://doi.org/10.1007/bf01386390

Dong, Y, Pi D (2018) Novel privacy-preserving algorithm based on frequent path for trajectory data publishing. Knowl.-Based Syst. 148:55–65. https://doi.org/10.1016/j.knosys.2018.01.007

Forsch A, Dehbi Y, Niedermann B, Oehrlein J, Rottmann P, Haunert J-H (2021) Multimodal travel-time maps with formally correct and schematic isochrones. Trans GIS 25(6):3233–3256. https://doi.org/10.1111/tgis.12821

Forsch A, Oehrlein J, Niedermann B, Haunert J-H (2022) Inferring routing preferences of cyclists from user-generated trajectories using a compression criterion. J Spatial Inform Sci. (manuscript submitted on 22nd September 2022)

Funke S, Laue S, Storandt S (2016) Deducing individual driving preferences for user-aware navigation. In: Ravada S, Ali ME, Newsam SD, Renz M, Trajcevski G (eds) Proc. 24th ACM SIGSPATIAL International Conference on Advances in Geographic Information Systems (ACM SIGSPATIAL GIS '16), pp 14:1–14:9. https://doi.org/10.1145/2996913.2997004

Funke S, Schnelle N, Storandt S (2017) Uran: a unified data structure for rendering and navigation. In: Web and Wireless Geographical Information Systems. Springer, Cham, pp 66–82. https://doi.org/10.1007/978-3-319-55998-8_5

Funke S, Rupp T, Nusser A, Storandt S (2019) PATHFINDER: storage and indexing of massive trajectory sets. In: Proc. 16th International Symposium on Spatial and Temporal Databases (SSTD '19), pp 90–99. https://doi.org/10.1145/3340964.3340978

Geisberger R, Sanders P, Schultes D, Vetter C (2012) Exact routing in large road networks using contraction hierarchies. Transport Sci 46(3):388–404. https://doi.org/10.1287/trsc.1110.0401

Haunert J-H, Budig B (2012) An algorithm for map matching given incomplete road data. In: Proc. 20th ACM SIGSPATIAL International Conference on Advances in Geographic Information Systems (ACM SIGSPATIAL GIS '12), pp 510–513. https://doi.org/10.1145/2424321.2424402

Haunert J-H, Schmidt D, Schmidt M (2021) Anonymization via clustering of locations in road networks. In: 11th International Conference on Geographic Information Science (GIScience 2021)—Part II Short Paper Proceedings. https://doi.org/10.25436/E2CC7P

Huo Z, Meng X, Hu H, Huang Y (2012) You can walk alone: trajectory privacy-preserving through significant stays protection. In: International Conference on Database Systems for Advanced Applications. Springer, Berlin, pp 351–366. https://doi.org/10.1007/978-3-642-29038-1_26

Imielińska C, Kalantari B, Khachiyan L (1993) A greedy heuristic for a minimum-weight forest problem. Oper Res Lett 14(2):65–71. https://doi.org/10.1016/0167-6377(93)90097-Z

Koller H, Widhalm P, Dragaschnig M, Graser A (2015) Fast hidden Markov model map-matching for sparse and noisy trajectories. In: Proc. 18th IEEE International Conference on Intelligent Transportation Systems (ITSC '15). IEEE, pp 2557–2561. https://doi.org/10.1109/ITSC.2015.411

Krogh B, Jensen CS, Torp K (2016) Efficient in-memory indexing of network-constrained trajectories. In: Proc. 24th ACM SIGSPATIAL Int. Conf. on Advances in Geographic Information Systems. ACM, New York, pp 17:1–17:10. ISBN 978-1-4503-4589-7. https://doi.org/10.1145/2996913.2996972

Monreale A, Andrienko GL, Andrienko NV, Giannotti F, Pedreschi D, Rinzivillo S, Wrobel S (2010) Movement data anonymity through generalization. Trans Data Privacy 3(2):91–121.

Newson P, Krumm J (2009) Hidden markov map matching through noise and sparseness. In: Proceedings of the 17th ACM SIGSPATIAL International Conference on Advances in Geographic Information Systems (GIS '09). ACM, New York, pp 336–343. https://doi.org/10.1145/1653771.1653818

O'Sullivan D, Morrison A, Shearer J (2000) Using desktop GIS for the investigation of accessibility by public transport: an isochrone approach. Int J Geograph Inform Sci 14(1):85–104. https://doi.org/10.1080/136588100240976

Rissanen J (1978) Modeling by shortest data description. Automatica 14(5):465–471. https://doi.org/10.1016/0005-1098(78)90005-5

Rupp T, Baur L, Funke S (2022) PATHFINDER VIS (Demo Paper). In: SIGSPATIAL/GIS. ACM, New York, pp 11–14.

Sandu Popa I, Zeitouni K, Oria V, Barth D, Vial S (2011) Indexing in-network trajectory flows. J. Int J Very Large Data Bases 20(5):643–669. https://doi.org/10.1007/s00778-011-0236-8

Seybold MP (2017) Robust map matching for heterogeneous data via dominance decompositions. In: Proc. SIAM International Conference on Data Mining, pp 813–821. https://doi.org/10.1137/1.9781611974973.91

Song R, Sun W, Zheng B, Zheng Y (2014) Press: a novel framework of trajectory compression in road networks. Proc VLDB Endow 7(9):661–672. ISSN 2150-8097. https://doi.org/10.14778/2732939.2732940

Sweeney L (2002) k-anonymity: a model for protecting privacy. Int J Uncertainty Fuzziness Knowl.-Based Syst 10(05):557–570. https://doi.org/10.1142/S0218488502001648

Wang N, Kankanhalli MS (2020) Protecting sensitive place visits in privacy-preserving trajectory publishing. Comput Secur 97. https://doi.org/10.1016/j.cose.2020.101949

Yang X, Wang B, Yang K, Liu C, Zheng B (2018) A novel representation and compression for queries on trajectories in road networks. IEEE Trans Knowl Data Eng 30(4):613–629. https://doi.org/10.1109/TKDE.2017.2776927

Yarovoy R, Bonchi F, Lakshmanan LV, Wang WH (2009) Anonymizing moving objects: How to hide a mob in a crowd? In: Proceedings of the 12th International Conference on Extending Database Technology: Advances in Database Technology, pp 72–83. https://doi.org/10.1145/1516360.1516370

Zheng Y (2015) Trajectory data mining: an overview. ACM Trans Intell Syst Technol 6(3):29:1–29:41. ISSN 2157-6904. https://doi.org/10.1145/2743025

Chapter 4
Uncertainty-Aware Enrichment of Animal Movement Trajectories by VGI

Yannick Metz and Daniel A. Keim

Abstract Combining data from different sources and modalities can unlock novel insights that are not available by analyzing single data sources in isolation. We investigate how multimodal user-generated data, consisting of images, videos, or text descriptions, can be used to enrich trajectories of migratory birds, e.g., for research on biodiversity or climate change. Firstly, we present our work on advanced visual analysis of GPS trajectory data. We developed an interactive application that lets domain experts from ornithology naturally explore spatiotemporal data and effectively use their knowledge. Secondly, we discuss work on the integration of general-purpose image data into citizen science platforms. As part of inter-project cooperation, we contribute to the development of a classifier pipeline to semi-automatically extract images that can be integrated with different data sources to vastly increase the number of available records in citizen science platforms. These works are an important foundation for a dynamic matching approach to jointly integrate geospatial trajectory data and user-generated geo-referenced content. Building on this work, we explore the joint visualization of trajectory data and VGI data while considering the uncertainty of observations. *BirdTrace*, a visual analytics approach to enable a multi-scale analysis of trajectory and multimodal user-generated data, is highlighted. Finally, we comment on the possibility to enhance prediction models for trajectories by integrating additional data and domain knowledge.

Keywords Visualization · Trajectory data · User-generated con tent · Animal tracking

Y. Metz (✉) · D. A. Keim
Universität Konstanz, Konstanz, Germany
e-mail: yannick.metz@uni-konstanz.de; keim@uni-konstanz.de

4.1   Introduction

Our goal is to integrate and match the two heterogeneous large data sources to enrich the spatial databases with contextual information. Furthermore, we want to investigate in detail the uncertainty that is found in different VGI data sources. In this chapter, we describe our research toward this goal, by focusing on three core contributions:

1. Visual-interactive analysis of GPS trajectory data for domain experts
2. The usage of additional non-verified data sources for VGI in the context of birdwatching
3. A joint approach to analyze GPS trajectory data and VGI contributions considering the uncertainty that, e.g., was introduced by the previous extraction of non-verified data

Additionally, we present work on using visual analytics for the training of deep learning models for movement prediction, which can support future research in domains of geographic information science and forecasting for biologging data.

In the following, we first present the relevant background that motivates our research. Subsequently, we describe pursued research contributing to the aforementioned points, done in conjunction with domain experts or in close collaboration with other partners of the priority program. Finally, we give a brief outlook on possible future research directions.

4.2   Related Work

VGI contains insight that has the potential to solve fundamental and unsolved social and environmental challenges. The big scientific challenge consists of how to investigate and extract value from noisy VGI data sources. Coxen et al. (2017) stressed that there is a general concern about spatial biases in citizen science datasets. Geldmann et al. (2016) also underline that untrained volunteers might be introducing biases, leading to spatial biases toward densely populated areas and easy-to-watch observations. Integrating and assimilation of VGI into scientific models requires a change of paradigm that embraces uncertainty and bias. Some popular citizen science initiatives have already tried to assimilate VGI to improve the results of biodiversity models. Since 2002, the eBird project has been gathering bird observation records from volunteers around the world. The participation of volunteers has shown a rapid increase in recent years with millions of observations submitted each year. Since then, more than 500,000 users have visited the eBird website (Sullivan et al. 2009). Kelling et al. (2012) proposed a Human/Computer Learning Network for Biodiversity Conservation incorporating VGI coming from eBird using an active learning feedback loop to improve the results of the AI algorithms. Fink et al. (2010, 2013) introduced a spatiotemporal exploratory model,

STEM, and AdaSTEM, to study species distribution models. They used massively crowdsourced citizen science data to construct the corresponding models. The STEM model was afterward integrated into a visual analytics system, BirdVis (Ferreira et al. 2011), that allows the ornithologist to analyze abundance models to understand better bird populations. Coxen et al. (2017) compare two species distribution models using satellite tracking data vs. citizen science datasets coming from eBird datasets. Their results showed the effectiveness of citizen science datasets for this particular use. In other disciplines, such as atmospheric sciences, the shift to the user-generated information paradigm is even harder. Chapman et al. (2017) stated that in atmospheric science, high-quality and precise observation is deeply rooted in the essence of the discipline and other data sources with low-quality, bias, and imprecision are hard to accept. Other antecedents to understanding bird migration behavior and patterns are the work of Jain and Dilkina, who constructed a migration network using K-means clustering and Markov chain model (Jain and Dilkina 2015), and the tool of Wood et al., where they studied the seasonal behavior of bird species within a specific location (Wood et al. 2011). There is also previous work to understand the quality of the observations and uncertainty of the models, by characterizing bird watchers. Cole and Scott in 1999 used Texas Conservation Passport holders and members of the American Birding Association to categorize differences between two different groups of wildlife watching as casual wildlife watchers and serious birders (Cole and Scott 1999). These two groups were defined by their skill level at identifying birds, the frequency of participation, expenditures, and bird-watching behavior. Afterward, Scott conducted another study with Thigpen (Scott and Thigpen 2003) to understand bird watchers' behavior. Data were collected from the bird-watching festival in September 1995 at the Seventh Annual HummerBird Celebration in Rockport/Fulton, Texas. In Scott and Shafer (2001), specialization was measured with regard to birdwatcher behavior, level of skill, and commitment. Based on the this, they categorized birders into casual, interested, active, and skilled birders. The outcomes of the study by Scott and Thigpen have supported McFarlane's investigation of birders in Alberta (McFarlane 1994). She revealed that 80% of the general population in her example were casual or novice birders. This information could be very valuable to quantify the confidence of VGI observations. Previous work shows a glimpse of the unprecedented opportunity to materialize a change in the way scientific models treat data, changing the VGI data paradigm to embrace the uncertainty and bias of data provided by humans as part of the scientific investigation process. From now on, a fascinating endeavor comprises for us the development of new techniques that can bridge the gap between social, exact, and natural sciences by using VGI as an interlinkage among biodiversity and human behavior to provide effective and timely answers to societal calls such as climate change and nature preservation.

Previous work has shown how geo-tagged social media data reflects the spatiotemporal distribution of social groups (MacEachren et al. 2011). For example, scatterplots combine event detection and classification in investigating the geo-tagged social media, to enable situation awareness (Thom et al. 2012; Cao et al. 2012) for geospatial information diffusion. Our preliminary work in this area covers the dynamics of social groups and their expressions on social media. Our work on social media bubbles (Diehl et al. 2018) is the first step toward the structuring of complex social relationships on social media. The main connection point between the social group structure as the scaffolding of society and VGI is the uncertainty that humans introduce into the data and the trustworthiness of systems consumed by humans. This work addresses different aspects of uncertainty from a practical point of view. Our work on the visual assessment for visual abstractions (Sacha et al. 2017) addresses the trustworthiness of the users in the systems for the particular case of soccer data. The works above addressed the uncertainty from the perspective of the producer of VGI and the trustworthiness of the users on the systems. During the last few years, the study of uncertainty and its propagation through the visual analytics workflow have gained popularity. Early, in 2015, MacEachren proposed to consider the propagation of uncertainty through the whole VA workflow rather than just the visualization of the uncertainty at the end of the pipeline (MacEachren et al. 2011). He illustrated current challenges and possible approaches to tackle uncertainty using definitions from decision sciences. Kinkeldey et al. analyzed the impact of visually represented geodata uncertainty on decision-making and addressed possible approaches for the evaluation of uncertainty in visualizations. Previously, we tackled the uncertainty aspects of VGI using a theoretical framework (Diehl et al. 2018), which, for the first time, shapes the human factors of uncertainty in VGI and defines a new term "user uncertainty" to enclose them.

4.3 Analysis of GPS Trajectory Data

We start by looking at an analysis that is possible for stand-alone trajectory data in the biologging context. Analysis tools, both visually and algorithmically, build the foundation for our goal of enabling a joint approach for trajectory data and voluntary geographic information.

4.3.1 Motivation and Research Gap

Segmenting biologging time series of animals on multiple temporal scales is an essential step that requires complex techniques with careful parameterization and possibly cross-domain expertise. Yet, there is a lack of visual-interactive tools that strongly support such multi-scale segmentation. To close this gap, we present our MultiSegVA platform for interactively defining segmentation techniques and parameters on multiple temporal scales in our paper *MultiSegVA: Using Visual Analytics to Segment Biologging Time Series on Multiple Scales* (Meschenmoser et al. 2020). MultiSegVA primarily contributes tailored, visual-interactive means and visual analytics paradigms for segmenting unlabeled time series on multiple scales. Further, to flexibly compose the multi-scale segmentation, the platform contributes a new visual query language that links a variety of segmentation techniques. To illustrate our approach, we present a domain-oriented set of segmentation techniques derived in collaboration with movement ecologists. In the paper, the applicability and usefulness of MultiSegVA are demonstrated in two real-world use cases from movement ecology, related to behavior analysis after environment-aware segmentation and after progressive clustering. Expert feedback from movement ecologists shows the effectiveness of tailored visual-interactive means and visual analytics paradigms in segmenting multi-scale data, enabling them to perform semantically meaningful analyses. Here, we want to highlight two key aspects of the work, the characteristics of biologging time series data and the respective analysis, as well as how we can support this process within the visual analytics framework. For further details, we refer to the paper of Meschenmoser et al. (2020). For our work, we focus on biologging time series of moving animals: these time series have prototypical multi-scale characters and include widely unexplored behaviors, which are hidden in high resolutions and cardinalities. Additionally, biologging-driven movement ecology is an emerging field (Brown et al. 2013; Shepard et al. 2008), triggered by technical advances that enable academia to address open questions in innovative ways. The biologging time series stems from miniaturized tags and gives high-resolution information about, e.g., an animal's location, tri-axial acceleration, and heart rate. Here, semantics are typically distributed on diverse temporal scales, including life stages, seasons, days, day times, and (micro)movement frames. These temporal scales are complemented by spatial scales concerning, e.g., the overall migration range, migration stops, and foraging ranges. There are complex scale- and context-specific conditions (Benhamou 2014; Levin 1992), implying different energy expenditures, driving factors, and decisions for behavior. Hence, segmenting such time series on a single scale with global parameters does not sufficiently address their multi-scale character. The relevance of multi-scale segmentation can be further motivated by three reasons. First, analysts can deepen their understanding of how scales relate to each other: e.g., in terms of nesting relations, next to relative scale sizes and types. A multi-scale perspective can even enable one "to gain an insight on an entire knowledge domain or a relevant sub-part" (Nazemi et al. 2015). Second, even without labeled data or thoroughly parameterized single-scale techniques, it is possible to identify fine-grained patterns that are wrapped by lower-scale, context-yielding patterns. Such fine-grained and context-aware patterns are crucial to enriching existing classification and prediction models. Third, demands for more multi-scale analyses originate from domain literature. Such demands can be found in movement ecology and analysis (Andrienko and Andrienko 2013; Demšar et al. 2015), but also in, e.g., medical sciences (Alber et al. 2019) and social sciences (Cash et al. 2006). However, in practice, segmenting time series on multiple scales is often impeded by several factors. First, multi-scale techniques rely on more in-depth, theoretical foundations and inherent parameters that need to be carefully adapted. Therefore, analysts (e.g., movement ecologists) might require cross-domain expertise in statistical multi-scale time series analysis. Second, even with such expertise, it is difficult to decide on scale properties (e.g., size, dimension, number of scales) and further parameters. Third, we observe a lack of suitable visual-interactive approaches in related works (Sect. 3.2) that can strongly support and promote segmenting time series on multiple scales.

For MultiSegVa, we defined four requirements in cooperation with domain experts:

1. An application that integrates analysis tools at different time scales without the need to manually combine different algorithms or libraries.
2. Support time series segmentation by revealing the multi-scale structure and addressing its specifics.
3. The analysis should be able to flexibly parameterize segmentation algorithms to the specific context.
4. Visual-interactive features that can help the analysts' work.

4.3.2 Approach

To close the research gap of enabling multi-scale analysis of biologging time series, we present our web-based MultiSegVA platform that allows analysts to visually explore and refine a multi-scale segmentation, which results from a simple way of setting segmentation techniques on multiple scales. In the context of multi-scale segmentation, MultiSegVA primarily contributes to the use of tailored visual-interactive features and established VA paradigms. To flexibly configure segmentation techniques and parameters, MultiSegVA includes a new visual query language (VQL, C2) that links a variety of segmentation techniques across multiple scales. These techniques stem in the present case from a set that was derived together with movement ecologists and covers typical domain use cases. Figure 4.1 shows the main window of the *MultiSegVA* system. Here, the analyst can build visual queries and analyze the existing segmentation results in a hierarchy visualization,

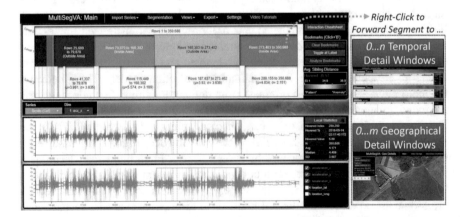

Fig. 4.1 A screenshot from **MultiSegVA**

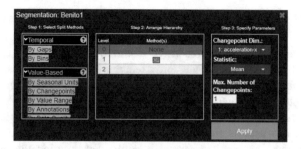

Fig. 4.2 The visual query language interface: In the left column, multiple time series segmentation methods can be selected. The hierarchy of applied methods can be changed via an interactive interface, shown in the second column. Finally, detailed settings for each method can be changed in the third column

which in turn is closely linked to one- and multidimensional time series plots. It is also possible to access additional details of segments via a temporal detail window or inspect the underlying trajectories on a map. MultiSegVA implements a feedback loop for iterative analysis and refinement of the segmentation. After importing a time series, e.g., GPS trajectory data of tracked animals via Movebank (Wikelski M 2023), the analyst can start to analyze the time series. After an initial visual inspection of the time series dimensions, an analyst can steer hierarchical time series segmentation using the visual query language (VQL). The interface is shown in Fig. 4.2. The VQL serves three purposes here: (1) Different types of segmentation techniques can be easily arranged across different time scales, (2) the hierarchical application order can be defined by manipulating with building blocks, and (3) the chosen techniques can be interactively parameterized. The query interface first provides a list of available techniques organized by category and can recommend appropriate techniques. To modify the hierarchical application of techniques, selected techniques are arranged as visual building blocks which can be modified by drag-and-drop interactions. This can alleviate some issues that arise with text-based queries. In particular, it avoids changing the ordering of nested queries, which can be tedious and error-prone in text-based queries. The query language also provides several selectors and operations to chain and link different techniques at the same or different scales. The finalized query is then processed in the backend of the application, and the results are visualized via the icicle hierarchy view; see the top of Fig. 4.1. The analyst can choose to adapt the query based on the achieved results to iteratively improve the segmentation results and get a detailed understanding of the time series data.

4.3.3 Results

MultiSegVA enables the comprehensive exploration and refining of multi-scale segmentation by tailored visual-interactive features and VA paradigms. MultiSegVA includes segment tree encoding, subtree highlighting, guidance, density-dependent features, adapted navigation, multi-window support, and a feedback-based work-flow. The VQL facilitates exploring and parameterizing different multi-scale structures. Still, a few aspects remain for further reflection. The icicle visualization meets expert requests and has several benefits. Yet, guiding the user by color to interesting parts of the segment tree is a challenging task. We tested global, level-based, and sibling-based guidance variants according to color fills. We chose sibling-based guidance (i.e., all siblings of one hovered segment are colored) that optimally captures local similarities while requiring more navigation effort across levels and nodes. Upcoming works will include an even more effective variant, i.e., guidance to local similarities with little interaction and one fixed color scale. Our VQL makes it trivial to build a multi-scale segmentation. Query building is a play with building blocks that benefits from strong abstraction and simple interactions. Rather, it is difficult to decide which multi-scale structure and building blocks are most appropriate: a decision that depends on data, analysis, and tasks. MultiSegVA facilitates this decision through extensive documentation, technique categorization, few technique parameters, and short processing times in a compact workflow. For further support, we plan predefined queries, and instant responses at query building, next to the parameter and technique suggestions. For suggesting parameters, we will apply estimators (Catarci et al. 1997; Yao 1988) for the number of change points as well as the elbow method for knn-searches. While motif length and HDBSCAN's minPts (Campello et al. 2013) optimally benefit from domain expertise, suggesting other parameters will simplify the interaction and can address another limitation. Now, a technique processes each segment of one scale with the same parameters; thus, slight data-dependent parameter modifications will be examined. For technique suggestions, we envision for each technique a scale-wise relevance score that reflects data properties and is part of a rule-based prioritization, shaped by domain expertise and meaningful hierarchies. It is essential to depict the semantics into which MultiSegVA can provide insights. First of all, MultiSegVA illuminates diverse multi-scale structures and gives insights into how scales relate to each other. Coarse behaviors can be distinguished by relatively simple techniques, motifs show repetitive behaviors, and knn-searches allow the matching with already explored segments. Segment lengths and similarities can be explored, next to local anomalies and spatial contexts. However, with the current techniques, it is difficult to broadly capture deeper, behavioral semantics (e.g., chew, scratch). Hereto more complex or learning techniques (e.g., HMMs, SVMs) will be needed that neither overfill the interface nor delimit generalizability due to the lack of learned patterns. The latter point goes hand in hand with our major limitation and the corresponding implication for upcoming work: integrating even more intelligent methods and automatism. These plans all relate to aspects from above, i.e., better guidance, technique, and more parameter suggestions, as well as techniques for deeper behavioral semantics. MultiSegVA relies on requirements from movement ecology experts and stands for an iterative, extensively collaborative, and interdisciplinary process. We can gather domain feedback on several stages, derive a domain-oriented set of techniques, and even link MultiSegVA to Movebank with >2.2 billion animal locations. With this application domain-focused, MultiSegVA underpins the value of multi-scale analyses and is certainly another step forward "to empower the animal tracking community and to foster new insight into the ecology and movement of tracked animals" (Spretke et al. 2011). Meanwhile, our third use case shows that MultiSegVA variants for other domains are conceivable, especially with tailored domain-oriented technique sets. This generalizability is promoted by the platform's I/O features and its ability to handle heterogeneous time series, with >1.2 million records.

4.4   Analysis of VGI Contributor Data

In joint collaborative work with partners of the priority program, we investigate the utility of using a novel pipeline based on a deep learning-based image classifier to integrate images from the social media platform *Flickr* with data from citizen science platforms: *A text and image analysis workflow using citizen science data to extract relevant social media records: combining red kite observations from Flickr, eBird and iNaturalist* (Hartmann et al. 2022). In our research agenda, this work serves a dual role: (1) We explore the characteristics of VGI image data in our chosen domain of migratory birds, as well as automated integration techniques, and (2) we integrate contributions from non-verified data sources, which directly connects to the research topic of uncertainty in data sources. Specifically, the confidence of the developed classification pipeline might be directly used as an uncertainty measure for the matching process we introduce in the following chapter.

4.4.1   Motivation and Research Gap

There is an urgent need to develop new methods to monitor the state of the environment. One potential approach is to use new data sources, such as user-generated content, to augment existing approaches. Despite a wide range of works discussing and demonstrating the potential of new data forms in the creation of indicators, we could not identify previous research which explicitly created a workflow designed to integrate data from different sources and of different modalities. Furthermore, although the properties of different forms of UGC are relatively well understood, they have not been effectively used to develop reproducible workflows. Finally, most studies evaluate the quality of extracted information in isolation through metrics such as precision and recall, but do not explore the added value of integrating data.

In the paper, we propose, implement, and evaluate a workflow taking advantage of citizen science data documenting and recording sightings of birds and more specifically red kites (*Milvus milvus*). Analyzing social media data until recently has often used simple keyword-based methods to perform an initial filtering or search step, meaning that content tagged in other ways was not found. However, improvements in content-based classification now mean that it is also possible to use off-the-shelf, pre-trained algorithms to reliably identify predefined classes such as the presence of buildings, people, or birds in image data with reasonable accuracy.

4.4.2 Approach

We take a new approach, using citizen science projects recording sightings of red kites (*Milvus milvus*) to train and validate a convolutional neural network (CNN) capable of identifying images containing red kites. This CNN is integrated into a sequential workflow that also uses an off-the-shelf bird classifier and text metadata to retrieve observations of red kites in the Chilterns, England. Our workflow reduces an initial set of more than 600,000 images to just 3065 candidate images. Manual inspection of these images shows that our approach has a precision of 0.658. A workflow using only text identifies 14% fewer images than that including image content analysis, and by combining image and text classifiers, we achieve an almost perfect precision of 0.992. Images retrieved from social media records complement those recorded by citizen scientists spatially and temporally, and our workflow is sufficiently generic that it can easily be transferred to other species.

Flickr is a social media site, where individuals can upload photographs and metadata, including tags and locations in the form of coordinates. Flickr's usage has declined in recent years, but it remains very popular in research, mostly because of its well-documented and easy-to-use API, which allows querying using search terms and bounding boxes. Our citizen scientist data came from two platforms: iNaturalist and eBird. iNaturalist allows participants to upload images of organisms such as plants and insects to the platform and use its community to crowdsource taxonomic identification. Currently, according to their website (https://inaturalist.org), iNaturalist hosts nearly 100 million observations of over 375000 species and is, therefore, one of the largest and most successful citizen science projects to date (Unger et al. 2020). eBird has similar features to iNaturalist but as a platform is exclusively specialized in bird observations. Their website states (https://ebird.org) that "eBird is among the world's largest biodiversity-related science projects, with more than 100 million bird sightings contributed annually." It predominantly hosts observation location data, but also corresponding bird images, as well as bird sounds (Sullivan et al. 2009; Wood et al. 2011).

Since our workflow is designed to be generic, take advantage of the text and image data, and combine records from citizen science reports with social media data, it uses a combination of a simple rule-based approach, existing pre-trained models, and a model trained specifically for our target species. Our approach is designed to take advantage of what we assume to be high-quality data collected by citizen scientists with an interest in ornithology, use off-the-shelf models where possible, and reduce the initial number of social media posts in a given region to a manageable size for manual verification. In the following, we want to give a summary of the proposed workflow:

1. We identify all geo-tagged social media records in a study area (in our case, the *Chiltern Hills* area in the UK).
2. Out of the identified records, we assign all records that contain the *Latin* name of our target species *Milvus milvus* to our result set. We assume that users familiar with the biological taxonomy are experts and thus treat these records as trustworthy.
3. We use a generic image classification model to identify images that contain birds (with a confidence threshold of p_B 0.5). These retained images are then processed further.
4. For these filtered images, we use metadata such as title or description to identify records that are highly likely Red Kites, e.g., because the description contains the common name in a European language (such as "Red Kite" or "Rotmilan" (German)). We include these images in the result set.
5. We use a secondary image classifier trained on citizen science data to identify images that likely are red kites (with a confidence of $p_{RK} > 50\%$). These are also added to the final set of candidate images.
6. As a final step, we assume that an expert can manually verify the extracted images. As the workflow significantly reduces the set of candidate images, this task becomes feasible and ensures high data quality.

The workflow creates a high-confidence dataset of images that can be integrated with existing citizen science platform data. As part of the paper, we ran a detailed study of the characteristics of the different data sources, namely, *Flickr, eBird*, and *iNaturalist*. In the chosen target area, we compare (1) spatial coverage, (2) temporal distribution, (3) contributor patterns, as well as (4) image data quality.

4.4.3   Results

Our workflow aimed to extract relevant images of red kites from Flickr data and to use these to complement citizen science records from eBird and iNaturalist. In the following, we, therefore, explore the following aspects of the results we obtained:

- How effective is our workflow at extracting relevant red kite images, and how much added value is obtained through the use of both text and image content?
- What are the properties of the extracted records within our study area, and do the social media data complement the citizen science platforms?

The workflow returned 3065 candidate images, downsampling the original dataset by 99.5%. These images were then individually inspected to identify true

Table 4.1 Precision using different combinations of the components in the workflow

approach	included posts	true positives	precision
only visual data	2763	1723	0.624
only text data	2215	1946	0.878
initial workflow	3065	2017	0.658
text + visual data (final workflow)	3559	2262	0.636

and false positives and allow us to calculate the precision. Images were marked as true positives if a red kite was identifiable in an image. This meant that images had to be sufficiently clear, such that distinctive features of red kites (e.g., their forked tails or red-brown coloring) were visible. Images where a bird was visible, but not unambiguously identifiable, images showing feathers or pellets, and images that were obviously irrelevant were all marked as false positives. A total of 2017 records were thus identified as true positives, with 1048 false positives and a resulting precision for the complete workflow of 0.658.

To understand the benefits of text and image analysis, we ran the components of the workflow individually and annotated any additional images extracted (Table 4.1).

1. In the textual workflow setting, records were returned if either the Latin name or a common name for red kite (in six language variations) were detected. This approach identified 2215 posts, of which 1946 were true positives and 269 false positives, resulting in a precision of 0.879.
2. In the visual workflow setting, only visual information was considered. A post was considered relevant and included if both the bird model and red kite model return a probability above 50% for the given image. This approach returned 2763 included posts, of which 1723 were true positives and 1040 were false positives, giving a precision of 0.624.

We found 1419 Flickr posts that were included by both settings, of which 1407 were true positives. This means that by only retaining candidate records identified by both textual and image-based information, we can achieve an almost perfect precision of 0.992. We then checked for records that were exclusively identified by either text or image analysis. Five hundred and thirty-nine posts were only detected by the textual analysis (point 1 in the list above), and 316 were only detected by the visual analysis (point 2 in the list above). Combining these results leads to a total of 3559 records, of which 2262 are true positives and 1297 are false positives, and a precision of 0.636. Looking back at the performance of our initial integrated workflow, we note that 245 (12%) additional true positive red kite posts were extracted by merging the results of separately performed textual and visual analysis. This increase in recall is at the cost of a very slight reduction in the precision of 0.02. Summarizing these findings, 62% of true positives were found using either text or image analysis. Twenty-four percent are only correctly classified by textual data, and 14% are missed if no visual analysis is performed.

We find that the workflow functions as a data filter, reducing the data volume by 99.5%. By reducing the data volume, it becomes realistic to analyze the remaining data manually to select true positives. The workflow thus addresses the research gap identified by Burke et al. (2022), using generalizable methods to extract target data from various unverified sources to enrich data.

We found that while keyword matching delivered high precision with little evidence of ambiguity, image analysis returned more potential candidates than textual analysis, but with lower precision. By only retaining posts identified by both textual and visual analysis, they were able to achieve almost perfect precision (0.992), at the cost of a lower recall. By combining the two approaches, they increased the extracted data volume by almost 14% while still downsampling the original dataset by around 99.5% and with a precision 0.636.

The visual distribution of points on the map in the article shows how different sources can complement one another when trying to determine patterns. The locations of Flickr posts tend to cluster around urban areas and points of interest along existing road networks, which suggests that the Flickr observations are often taken opportunistically. eBird and iNaturalist observations, on the other hand, are more heterogeneously allocated and show less obvious relationships to known spatial features, suggesting that birdwatchers go out with a clear intent to observe birds and seek a variety of locations for that purpose.

The study found that the temporal coverage of red kite observations in the Chilterns was different on a yearly and monthly scale. Aggregating data over years showed that the pattern shown by Flickr was different from the ones of eBird and iNaturalist. Year-on-year changes appeared to be more driven by underlying platform dynamics, such as user base and popularity changes. The study found that the rapid drop in Flickr observations from 2012 onward represented a decrease in Flickr popularity rather than a decline in red kites in the Chilterns. On the other hand, the study found that there was a strong increase for eBird and iNaturalist from the year 2016 onward. This could be the result of increased popularity, increased interest in red kites, or increased visits to the study region. Looking at monthly temporal scales showed a trend toward the warmer spring and summer months between March and June. These results may suggest higher visitation rates to the Chilterns in warmer periods, but could also be influenced by specific red kite behavioral patterns. Investigating the number of unique users per data source revealed that representativeness varies between platforms. eBird data was contributed by the fewest individuals, whereas Flickr and iNaturalist offered a more diverse user base. This observation could be attributed to higher platform popularity and an overall larger user base of the latter. Knowing the share of the population represented by a UGC-based analysis is crucial for policymakers to make adequate decisions that reflect the people's opinion (Wang et al. 2019b).

The image quality analysis revealed clear differences between social media data on Flickr and citizen science data in eBird and iNaturalist. Flickr's users are interested in capturing scenic and visually pleasing images, while eBird and iNaturalist users are more concerned about capturing the target species itself as proof of observation and less about the image quality. This discovery may point to the potential usefulness of social media data for the identification and tracking of individuals.

4.5 BirdTrace: A Visual Analytics Application to Jointly Analyze Trajectory Data and VGI

4.5.1 Motivation and Research Gap

BirdTrace makes use of two primary data sources: GPS data from tagged birds and user-generated content from birders. GPS data typically is of high quality but is only available on a small scale, while user-generated data is more abundant but of variable quality. By combining these two data sources, BirdTrace can provide a more complete picture of bird populations. The system uses a dynamic matching approach to semantically enrich trajectory data with geo-referenced data like images or textual descriptions.

4.5.2 Automated Matching

A key step was the development of semi-automatic methods to extract, integrate, and match data from VGI and tracked spatiotemporal datasets. This has allowed for further knowledge to be gained about individuals and populations of animals, including information about local animal habitats, animal migrations across continents, land-use change, biodiversity loss, invasive species, the spread of diseases, and climate change. There are yet no existing methods and systems that integrate and fuse mixed VGI data from birdwatchers and tracked trajectories of wildlife animals from the ICARUS (Movebank), so we developed them as part of the project.

We have already described the analysis of the trajectories, but now want to find relevant VGI contributions (e.g., images, video, audio, or text descriptions) for trajectories. Here, *relevance* refers to "how well a domain expert can use the found VGI contributions to answer specific questions." As this implies, the criteria for relevance might therefore depend on the problem. We tackle this problem by giving users the possibility to choose between different matching criteria, e.g., based on spatial or temporal distance, and potential classified behaviors like breeding. By including additional data sources, one might increase the number of possible matching criteria in the future. Let's look at the way we enable automated matching; see Fig. 4.3:

1. We assume the availability of GPS trajectory data for individuals of a species of interest. As our data source, we utilize Movebank.
2. Secondly, collect and locally store multimedia VGI data from citizen science platforms and potentially Flickr (as described in the previous chapter).

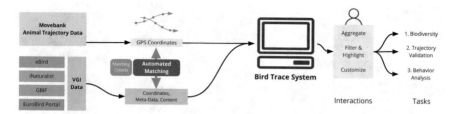

Fig. 4.3 The pipeline and workflow of the BirdTrace system. We combine animal movement trajectory data with VGI data from different citizen science portals. An automated matching approach is used to filter relevant VGI contributions to respective movement trajectories. Both trajectories and VGI data points are then jointly visualized in a shared visual interface. The interface enables aggregating, filtering, or annotating the given data

3. Based on a user query and selected matching criteria, we match individual VGI contributions with GPS trajectory data.
4. Trajectories and VGI contributions are jointly visualized in an interactive visual analytics application, which facilitates analysis by a domain expert.
5. The user can use additional interactive tools to search, filter, and highlight the matched contributions.

To facilitate the matching process, we implemented a data processing pipeline, which applies appropriate preprocessing to both the GPS trajectory data and the VGI contributions. We apply steps like line simplification, motif discovery, and outlier detection to the trajectory data to reduce the size of the data and to simplify the matching computations. VGI contributions from VGI portals like eBird, iNaturalist, and GBIF are collected and processed. To simplify analysis, we use precomputing and caching of data. This enables the efficient clustering of VGI contributions and fast matching of VGI contributions with trajectory data.

4.5.3   A Joint Visual Analytics Workspace

We developed **BirdTrace**,[1] a novel visual analytics method and interfaces to support the semantic annotation of the integrated database consisting of the VGI and the tracked trajectory data. The goal of the application is to add context information semantically from a domain expert (ornithologist) to increase the quality and enrich the integrated data sources previously discussed. The primary challenge is to reduce the uncertainty of the combined data sources and to raise awareness of the remaining uncertainty in our resulting database. Specifically, the tool allows annotating spatiotemporal databases and VGI in their semantic context. We will further support the annotation process with a semi-automatic process to enable the fast and reliable annotation of large datasets. We have to add semi-automatically

---

[1] Available at https://birdtrace.dbvis.de.

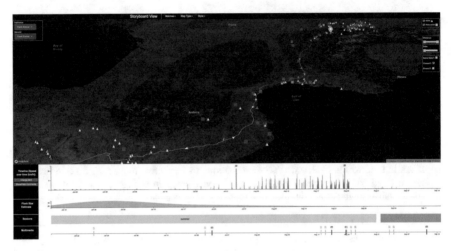

Fig. 4.4 A screenshot from **BirdTrace**

annotation to the dataset, as the domain experts do not have the time to review a large number of matchings.

Using the semantic annotations, we will be able to enrich our joined database with more knowledge from domain experts to increase the data quality of our integrated database. For instance, an ornithologist can verify or oppose the VGI information. The semantic annotation will also assist to clean and prune possibly incorrect merged data records and increase the awareness of uncertainty. Figure 4.4 shows the user interface of BirdTrace, showing a spatial map view on top, and a temporal "timeline" view on the bottom.

4.6 Data-Driven Modeling of Tracked and Observed Animal Behavior

Finally, we explored the challenging tasks of training prediction models, applicable, e.g., to animal trajectory forecasting. To improve uncertainty-aware prediction models for animal trajectories, we explored techniques from deep imitation and reinforcement learning. Specifically, we explored how to use data-driven deep learning methods to predict the movement of fish swarms. A complete presentation of results on predictive models would be beyond the scope of this chapter. We, therefore, want to focus on the workflow and, specifically, how concepts from visual analytics can be used here. We leave a discussion on model implementations for future work. In general, although deep learning-based approaches for related tasks are very promising, we still observe low adoption. Multiple challenges hinder the application of reinforcement learning algorithms in experimental and real-world use cases. Such challenges occur at different stages of the development and deployment of such models. While reinforcement learning workflows share similarities with machine learning approaches, we argue that distinct challenges can be tackled and overcome using visual analytic concepts. Thus, we propose a comprehensive workflow for reinforcement learning and present an implementation of our workflow incorporating visual analytic concepts integrating tailored views and visualizations for different stages and tasks of the workflow (Metz et al. 2022). In this final section, we would like to shine a light on how our workflow supports experimentation in this space and encourage future research in the application of novel RL-based methods for a wide range of problems in the context of geoinformatics and VGI, e.g., trajectory forecasting.

4.6.1 Motivation and Research Gap

Recently, there have been notable examples of the capabilities of reinforcement learning (RL) in diverse fields like robotics (Nguyen and La 2019), physics (Martín-Guerrero and Lamata 2021), or even video compression (Mandhane et al. 2022). Despite these successes, the application and evaluation of recent deep reinforcement and imitation learning techniques in real-world scenarios are still limited. Existing research almost exclusively focuses on synthetic benchmarks and use cases (Bellemare et al. 2013). We argue that the usage and evaluation in realistic scenarios is a mandatory step in assessing the capabilities of current approaches and identifying existing weaknesses and possibilities for further development. In this chapter, we present a visual analytics workflow and an instantiation of the approach that facilitates the application of state-of-the-art algorithms to various scenarios. Our presented approach is designed specifically to support domain experts, with basic knowledge of core concepts in reinforcement learning, who are interested in applying RL algorithms to domain-specific sequential decision-making tasks. The goal is to enable the effective application of their knowledge to (1) design agents and simulation environments including reward functions and (2) a detailed assessment of trained agents' capabilities in terms of performance, robustness, and traceability. A structured and well-defined approach can also help to critically investigate and combat some fundamental difficulties of reinforcement learning like *brittleness*, *generalization* to new tasks and environments, and issues of *reproducibility* (Dulac-Arnold et al. 2020; Henderson et al. 2017).

Outside of reinforcement and imitation learning, there exists a wide range of workflows and interactive visual analytics (VA) tools for the training and evaluation of ML models (Amershi et al. 2015; Endert et al. 2018; Spinner et al. 2020). Compared to other fields of machine learning, there has been less work on applying visual analytics in the space of reinforcement and especially imitation learning. A large number of necessary decisions and the existence of interconnected tasks make the application of interactive machine learning, with a close coupling of model and human, especially valuable for reinforcement learning.

Existing work such as *DQNViz* by Wang et al. (2019a) enables the analysis of spatial behavior patterns of agents in Atari environments like breakout (see arcade learning environment (Bellemare et al. 2013)) using visual analytics. He et al.

present *DynamicsExplorer* (He et al. 2020) to evaluate and diagnose a trained policy in a robotics use case, which incorporates views to track the trajectories of a ball in the maze during episodes. The application enables the inspection of the effect of real-world conditions for trained agents. Saldanha et al. (2019) showcase an application that supports data scientists during experimentation by increasing situational awareness. Key elements are thumbnails summarizing agent performance during episodes and specialized views to understand the connection between particular hyperparameter settings and training performance.

Compared to the existing approaches, we (1) extend the existing frameworks to encompass a holistic view of the relevant stages of the reinforcement learning process instead of just sub-tasks; (2) present a generic, easily adaptable application, which can be instantiated to specific use cases; (3) explicitly consider imitation learning, due to the frequent use in conjunction with reinforcement learning; and (4) apply our framework in a novel, custom real-world use case instead of an existing benchmark environment.

4.6.2 Approach

There has not been a comprehensive workflow for the experimentation and application of reinforcement learning tightly incorporating users. This leaves both researchers and practitioners to loosely defined best practices. In the following chapter, we outline a conceptual workflow for developers and researchers, which we base on guides, projects, and popular open-source libraries. We follow the terminology used, e.g., in the *Gym* package (Brockman et al. 2016). As a starting point, we consider the fundamental workflow from Sacha et al. (2019) that is aimed at generic ML tasks:

1. Prepare-Data: Data selection, cleaning, and transformations; detection of faulty or missing data
2. Prepare-Learning: Specification of an initial model, preparation of training, selection of algorithms, and training parameters
3. Model-Learning: Training of the actual model, monitoring, and supervision
4. Evaluate-Model: Apply the model to testing data, selecting and analyzing quality metrics, and understanding the model

We are interested in highlighting steps and tasks that are specific and critical to reinforcement and imitation learning and which have not been captured previously by more generic workflows. Figure 4.5 summarizes our proposed workflow described in this chapter. In the paper, we discuss the specific stages for imitation and reinforcement learning mirroring the workflow. Specifically, we highlight user tasks during (1) setup and design of the environment which corresponds to mapping a domain-specific problem to a setup applicable to RL algorithms, (2) model training and supervision, and finally (3) evaluation and understanding of trained models. For each of these steps, we present further detailed user tasks and highlight how

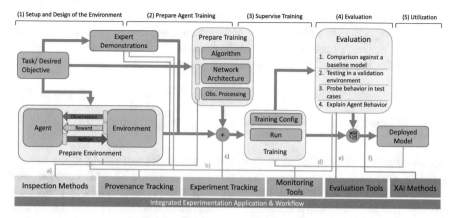

Fig. 4.5 Overview of the **RIVA** (**R**einforcement and **I**mitation Learning with **V**isual **A**nalytics) workflow. **RIVA** is an integrated experimentation workflow and application that provides a range of tools to support all major critical steps: (**a**) inspecting observations, actions, and rewards and ensuring matching values between simulation, expert demonstrations, and architectures, (**b**) provenance tracking of interesting states to enable targeted case-based evaluation, (**c**) tracking of parameters and settings to ensure reproducibility and understand the effect of design decision, (**d**) interactively monitor training and final performance beyond reward, (**e**) enable effective evaluation by integrating multiple evaluation tools, and (**f**) explain behavior by natively integrating XAI methods like input attribution techniques

visual analytics concepts are applicable. We apply the proposed framework and developed an application in the use case of imitation and reinforcement learning for collective behavior: data-driven learning of the behavior of fish schools (collective movement of fish swarms). We cooperated with a domain expert throughout the entire process, from designing custom environments and agents, training, to final evaluation. Modeling the behavior of individual actors in swarm systems has been a long-standing problem in biology (Reynolds 1987; Sumpter 2006; Calovi et al. 2013). Learning individual policies that lead to coordinated collective behavior via both reinforcement learning and imitation learning from recorded trajectories is an exciting application that promises to overcome existing simplifications in hand-crafted models.

4.6.3   Results and Discussion

The use case can be well integrated into our workflow and application with minimal modifications. Noticeably, a custom interactive rendering of the environment was added. We utilize the modularity of the software to integrate additional components like custom visualizations. During the design phase, the inspection views were used to ensure consistency between environment, agent, and dataset, e.g., to spot premature episode termination. Our workflow was highly effective in maintaining a high level of productivity and consistency through an iterative design process, in which we experimented with different observation space designs, reward functions, network types, and hyperparameter configurations. The set of evaluation tools is used both for internal evaluation and external presentation.

4.7 Discussion and Conclusion

In this chapter, we have given an overview of research contributing to the overall goal of enriching high-quality sparse and curated data with VGI contributions to enable different applications like analysis of species distribution or prediction modeling of movement. In particular, we highlighted the potential of visual analytics solutions for different stages of curation, analysis, and model building using VGI data. Visualizations can be especially suited to present data of varying quality and express uncertainty. Both our joint collaborative work on integrating Flickr images with citizen science data and the BirdTrace platform highlight the potential of integrating different data sources, ranking from citizen science platforms, social media, to professional data collection efforts.

Acknowledgments This research was supported by the German Research Foundation DFG within Priority Research Program 1894 *Volunteered Geographic Information: Interpretation, Visualization, and Social Computing* (VGIscience). In this chapter, results from the project *Uncertainty-Aware Enrichment of Animal Movement Trajectories by VGI* (BirdTrace, Project number 314671965) are reviewed.

References

Alber M, Buganza Tepole A, Cannon WR, De S, Dura-Bernal S et al (2019) Integrating machine learning and multiscale modeling–perspectives, challenges, and opportunities in the biological, biomedical, and behavioral sciences. NPJ Digit Med 2(1):1–11

Amershi S, Chickering M, Drucker SM, Lee B, Simard P et al (2015) Modeltracker: redesigning performance analysis tools for machine learning. In: Proceedings of the 33rd Annual ACM Conference on Human Factors in Computing Systems, CHI '15, New York, NY, USA, 2015. Association for Computing Machinery, pp 337–346. ISBN 9781450331456. https://doi.org/10.1145/2702123.2702509

Andrienko N, Andrienko G (2013) Visual analytics of movement: an overview of methods, tools and procedures. Inform Visualization 12(1):3–24

Bellemare MG, Naddaf Y, Veness J, Bowling M (2013) The arcade learning environment: An evaluation platform for general agents. J Artif Intell Res 47:253–279

Benhamou S (2014) Of scales and stationarity in animal movements. Ecol Lett 17(3):261–272

Brockman G, Cheung V, Pettersson L, Schneider J, Schulman J et al (2016) Openai gym

Brown DD, Kays R, Wikelski M, Wilson R, Klimley AP (2013) Observing the unwatchable through acceleration logging of animal behavior. Animal Biotelem 1(1):1–16

Burke HM, Tingley R, Dorin A (2022) Tag frequency difference: rapid estimation of image set relevance for species occurrence data using general-purpose image classifiers. Ecol Inform 69:101598. ISSN 1574-9541. https://doi.org/10.1016/j.ecoinf.2022.101598

Calovi D, Lopez U, Ngo S, Sire C, Chaté H et al (2013) Swarming, schooling, milling: phase diagram of a data-driven fish school model. New J Phys 16:015026. https://doi.org/10.1088/1367-2630/16/1/015026

Campello RJ, Moulavi D, Sander J (2013) Density-based clustering based on hierarchical density estimates. In: Advances in Knowledge Discovery and Data Mining: 17th Pacific-Asia Conference, PAKDD 2013, Gold Coast, Australia, April 14–17, 2013, Proceedings, Part II 17. Springer, Berlin, pp 160–172

Cao N, Lin Y-R, Sun X, Lazer D, Liu S et al (2012) Whisper: tracing the spatiotemporal process of information diffusion in real time. IEEE Trans Visualization Comput Graph 18(12):2649–2658

Cash DW, Adger WN, Berkes F, Garden P, Lebel L et al (2006) Scale and cross-scale dynamics: governance and information in a multilevel world. Ecol Soc 11(2)

Catarci T, Costabile MF, Levialdi S, Batini C (1997) Visual query systems for databases: a survey. J Visual Lang Comput 8(2):215–260

Chapman L, Bell C, Bell S (2017) Can the crowdsourcing data paradigm take atmospheric science to a new level? A case study of the urban heat island of London quantified using Netatmo weather stations. Int J Climatol 37(9):3597–3605

Cole JS, Scott D (1999) Segmenting participation in wildlife watching: a comparison of casual wildlife watchers and serious birders. Hum Dimens Wildl 4(4):44–61

Coxen CL, Frey JK, Carleton SA, Collins DP (2017) Species distribution models for a migratory bird based on citizen science and satellite tracking data. Global Ecol Conserv 11:298–311

Demšar U, Buchin K, Cagnacci F, Safi K, Speckmann B et al (2015) Analysis and visualisation of movement: an interdisciplinary review. Movement Ecol 3(1):1–24

Diehl A, Yang B, Das RD, Chen S, Andrienko G et al (2018) User-uncertainty: a human-centred uncertainty taxonomy for VGI through the visual analytics workflow

Dulac-Arnold G, Levine N, Mankowitz DJ, Li J, Paduraru C et al (2020) An empirical investigation of the challenges of real-world reinforcement learning. arXiv e-prints, art. arXiv:2003.11881

Endert A, Ribarsky W, Turkay C, Wong W, Nabney I et al (2018) The state of the art in integrating machine learning into visual analytics. arXiv e-prints, art. arXiv:1802.07954

Ferreira N, Lins L, Fink D, Kelling S, Wood C et al (2011) Birdvis: visualizing and understanding bird populations. IEEE Trans Visualization Comput Graph 17(12):2374–2383

Fink D, Hochachka WM, Zuckerberg B, Winkler DW, Shaby B et al (2010) Spatiotemporal exploratory models for broad-scale survey data. Ecol Appl 20(8):2131–2147

Fink D, Damoulas T, Dave J (2013) Adaptive spatio-temporal exploratory models: Hemisphere-wide species distributions from massively crowdsourced ebird data. In: Proceedings of the AAAI Conference on Artificial Intelligence, vol 27, pp 1284–1290

Geldmann J, Heilmann-Clausen J, Holm TE, Levinsky I, Markussen B et al (2016) What determines spatial bias in citizen science? Exploring four recording schemes with different proficiency requirements. Diversity Distrib 22(11):1139–1149

Hartmann MC, Schott M, Dsouza A, Metz Y, Volpi M et al (2022) A text and image analysis workflow using citizen science data to extract relevant social media records: combining red kite observations from flickr, ebird and inaturalist. Ecol Inform 71:101782

He W, Lee T, van Baar J, Wittenburg K, Shen H (2020) Dynamicsexplorer: visual analytics for robot control tasks involving dynamics and LSTM-based control policies. In: 2020 IEEE Pacific Visualization Symposium (PacificVis), pp 36–45. https://doi.org/10.1109/PacificVis48177.2020.7127

Henderson P, Islam R, Bachman P, Pineau J, Precup D et al (2017) Deep reinforcement learning that matters. In: 32nd AAAI Conference on Artificial Intelligence, AAAI 2018, pp 3207–3214. http://arxiv.org/abs/1709.06560.

Jain N, Dilkina B (2015) Coarse models for bird migrations using clustering and non-stationary Markov chains. In: AAAI Workshop: Computational Sustainability

Kelling S, Gerbracht J, Fink D, Lagoze C, Wong W-K et al (2012) ebird: a human/computer learning network for biodiversity conservation and research. In: Twenty-Fourth IAAI Conference

Levin SA (1992) The problem of pattern and scale in ecology: the robert h. macarthur award lecture. Ecology 73(6):1943–1967

MacEachren AM, Jaiswal A, Robinson AC, Pezanowski S, Savelyev A et al (2011) Senseplace2: Geotwitter analytics support for situational awareness. In: 2011 IEEE conference on visual analytics science and technology (VAST). IEEE, pp 181–190

Mandhane A et al (2022) MuZero with self-competition for rate control in VP9 video compression. arXiv e-prints, art. arXiv:2202.06626

Martín-Guerrero, JD, Lamata L (2021) Reinforcement learning and physics. Appl Sci 11(18). ISSN 2076-3417. https://doi.org/10.3390/app11188589

McFarlane BL (1994) Specialization and motivations of birdwatchers. Wildl Soc Bull 361–370

Meschenmoser P, Buchmüller JF, Seebacher D, Wikelski M, Keim DA (2020) Multisegva: using visual analytics to segment biologging time series on multiple scales. IEEE Trans Visualization Comput Graph 27(2):1623–1633

Metz Y, Schlegel U, Seebacher D, El-Assady M, Keim DA (2022) A comprehensive workflow for effective imitation and reinforcement learning with visual analytics. In: 13th International EuroVis Workshop on Visual Analytics (EuroVA 2022), pp 19–23

Nazemi K, Burkhardt D, Ginters E, Kohlhammer J (2015) Semantics visualization–definition, approaches and challenges. Proc Comput Sci 75:75–83

Nguyen H, La H (2019) Review of deep reinforcement learning for robot manipulation. In: 2019 Third IEEE International Conference on Robotic Computing (IRC), pp 590–595. https://doi.org/10.1109/IRC.2019.00120

Reynolds CW (1987) Flocks, herds and schools: a distributed behavioral model. In: Proceedings of the 14th Annual Conference on Computer Graphics and Interactive Techniques, SIGGRAPH '87, New York, NY, USA, 1987. Association for Computing Machinery, pp 25–34. ISBN 0897912276. https://doi.org/10.1145/37401.37406

Sacha D, Al-Masoudi F, Stein M, Schreck T, Keim DA et al (2017) Dynamic visual abstraction of soccer movement. In: Computer Graphics Forum, vol 36. Wiley Online Library, pp 305–315

Sacha D, Kraus M, Keim DA, Chen M (2019) VIS4ML: an ontology for visual analytics assisted machine learning. IEEE Trans Vis Comput Graph 25(1):385–395. https://doi.org/10.1109/TVCG.2018.2864838

Saldanha E, Praggastis B, Billow T, Arendt D (2019) ReLVis : visual analytics for situational awareness during reinforcement learning experimentation. In: EuroVis (Short Papers). Eurographics Association, pp 43–47. https://doi.org/10.2312/evs.20191168

Scott D, Shafer CS (2001) Recreational specialization: a critical look at the construct. J Leisure Res 33(3):319–343

Scott D, Thigpen J (2003) Understanding the birder as tourist: segmenting visitors to the Texas hummer/bird celebration. Hum Dimensions Wildl 8(3):199–218

Shepard EL, Wilson RP, Quintana F, Laich AG, Liebsch N et al (2008) Identification of animal movement patterns using tri-axial accelerometry. Endangered Species Res 10:47–60

Spinner T, Schlegel U, Schäfer H, El-Assady M (2020) explainer: a visual analytics framework for interactive and explainable machine learning. IEEE Trans Vis Comput Graph 26(1):1064–1074. https://doi.org/10.1109/TVCG.2019.2934629

Spretke D, Bak P, Janetzko H, Kranstauber B, Mansmann F, Davidson S (2011) Exploration through enrichment: a visual analytics approach for animal movement. In: Proceedings of the 19th ACM SIGSPATIAL International Conference on Advances in Geographic Information Systems, pp 421–424.

Sullivan BL, Wood CL, Iliff MJ, Bonney RE, Fink D et al (2009) ebird: a citizen-based bird observation network in the biological sciences. Biol Conserv 142(10):2282–2292

Sumpter D (2006) The principles of collective animal behavior. Philos Trans R Soc Lond Series B: Biol Sci 361:5–22. https://doi.org/10.1098/rstb.2005.1733

Thom D, Bosch H, Koch S, Wörner M, Ertl T (2012) Spatiotemporal anomaly detection through visual analysis of geolocated twitter messages. In: 2012 IEEE Pacific Visualization Symposium. IEEE, pp 41–48

Unger S, Rollins M, Tietz A, Dumais H (2020) inaturalist as an engaging tool for identifying organisms in outdoor activities. J Biol Educ 0(0):1–11. https://doi.org/10.1080/00219266.2020.1739114

Wang J, Gou L, Shen HW, Yang H (2019a) DQNViz: a visual analytics approach to understand deep Q-networks. IEEE Trans Vis Comput Graph 25(1):288–298. ISSN 19410506. https://doi.org/10.1109/TVCG.2018.2864504

Wang Z, Hale S, Adelani DI, Grabowicz P, Hartman T et al (2019b) Demographic inference and representative population estimates from multilingual social media data. In: The World Wide Web Conference, WWW '19, New York, NY, USA, 2019b. Association for Computing Machinery, pp 2056–2067. ISBN 9781450366748. https://doi.org/10.1145/3308558.3313684

Wikelski M, Davidson SC (2023) Movebank: archive, analysis and sharing of animal movement data. Hosted by the max planck institute of animal behavior. www.movebank.org

Wood C, Sullivan B, Iliff M, Fink D, Kelling S (2011) ebird: engaging birders in science and conservation. PLoS Biol 9(12):e1001220

Yao Y-C (1988) Estimating the number of change-points via Schwarz'criterion. Stat Probab Lett 6(3):181–189

Chapter 5
Two Worlds in One Network: Fusing Deep Learning and Random Forests for Classification and Object Detection

Christoph Reinders, Michael Ying Yang, and Bodo Rosenhahn

Abstract Neural networks have demonstrated great success; however, large amounts of labeled data are usually required for training the networks. In this work, a framework for analyzing the road and traffic situations for cyclists and pedestrians is presented, which only requires very few labeled examples. We address this problem by combining convolutional neural networks and random forests, transforming the random forest into a neural network, and generating a fully convolutional network for detecting objects. Because existing methods for transforming random forests into neural networks propose a direct mapping and produce inefficient architectures, we present neural random forest imitation—an imitation learning approach by generating training data from a random forest and learning a neural network that imitates its behavior. This implicit transformation creates very efficient neural networks that learn the decision boundaries of a random forest. The generated model is differentiable, can be used as a warm start for fine-tuning, and enables end-to-end optimization. Experiments on several real-world benchmark datasets demonstrate superior performance, especially when training with very few training examples. Compared to state-of-the-art methods, we significantly reduce the number of network parameters while achieving the same or even improved accuracy due to better generalization.

Keywords Classification · Neural networks · Random forests · Object detection · Localization · Imitation learning

C. Reinders (✉) · B. Rosenhahn
Institute for Information Processing, L3S/Leibniz University Hannover, Hannover, Germany
e-mail: reinders@tnt.uni-hannover.de; rosenhahn@tnt.uni-hannover.de

M. Y. Yang
Scene Understanding Group, University of Twente, Enschede, Netherlands
e-mail: michael.yang@utwente.nl

D. Burghardt et al. (eds.), *Volunteered Geographic Information*,
https://doi.org/10.1007/978-3-031-35374-1_5

5.1   Introduction

During the last few years, the availability of spatial data has rapidly developed. An essential aspect of this development is the involvement of a large number of users, who often use smartphones and mobile devices, to generate and make freely available volunteered geographic information (VGI). For example, apps like *Waze* combine the local velocities of smartphones (in cars) to predict the flow velocities (and time delay) of traffic jams. Users can recommend and comment on specific traffic situations. Although GPS and gyroscope data (e.g., in fitness straps) are common, images allow a comprehensive scene understanding. The collection of large amounts of unlabeled images is easy; however, the development of machine learning methods for scene analysis with limited amounts of labeled data is challenging.

Neural networks have become very popular in many areas, such as computer vision (Krizhevsky et al. 2012; Reinders et al. 2022; Ren et al. 2015; Simonyan and Zisserman 2015; Zhao et al. 2017; Qiao et al. 2021; Rudolph et al. 2022; Sun et al. 2021a), speech recognition (Graves et al. 2013; Park et al. 2019; Sun et al. 2021a), automated game-playing (Mnih et al. 2015; Dockhorn et al. 2017), or natural language processing (Collobert et al. 2011; Sutskever et al. 2014; Otter et al. 2021). Researchers have published many datasets for training neural networks and put enormous effort into providing labels for each data sample. For real-world applications, the dependency on large amounts of labeled data represents a significant limitation (Breiman et al. 1984; Hekler et al. 2019; Barz and Denzler 2020; Qi and Luo 2020; Phoo and Hariharan 2021; Wang et al. 2021). Frequently, there is little or even no labeled data for a particular task, and hundreds or thousands of examples have to be collected and annotated. This particularly affects new applications and rare labels (e.g., detecting rare diseases or defects in manufacturing). Transfer learning and regularization methods are usually applied to reduce overfitting. However, for training with little data, the networks still have a considerable number of parameters that have to be fine-tuned—even if just the last layers are trained.

In contrast to neural networks, random forests are very robust to overfitting due to their ensemble of multiple decision trees. Each decision tree is trained on randomly selected features and samples. Random forests have demonstrated remarkable performance in many domains (Fernández-Delgado et al. 2014). While the generated decision rules are simple and interpretable, the orthogonal separation of the feature space can also be disadvantageous on other datasets, especially with correlated features (Menze et al. 2011). Additionally, random forests are not differentiable and cannot be fine-tuned with gradient-based optimization.

In this research project *Comprehensive Conjoint GPS and Video Data Analysis for Smart Maps* (COVMAP), we are interested in combining GPS, gyroscope, and image data to analyze road and traffic situations for cyclists and pedestrians. Our standard setting is a smartphone attached to a bicycle, which records the GPS coordinates, images, motion information, local weather information, and time. We present a framework for detecting traffic signs that are of interest for cyclists and pedestrians. Related to this work, Chap. 3 introduces methods for anonymizing and map-matching trajectories, and Chap. 1 presents a geographic knowledge graph for a semantic representation of geographic entities in OSM. The goal of this work is to minimize the costs of annotating a dataset and enable the detection of objects with only a handful of examples per class. For that, we combine neural networks and random forests and bring both worlds together. After generating a classifier for image patches, the random forest is mapped to a neural network to combine all modules in a single pipeline, and a fully convolutional network is created for object detection.

Mapping random forests into neural networks is already used in many applications such as network initialization (Humbird et al. 2019), camera localization (Massiceti et al. 2017), object detection (Reinders et al. 2018), or semantic segmentation (Richmond et al. 2016). State-of-the-art methods (Massiceti et al. 2017; Sethi 1990; Welbl 2014) create a two-hidden-layer neural network by adding a neuron for each split node and each leaf node of the decision trees. The number of parameters of the networks becomes enormous as the number of nodes grows exponentially with the increasing depth of the decision trees. Additionally, many weights are set to zero so that an inefficient representation is created. Due to both reasons, the mappings do not scale and are only applicable to simple random forests.

In this work, we present an imitation learning approach to generate neural networks from random forests, which results in very efficient models. We introduce a method for generating training data from a random forest that creates any amount of input-target pairs. With this data, a neural network is trained to imitate the random forest. Experiments demonstrate that the accuracy of the imitating neural network is equal to the original accuracy or even slightly better than the random forest due to better generalization while being significantly smaller. To summarize, our **contributions** are:

- We present a pipeline for detecting and localizing traffic signs for cyclists and pedestrians with very few labeled training examples by combining convolutional neural networks and random forests.
- We propose a novel method for implicitly transforming random forests into neural networks by generating data from a random forest and training an random forest-imitating neural network. Labeled data samples are created by evaluating the decision boundaries and guided routing to selected leaf nodes.
- In contrast to direct mappings, our imitation learning approach is scalable to complex classifiers and deep random forests.
- We enable learning and initialization of neural networks with very little data.
- Neural networks and random forests can be combined in a fully differentiable, end-to-end pipeline for acceleration and further fine-tuning.

5.2   Related Work

Many deep learning-based methods have been presented for object detection in recent years. Two-stage methods like R-CNN (Girshick et al. 2014), Fast R-CNN (Girshick 2015), and Faster R-CNN (Ren et al. 2015) include a region proposal mechanism and predict the object scores and boundaries based on the pooled features. Cascade R-CNN (Cai and Vasconcelos 2018) consists of multiple R-CNN stages that progressively refine the predicted bounding boxes. Sparse R-CNN (Sun et al. 2021b) learns a fixed set of bounding box candidates. One-stage methods achieve great performance by regressing and classifying candidate bounding boxes of a predefined set of anchor boxes. Well-known methods are SSD (Liu et al. 2016), YOLO (Redmon and Farhadi 2016), and RetinaNet (Lin et al. 2017). CenterNet (Duan et al. 2019) introduces a triplet representation, including one center keypoint and two corners. FCOS (Tian et al. 2019) presents a center-ness branch for anchor-free detection. YOLOF (Chen et al. 2021) uses a single-scale feature map without feature pyramid network. DETR (Carion et al. 2020) models object detection as a set prediction problem and introduces a vision transformer architecture. $R(Det)^2$ (Li and Wang 2022) presents a combination of soft decision trees and neural networks for randomized decision routing. All the presented methods have a huge number of trainable parameters and require large amounts of labeled data for training.

Random forests and neural networks share some similar characteristics, such as the ability to learn arbitrary decision boundaries; however, both methods have different advantages. Random forests are based on decision trees. Various tree models have been presented—the most well known are C4.5 (Quinlan 1993) and CART (Breiman et al. 1984). Decision trees learn rules by splitting the data. The rules are easy to interpret and additionally provide an importance score of the features. Random forests (Breiman 2001) are an ensemble method consisting of multiple decision trees, with each decision tree being trained using a random subset of samples and features. Fernández-Delgado et al. (2014) conduct extensive experiments comparing 179 classifiers on 121 UCI datasets (Dua and Graff 2017). The authors show that random forests perform best, followed by support vector machines with a radial basis function kernel. Therefore, random forests are often considered as a reference for new classifiers.

Neural networks are universal function approximators. The generalization performance has been widely studied. Zhang et al. (2017) demonstrate that deep neural networks are capable of fitting random labels and memorizing the training data. Bornschein et al. (2020) analyze the performance across different dataset sizes. Olson et al. (2018) evaluate the performance of modern neural networks using the same test strategy as Fernández-Delgado et al. (2014) and find that neural networks achieve good results but are not as strong as random forests.

Sethi (1990) presents a mapping of decision trees to two-hidden-layer neural networks. In the first hidden layer, the number of neurons equals the number of split nodes in the decision tree. Each of these neurons implements the decision function of the split nodes and determines the routing to the left or right child node. The second hidden layer has a neuron per leaf node in the decision tree. Each of the neurons is connected to all split nodes on the path from the root node to the leaf node to evaluate if the data is routed to the respective leaf node. Finally, the output layer is connected to all leaf neurons and aggregates the results by implementing the leaf votes. By using hyperbolic tangent and sigmoid functions, respectively, as activation functions between the layers, the generated network is differentiable and, thus, trainable with gradient-based optimization algorithms. The method can be easily extended to random forests by mapping all trees.

Welbl (2014) and Biau et al. (2019) follow a similar strategy. The authors propose a method that maps random forests into neural networks as a smart initialization and then fine-tunes the networks by backpropagation. Two training modes are introduced: *independent* and *joint*. Independent training fits all networks one after the other and creates an ensemble of networks as a final classifier. Joint training concatenates all tree networks into one single network so that the output layer is connected to all leaf neurons in the second hidden layer from all decision trees and all parameters are optimized together. Additionally, the authors evaluate sparse and full connectivity. Sparse connectivity maintains the tree structures and has fewer weights to train. In practice, sparse weights require a special differentiable implementation, which can drastically decrease performance, especially when training on a GPU. Full connectivity optimizes all parameters of the fully connected network. Massiceti et al. (2017) extend this approach and introduce a network splitting strategy by dividing each decision tree into multiple subtrees. The subtrees are mapped individually and share common neurons for evaluating the split decision.

These techniques, however, are only applicable to trees of limited depth. As the number of nodes grows exponentially with the increasing depth of the trees, inefficient representations are created, causing extremely high memory consumption. In this work, we address this issue by proposing an imitation learning-based method that results in much more efficient models.

5.3 Traffic Sign Recognition

In the first part of this chapter, we present a framework for object detection and localization that is able to recognize traffic signs for cyclists and pedestrians with very few labeled examples. While there are a lot of datasets for cars, the amount of labeled data for cyclists and pedestrians is very limited. Therefore, the advantages of convolutional neural networks and random forests are combined to build a robust object detector. After the detection of the objects, the image, GPS, and motion data are fused to localize the traffic signs on the map. We introduce an app for collecting and synchronizing data with a customary smartphone and present the captured dataset. Finally, experiments are performed to analyze the recognition performance. All details and further evaluations can be found in Reinders et al. (2018) and Reinders et al. (2019).

5.3.1 Framework

The framework consists of three modules. First, a system for object detection based on convolutional neural networks and random forests is presented. Afterward, the detected traffic signs are localized on the map by integrating GPS and motion information. Lastly, multiple observations are clustered to improve the precision.

5.3.1.1 Object Detection

In the first step, we train a convolutional neural network for representation learning on a related task where large amounts of data are available. In this application, the GTSRB (Stallkamp et al. 2012) dataset is selected, which consists of images of traffic signs for cars. The architecture of the network is a standard backbone (Springenberg et al. 2015) with multiple convolutional layers and a global average pooling. For generating the feature representations, the output of the last layer before the final classification layer is calculated.

On the downstream task, we start with a classifier for image patches. The feature representations of all patches and a fixed number of background patches are extracted. Because only a few number of labeled examples are available, we train a random forest to classify the image features and predict one of the C classes or background. The ensemble of multiple decision trees trained on different subsets of features and samples is very robust to overfitting (Breiman 2001).

Afterward, the convolutional neural network for feature generation and random forest for classification are combined in one pipeline. For that, we transform the random forest into a neural network using a method presented by Sethi (1990) and Welbl (2014). The method creates a two-hidden-layer neural network by mapping each decision tree of the random forest. An example of mapping a decision tree into a neural network is visualized in Fig. 5.1. For each split node in the decision tree, a neuron is created in the first hidden layer. The neurons are connected to the respective split features (all other weights are set to zero if no sparse architecture is used) and evaluate the split decisions, i.e., the routing to the left or right child node. In the second hidden layer, a neuron is created for each leaf node in the decision tree. The neurons combine the split decisions from the previous layer and determine whether the sample is routed to the respective leaf. In the output layer, the number of neurons corresponds to the number of classes. Each neuron stores the class votes from the leafs. Mapping a random forest, i.e., multiple decision trees, is done by mapping each decision tree and combining the neural networks. Now, we are able to create a fully convolutional network (Shelhamer et al. 2017) by replacing the fully connected layers with convolutional layers that perform the identical operation. Due to the shared features, the processing of the images is significantly accelerated. The images are analyzed by the fully convolutional network at multiple scales, and the output predicts the probability of each traffic sign class at each spatial position. In a post-processing, all detections with a probability larger than a defined threshold are extracted, and a non-maximum suppression is performed.

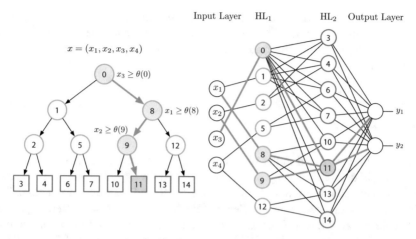

Fig. 5.1 A decision tree (left) can be mapped to a two-hidden-layer neural network (right). For each *split node* (green circle) in the decision tree, a neuron in the first hidden layer is created which evaluates the split rule. For each *leaf node* (blue rectangle), a neuron in the second hidden layer is created which determines the leaf membership. A routing to leaf node 11, for example, involves the split nodes 0, 8, and 9. The relevant connections for the corresponding calculation in the neural network are highlighted

5.3.1.2 Localization

The detected 2D bounding boxes are localized on the map by integrating GPS and heading information. Each image is associated with a GPS position and a heading. The heading points in the direction in which the device is oriented. For each bounding box, the depth is estimated by assuming a simple pinhole camera model, and the relative heading is determined based on the horizontal position in the image. Afterward, the information can be combined with the GPS position and heading of the image to generate the latitude, longitude, and heading of the traffic sign.

5.3.1.3 Clustering

After localizing the traffic signs, we merge multiple observations of the same traffic sign. Clustering algorithms (MacQueen et al. 1967; Fukunaga and Hostetler 1975; Dockhorn et al. 2015, 2016; Schier et al. 2022) automatically discover natural groupings in the data. If multiple detections exist in an image, we can generate additional constraints because we know that multiple traffic signs exist and the respective traffic signs should not be grouped. The additional information is represented as cannot-link constraints. For weakly supervised clustering with an unknown number of clusters, constrained mean shift (CMS) (Schier et al. 2022) clustering is performed to merge the detections. CMS is a density-based clustering algorithm that extends mean shift clustering (Fukunaga and Hostetler 1975) by enabling sparse supervision using cannot-link constraints. The clustering of the detections improves the localization accuracy and makes the position estimation more robust.

5.3.2 Dataset

To analyze the road and traffic situations for cyclists and pedestrians, we collected a real-world dataset. For that, we developed an app for capturing and synchronizing images and data from other sensors, like GPS and motion information. The smartphone is attached to the handlebar of the bicycle so that the camera is pointed in the direction of travel. Because monotonous routes, e.g., in rural areas, produce many similar images, we therefore introduce an adaptive filtering of the images to automatically detect points of interest. For that, we integrate motion information and apply a twofold filtering strategy based on decreases in speed and acceleration: (i) Decreases in speed indicate situations where the cyclist has to slow down because of potential traffic obstructions such as traffic jams, construction works, or other road users. (ii) Acceleration is used to analyze the road conditions and to detect, for example, potholes.

The collected dataset consists of 500 tours with a total riding time of 6 days in different cities. A visualization of the collected tours in Hanover is shown in Fig. 5.2. After filtering, the dataset has 56000 images with a size of 1080×1920 pixels. For the detection of traffic signs, we selected ten traffic signs that are of interest for

Fig. 5.2 Example tracks collected around Hanover. In total, we collected in Hanover, 450 tours, >47K images, 5.4 days of riding; Enschede, 40 tours, >8K images, 18 hours of riding; and Heidelberg, 11 tours, 1000 images, several hours of riding

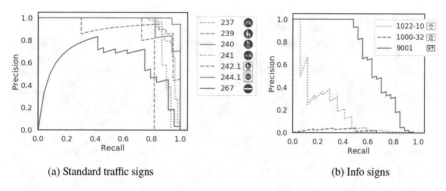

(a) Standard traffic signs   (b) Info signs

Fig. 5.3 Precision-recall curve for each class to analyze the recognition performance on the test set. (a) Standard traffic signs. (b) Info signs

cyclists and pedestrians and manually annotated the ground truth for a set of images to have data for training and testing. Overall, 524 bounding boxes are annotated in the images and split 50/50 in training and testing. The splitting is repeated multiple times with different seeds.

5.3.3 Experiments

The framework is evaluated on the presented dataset to analyze the recognition performance. For that, all bounding boxes are predicted at multiple scales and assigned to the ground truth bounding box with the highest overlap if the IoU is greater or equal than 0.5. The resulting precision-recall curve for each class is presented in Fig. 5.3. While the performance of the standard traffic signs is good, the more inconspicuous traffic signs are detected worse. The recognition performance of the latter correlates with the number of examples that are available for training. Qualitative examples are shown in Fig. 5.4. For more details and further analyses, please see Reinders et al. (2018) and Reinders et al. (2019).

5.4 Neural Random Forest Imitation

We propose a novel method, called *neural random forest imitation* (NRFI), for implicitly transforming random forests into neural networks that learns the decision boundaries and generates efficient representations. The advantages of our approach for mapping random forests into neural networks are threefold: (1) We enable the generation of neural networks with very few training examples. (2) The resulting network can be used as a warm start, is fully differentiable, and allows further end-to-end fine-tuning. (3) The generated network can be easily integrated into

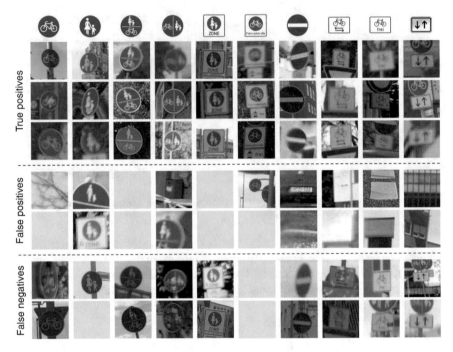

Fig. 5.4 Qualitative results of randomly selected examples on the test set. True positives, false positives, and false negatives are shown for each class. Some classes have less than two false positives or false negatives, respectively

any trainable pipeline (e.g., jointly with feature extraction), and existing high-performance deep learning frameworks can be used directly. This accelerates the process and enables parallelization via GPUs. In the following, we evaluate on standard benchmark datasets to present a general approach for various domains. While we focus on classification tasks in this work, NRFI can be simply adapted for regression tasks.

5.4.1 Background and Notation

In this section, we briefly describe decision trees (Breiman et al. 1984), random forests (Breiman 2001), and the notation used throughout this work. Decision trees consist of *split nodes* N^{split} and *leaf nodes* N^{leaf}. Each split node $s \in N^{\text{split}}$ performs a *split decision* and routes a data sample x to the left or right child node, denoted as $c_{\text{left}}(s)$ and $c_{\text{right}}(s)$, respectively. When using binary, axis-aligned split decisions, a single feature $f(s) \in \{1, \ldots, N\}$ and a threshold $\theta(s) \in \mathbb{R}$ are the basis for the split, where N is the number of features. If the value of feature $f(s)$ is smaller than

$\theta(s)$, the data sample is routed to the left child node and otherwise to the right child node, denoted as

$$x \in c_{\text{left}}(s) \iff x_{f(s)} < \theta(s) \tag{5.1}$$

$$x \in c_{\text{right}}(s) \iff x_{f(s)} \geq \theta(s). \tag{5.2}$$

Data samples are routed through a decision tree until a leaf node $l \in N^{\text{leaf}}$ is reached which stores the target value. For the classification task, these are the estimated class probabilities $P_{\text{leaf}}(l) = (p_1^l, \ldots, p_C^l)$, where C is the number of classes. Decision trees are trained by creating a root node and consecutively finding the best split of the data based on a criterion. The resulting subsets are assigned to the left and right child node, and the subtrees are processed further. Commonly used criteria are the *Gini impurity* or *entropy*.

A single decision tree is very fast and operates on high-dimensional data. However, it tends to overfit the training data by constructing a deep tree that separates perfectly all training examples. While having a very small training error, this easily results in a large test error. Random forests address this problem by learning an ensemble of n_T decision trees. Each tree is trained with a random subset of training examples and features. The prediction $RF(x)$ of a random forest is calculated by averaging the predictions of all decision trees.

5.4.2 Methodology

Our proposed neural random forest imitation approach implicitly transforms random forests into neural networks. The main concept includes (1) generating training data from decision trees and random forests, (2) adding strategies for reducing conflicts and increasing the variety of the generated examples, and (3) training a neural network that imitates the random forest by learning the decision boundaries. As a result, NRFI enables the transformation of random forests into efficient neural networks. An overview of the proposed method is shown in Fig. 5.5.

5.4.2.1 Data Generation

First, we propose a method for generating data from a given random forest. A data sample $x \in \mathbb{R}^N$ is an N-dimensional vector, where N is the number of features. We select a target class $t \in [1, \ldots, C]$ from C classes and generate a data sample for the selected class.

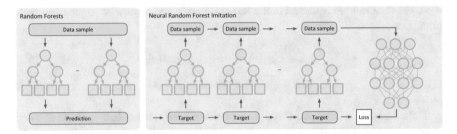

Fig. 5.5 Neural random forest imitation enables an implicit transformation of random forests into neural networks. Usually, data samples are propagated through the individual decision trees, and the split decisions are evaluated during inference. We propose a method for generating input-target pairs by reversing this process and training a neural network that imitates the random forest. The resulting network is much smaller compared to current state-of-the-art methods, which directly map the random forest

Data Initialization

A data sample x is initialized randomly. In the following, the feature-wise minimum and maximum of the training samples will be denoted as $f_{\min}, f_{\max} \in \mathbb{R}^N$. To initialize x, we sample $x \sim U(f_{\min}, f_{\max})$. In the next step, we will present a method for adapting the data sample to obtain characteristics of the target class.

Data Generation from Decision Trees

A decision tree processes an input vector x by routing the data through the tree until a leaf is reached. At each node, a split decision is evaluated, and the input is passed to the left child node or the right child node. Finally, a leaf l is reached which stores the estimated probabilities $P_{\text{leaf}}(l) = (p_1^l, \ldots, p_C^l)$ for each class.

We reverse this process and present a method for generating training data from a decision tree. An overview of the proposed data generation process is shown in Fig. 5.6. First, the class distribution information is propagated bottom-up from the leaf nodes to the split nodes (see Fig. 5.6a), and we define the class weights $W(n) = (w_1^n, \ldots, w_C^n)$ for every node n as follows:

$$W(n) = \begin{cases} P_{\text{leaf}}(n) & \text{if } n \in N^{\text{leaf}} \\ W(c_{\text{left}}(n)) + W(c_{\text{right}}(n)) & \text{if } n \in N^{\text{split}} \end{cases} \tag{5.3}$$

For every leaf node, the class weights are equal to the stored probabilities in the leaf. For every split node, the class weights in the child nodes are summed up.

After preparation, data samples for a target class t are generated (see Fig. 5.6b). For that, characteristics of the target class are successively added to the data sample. Starting at the root node, we modify the input data so that it is routed through

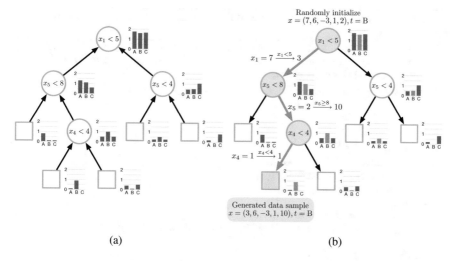

Fig. 5.6 Overview of the data generation process from a decision tree. First, the class distribution information is propagated from the leaf nodes to the split nodes (**a**). Afterward, data samples are generated by guided routing (Sect. 5.4.2.1) and modifying the data based on the split decisions (**b**). The weights for sampling the left or right child node are highlighted in orange

selected split nodes until a leaf node is reached. The pseudocode is presented in Algorithm 1.

The routing is guided based on the weights for the target class in the left child node $w_{\text{left}} = w_t^{c_{\text{left}}(n)}$ and right child node $w_{\text{right}} = w_t^{c_{\text{right}}(n)}$. The weights are normalized by their L2-norm, denoted as $\hat{w}_{\text{left}}$ and $\hat{w}_{\text{right}}$. Afterward, the left or right child node is randomly selected as next child node n_{next} depending on $\hat{w}_{\text{left}}$ and $\hat{w}_{\text{right}}$.

In the next step, the data sample is updated. We verify that the data sample is routed to the selected child node by evaluating the split decision. A split node s routes the data to the left or right child node based on a split feature $f(s)$ and a threshold $\theta(s)$. If the value of the split feature $x_{f(s)}$ is smaller than $\theta(s)$, the data sample is routed to the left child node and otherwise to the right child node. The data sample is modified if it is not already routed to the selected child node by assigning a new value. If the selected child node is the left child node, the value has to be smaller than the threshold $\theta(s)$, and a new value within the minimum feature value $f_{\text{min},f(s)}$ and $\theta(s)$ is randomly sampled:

$$x_{f(s)} \sim U(f_{\text{min},f(s)}, \theta(s)). \tag{5.4}$$

If the data sample is supposed to be routed to the right child node, the new value is randomly sampled between $\theta(s)$ and the maximum feature value $f_{\text{max},f(s)}$:

$$x_{f(s)} \sim U(\theta(s), f_{\text{max},f(s)}). \tag{5.5}$$

Algorithm 1 DATAGENERATIONFROMTREE Generate data samples from a decision tree

---

Input: Decision tree split features $f(n)$ and thresholds $\theta(n)$, target class t, feature minimum f_{min} and maximum f_{max}, class weights $W(n) = (w_1^n, \ldots, w_C^n)$ for all nodes $n \in N^{split} \cup N^{leaf}$

Output: Data sample for target class t

1: Sample $x \sim U(f_{min}, f_{max}) \in \mathbb{R}^N$
2: $n \leftarrow$ root node
3: **while** $n \notin N^{leaf}$ **do**
4: 　　　$w_{left} \leftarrow w_t^{c_{left}(n)}$
5: 　　　$w_{right} \leftarrow w_t^{c_{right}(n)}$
6: 　　**if** feature $f(n)$ is already used **then**
7: 　　　　weight current route with w_{path}
8: 　　　　$w_{current} \leftarrow w_{current} \cdot w_{path}$
9: 　　**end if**
10: 　　$\hat{w}_{left}, \hat{w}_{right} \leftarrow$ normalize w_{left} and w_{right}
11: 　　$n_{next} \leftarrow$ randomly select left or right child node with probability of $\hat{w}_{left}$ and $\hat{w}_{right}$, respectively
12: 　　**if** $n_{next} = c_{left}(n)$ **then**
13: 　　　　**if** $x_{f(n)} \geq \theta(n)$ **then**
14: 　　　　　　$x_{f(n)} \sim U(f_{min,f(n)}, \theta(n))$
15: 　　　　**end if**
16: 　　**else**
17: 　　　　**if** $x_{f(n)} < \theta(n)$ **then**
18: 　　　　　　$x_{f(n)} \sim U(\theta(n), f_{max,f(n)})$
19: 　　　　**end if**
20: 　　**end if**
21: 　　mark feature $f(n)$ as used
22: 　　$n \leftarrow n_{next}$
23: **end while**
24: **return** x

This process is repeated until a leaf node is reached. In each node, characteristics are added that classify the data sample as the target class.

During this process, modifications can conflict with previous decisions because features are used multiple times within a decision tree or across multiple decision trees. Therefore, the current routing is weighted with a factor $w_{path} \geq 1$ to prioritize the path and not change the data sample if possible. Overall, the presented method enables the generation of data samples and corresponding labels from a decision tree without adding any further data.

Data Generation from Random Forests

In the next step, we extend the method to generate data from random forests. Random forests consist of n_T decision trees $RF = \{T_1, \ldots, T_{n_T}\}$. For generating a data sample x, the presented method for a single decision tree is applied to multiple decision trees consecutively. The initialization is performed only once, and the visited features are shared. In each decision tree, the data sample is modified and routed to selected nodes based on the target class t. When using all decision trees, data samples are created where all trees agree with a high probability. For generating examples with varying confidence, i.e., the predictions of the individual decision trees diverge, we select a subset of n_{sub} decision trees $RF_{sub} \subseteq RF$.

All decision trees in RF_{sub} are processed in random order to generate a data sample. For each decision tree, the presented method modifies the data sample based on the target class. Finally, the output of the random forest $y = \mathrm{RF}(x)$ is predicted. In most cases, the prediction matches the target class. Due to factors such as the stochastic process, a small subset size, or varying predictions of the decision trees, it can be different occasionally. Thus, an input-target pair (x, y) has been created, showing similar characteristics as the target class and any amount of data can be generated by repeating this process.

Automatic Confidence Distribution

The number of decision trees n_{sub} can be set to a fixed value or sampled uniformly. Alternatively, we will present an automatic process for determining an optimal distribution of the confidences for generating a wide variety of different examples. The strategy is motivated by *importance weighting* (Fang et al. 2020). We generate n data samples (n is empirically set to 1000) for each number of decision trees $j \in [1, n_T]$. The respective generated datasets will be denoted as D_j.

An optimal sampling process generates highly diverse data samples with different confidences. To achieve that, an automated balancing of the distributions is determined. A histogram with H bins is calculated for each D_j, where h_i^j denotes the number of generated examples in the ith interval (equally distributed) from the distribution with j decision trees. In the next step, a weight w_j^D is defined for each distribution, and we optimize w^D as follows:

$$
\min_{w^D} \left\| \begin{bmatrix} \sum_{j=1}^{n_T} w_j^D h_1^j & \cdots & \sum_{j=1}^{n_T} w_j^D h_H^j \end{bmatrix}^T - \begin{bmatrix} 1 \\ \vdots \\ 1 \end{bmatrix} \right\|^2 \quad \text{s.t.} \quad \forall_j \; 0 \le w_j^D, \tag{5.6}
$$

where $w^D \in \mathbb{R}^{n_T}$. This optimization finds a weighting of the number of decision trees so that the generated confidences cover the full range equally. For that, the number of samples per bin h_i^j is summed up, weighted over all numbers of decision trees. After determining w^D, the number of decision trees can be sampled depending on w_j^D. An analysis of different sampling methods will be presented in Sect. 5.4.3.4. Automatically balancing the number of decision trees generates data samples with low and high confidence very equally distributed. The process does not require training data and provides a universal solution.

5.4.2.2   Imitation Learning

Finally, a neural network that imitates the random forest is trained. The network learns the decision boundaries from the generated data and approximates the same function as the random forest. The network architecture is based on a fully connected network with one or multiple hidden layers. The data dimensions are the same as those of the random forest, i.e., an N-dimensional input and C-dimensional output. Each hidden layer is followed by a ReLU activation (Nair and Hinton 2010). The last fully connected layer is using a softmax activation.

For training, we generate input-target pairs (x, y) as described in the last section. These training examples are fed into the training process to teach the network to predict the same results as the random forest. To avoid overfitting, the data is generated on-the-fly so that each training example is unique. In this way, we learn an efficient representation of the decision boundaries and are able to transform random forests into neural networks implicitly. In addition to that, the training is performed end to end on the generated data, and we can easily integrate the original training data.

5.4.3   Experiments

In this section, we perform several experiments to analyze the performance of neural random forest imitation and compare our method to state-of-the-art methods.

5.4.3.1   Datasets

The experiments are evaluated on nine classification datasets from the UCI Machine Learning Repository (Dua and Graff 2017) (*Car, Connect-4, Covertype, German Credit, Haberman, Image Segmentation, Iris, Soybean,* and *Wisconsin Breast Cancer (Original)*). The datasets cover many real-world problems in different areas, such as finance, computer vision, games, or medicine.

Following Fernández-Delgado et al. (2014), each dataset is split into a training and a test set using a 50/50 split while maintaining the label distribution. Afterward, the number of training examples is limited to n_{limit} examples per class. We evaluate the training with 5, 10, 20, and 50 examples per class. In contrast to Fernández-Delgado et al. (2014), we extract validation sets from the training set (e.g., for hyperparameter tuning). This ensures that the training and validation data are not mixed with the test data. For some datasets which provide a separate test set, the test accuracy is evaluated on the respective set. Missing values are set to the mean value of the feature. All experiments are repeated ten times with randomly sampled splits. The methods are repeated additionally four times with different seeds on each split.

5.4.3.2   Implementation Details

In all our experiments, stochastic gradient descent with Nesterov momentum as optimizer and cross-entropy loss are used. The initial learning rate is set to 0.1, momentum to 0.9, and weight decay to 0.0005. The batch size is set to 128 and 512, respectively, for generated data. The input data is normalized to $[-1, 1]$. For generating a wide variety of data, the prioritization of the current path $w_{\text{path}} \sim 1 + |\mathcal{N}(0, 5)|$ is sampled for each data sample individually. A new random forest is trained every 100 epochs to average the influence of the stochastic process, and the generated data samples are mixed. In the following, training on generated data will be denoted as *NRFI (gen)* and training on generated and original data as *NRFI (gen+ori)*. The fraction of NRFI data is set to 0.9. Random forests are trained with 500 decision trees, which are commonly used in practice (Fernández-Delgado et al. 2014; Olson et al. 2018). The decision trees are constructed up to a maximum depth of 10. For splitting, the Gini impurity is used, and $\sqrt{N}$ features are randomly selected, where N is the number of features.

5.4.3.3   Results

The proposed method generates data from a random forest and trains a neural network that imitates the random forest. The goal is that the neural network approximates the same function as the random forest. This also implies that the network reaches the same accuracy if successful.

We analyze the performance by training random forests for each dataset and evaluating neural random forest imitation with different network architectures. A variety of network architectures with different depths, widths, and additional layers such as dropout have been studied. In this work, we focus on two-hidden-layer networks with an equal number of neurons in both layers for clarity. The results are shown in Fig. 5.7 exemplarily for the *Car, Covertype,* and *Wisconsin Breast Cancer (Original)* dataset. The other datasets show similar characteristics. The overall evaluation on all datasets is presented in the next section. The number of training examples per class is shown in parentheses and increases in each row from left to right. For each setting, the test accuracy of the random forest is indicated by a red dashed line. The average test accuracy and standard deviation depending on the network architecture, i.e., the number of neurons in the first and second hidden layer, are plotted for different architectures. NRFI (gen), which is trained with generated data only, is shown in orange, and NRFI (gen+ori), which is trained with generated and original data, is shown in blue.

The analysis shows that the accuracy of the neural networks trained by NRFI reaches the accuracy of the random forest for all datasets. Only very small networks do not have the required capacity. The proposed method for generating labeled data from random forests by analyzing the decision boundaries enables training neural networks that imitate the random forests. For instance, in the case of 5 training examples per class, a two-hidden-layer network with 16 neurons in both

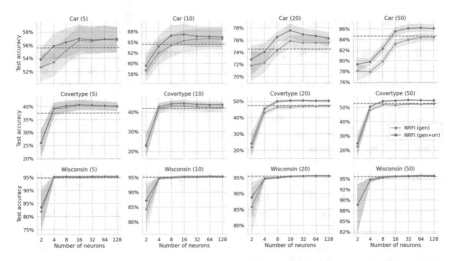

Fig. 5.7 Test accuracy depending on the network architecture (i.e., number of neurons in both hidden layers). Different datasets are shown per row, with an increasing number of training examples per class from left to right (indicated in parentheses). The red dashed line shows the accuracy of the random forest. NRFI with generated data is shown in orange and NRFI with generated and original data in blue. With increasing network capacity, NRFI is capable of imitating and even outperforming the random forest

layers already achieves the same accuracy as the random forest across all 3 datasets in Fig. 5.7. Additionally, the experiment shows that the training is very robust to overfitting even when the number of parameters in the network increases. When combining the generated data and original data, the accuracy on *Car* and *Covertype* improves with an increasing number of training examples.

Overall, the experiment shows that the accuracy increases with an increasing number of neurons in both layers and NRFI is robust to different network architectures. NRFI is capable of generating a large variety of unique examples from random forests which have been initially trained on a limited amount of data.

Comparison to State of the Art

We now compare the proposed method to state-of-the-art methods for mapping random forests into neural networks and classical machine learning classifiers such as random forests and support vector machines with a radial basis function kernel that have shown to be the best two classifiers across all UCI datasets (Fernández-Delgado et al. 2014). In detail, we will evaluate the following methods:

- DT: A decision tree (Breiman et al. 1984) learns simple and interpretable split decisions to classify data. The Gini impurity is used for splitting.
- SVM: Support vector machine (Chang and Lin 2011) is a popular classifier that tries to find the best hyperplane that maximizes the margin between the

classes. As evaluated by Fernández-Delgado et al. (2014), the best performance is achieved with a radial basis function kernel.

- RF: Random forest (Breiman 2001) is an ensemble-based method consisting of multiple decision trees. Each decision tree is trained on a different randomly selected subset of features and samples. The classifier follows the same overall setup, i.e., 500 decision trees and a maximum depth of 10.

- NN: A neural network (Rumelhart et al. 1988) with two hidden layers is trained using ReLU activation and cross-entropy loss. Possible values for the initial learning rate are $\{0.1, 0.01, 0.001, 0.0001, 0.00001\}$ and $\{2, 4, 8, 16, 32, 64, 128\}$ for the number of neurons in both hidden layers. The best hyperparameters are selected by performing a fourfold cross-validation.

- Sethi: The method proposed by Sethi (1990) maps a random forest into a two-hidden-layer neural network by adding a neuron for each split node and each leaf node. The weights are set corresponding to the split decisions.

- Welbl: Welbl (2014) and Biau et al. (2019) present a similar mapping with subsequent fine-tuning. The authors introduce two training modes: *independent* and *joint*. The first optimizes each small network individually, while the latter joins all mapped decision trees into one network. Additionally, the authors evaluate a network with sparse connections and regular fully connected networks (denoted as *sparse* and *full*).

- Massiceti: Massiceti et al. (2017) present a network splitting strategy to reduce the number of network parameters. The decision trees are divided into subtrees and mapped individually while sharing common split nodes. The optimal depth of the subtrees is determined by evaluating all possible values.

First, we analyze the performance of state-of-the-art methods for mapping random forests into neural networks and neural random forest imitation. The results are shown in Fig. 5.8 for different numbers of training examples per class. For each method, the average number of parameters of the generated networks across all datasets is plotted depending on the test error. That means that the methods aim for the lower-left corner (smaller number of network parameters and higher accuracy). Please note that the y-axis is shown on a logarithmic scale. The average performance of the random forests is indicated by a red dashed line.

The analysis shows that Sethi, Welbl (ind-full), and Welbl (joint-full) generate the largest networks. Network splitting (Massiceti et al. 2017) slightly improves the number of parameters of the networks. Using a sparse network architecture reduces the number of parameters. However, it should be noted that this requires special operations. NRFI with and without the original data is shown for different network architectures. The smallest architecture has 2 neurons in both hidden layers and the largest 128. For NRFI (gen-ori), we can see that a network with 16 neurons in both hidden layers (NN-16-16) is already sufficient to learn the decision boundaries of the random forest and achieve the same accuracy. When fewer training samples are available, NN-8-8 already has the required capacity. In the following, we will further analyze the accuracy and number of network parameters.

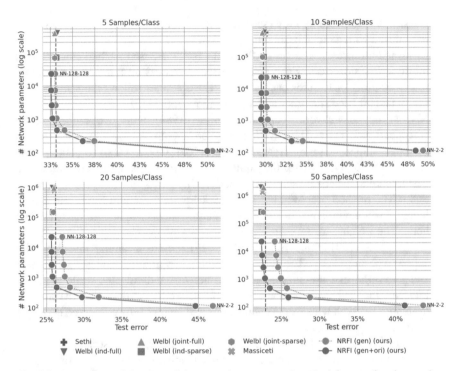

Fig. 5.8 Comparison of the state-of-the-art and our proposed method for transforming random forests into neural networks. The closer a method is to the lower-left corner, the better it is (fewer number of network parameters and lower test error). For neural random forest imitation, different network architectures are shown. Note that the number of network parameters is shown on a logarithmic scale

Accuracy

The average test accuracy and standard deviation for all methods are shown in Table 5.1. Here, we additionally include decision trees, support vector machines, random forests, and neural networks in the comparison. The evaluation is performed on all nine datasets, and results for different numbers of training examples are shown (increasing from left to right). The overall performance of each method is summarized in the last column. For neural random forest imitation, a network architecture with 128 neurons in both hidden layers is used. From the analysis, we can make the following observations: (1) When training neural random forest imitation with generated data only, the method achieves 99.18% of the random forest accuracy (71.44% compared to 72.03%). This shows that NRFI is capable of learning the decision boundaries. (2) Overall, NRFI trained with generated and original data reaches state-of-the-art performance (50 samples per class) or outperforms the other methods (5, 10, and 20 samples per class).

Table 5.1 Average test accuracy [%] and standard deviation on all nine datasets for different numbers of training examples per class. The overall performance of each method is summarized in the last column. The best methods are highlighted in bold

Method	Samples per class				
	5	10	20	50	mean
DT	62.95 ± 5.41	66.89 ± 4.18	70.82 ± 2.93	73.66 ± 2.20	68.58 ± 3.68
SVM	65.21 ± 4.81	68.15 ± 4.44	71.91 ± 3.33	75.96 ± 2.22	70.31 ± 3.70
RF	66.91 ± 4.01	70.31 ± 3.86	73.81 ± 2.46	77.08 ± 1.90	72.03 ± 3.06
NN	65.50 ± 5.15	69.89 ± 4.13	73.11 ± 3.19	76.50 ± 2.53	71.25 ± 3.75
Sethi	66.93 ± 4.01	70.06 ± 4.28	74.00 ± 3.00	77.50 ± 2.23	72.12 ± 3.38
Welbl (ind-full)	66.72 ± 4.04	70.21 ± 3.91	74.19 ± 2.50	**77.63 ± 1.81**	72.19 ± 3.06
Welbl (joint-full)	67.01 ± 4.14	70.42 ± 4.07	74.02 ± 2.80	77.31 ± 1.76	72.19 ± 3.19
Welbl (ind-sparse)	66.81 ± 4.07	70.27 ± 4.15	74.14 ± 2.58	77.60 ± 1.82	72.20 ± 3.15
Welbl (joint-sparse)	67.02 ± 4.17	70.41 ± 4.11	74.09 ± 2.77	77.36 ± 1.61	72.22 ± 3.17
Massiceti	66.97 ± 4.05	70.07 ± 4.28	73.98 ± 3.05	77.45 ± 2.26	72.12 ± 3.41
NRFI (gen) (ours)	66.99 ± 4.09	69.95 ± 4.21	72.90 ± 2.67	75.90 ± 2.22	71.44 ± 3.30
NRFI (gen+ori) (ours)	**67.42 ± 4.15**	**70.57 ± 4.05**	**74.36 ± 2.44**	77.62 ± 1.90	**72.49 ± 3.14**

Table 5.2 Comparison to state-of-the-art methods. For each method, the average number of parameters of the generated neural networks is shown. While achieving the same or even slightly better accuracy, neural random forest imitation generates much smaller models, enabling the mapping of complex random forests

Method	Samples per class				
	5	10	20	50	mean
	Number of network parameters				
Sethi	374299	592384	985294	1973341	981330
Welbl (ind-full)	374729	592147	984626	1972604	981027
Welbl (joint-full)	371965	589220	981816	1968118	977780
Welbl (ind-sparse)	70070	102895	154740	254344	145512
Welbl (joint-sparse)	67344	100131	151944	251598	142754
Massiceti	348972	522640	792410	1328731	748188
NRFI (ours)	**2676**	**2676**	**2676**	**2676**	**2676**

Network Parameters

Finally, we will analyze the number of parameters of the generated networks in detail. The results are shown in Table 5.2. Current state-of-the-art methods directly map random forests into neural networks. The number of parameters of the resulting network is evaluated on all datasets with different numbers of training examples. The overall performance is shown in the last column. Due to the stochastic process when training the random forests, the results can vary marginally.

Sethi, Welbl (ind-full), and Welbl (joint-full) generate networks with around 980 000 parameters on average. Of the four variants proposed by Welbl, joint training has a slightly smaller number of parameters compared to independent training because of shared neurons in the output layer. Network splitting proposed by Massiceti et al. (2017) maps multiple subtrees while sharing common split nodes and reduces the average number of network parameters to 748 000. Using sparse network architectures additionally reduces the number of network parameters to about 142 000; however, this requires a special implementation for sparse matrix multiplication. All of the methods show a drastic increase with the growing complexity of the classifiers. Sethi, for example, generates networks with 374 000 parameters when training with 5 examples per class. The average number of network parameters increases to 1.9 million when training with 50 examples per class.

NRFI introduces imitation instead of direct mapping. In the following, a network architecture with 32 neurons in both hidden layers is selected. The previous analysis has shown that this architecture is capable of imitating the random forests (see Fig. 5.8 for details) across all datasets and different numbers of training examples. Our method significantly reduces the number of parameters of the generated networks while reaching the same or even slightly better accuracy. The current best-performing methods generate networks with an average number of parameters of either 142 000, if sparse processing is available, or 748 000 when using usual fully connected neural networks. In comparison, neural random forest imitation requires only 2676 parameters. Another advantage is that the proposed method does not create a predefined architecture but enables arbitrary network architectures. As a result, NRFI enables the transformation of very complex classifiers into neural networks.

5.4.3.4   Analysis of the Generated Data

To study the sampling process, we analyze the variability of the generated data as well as different sampling modes in the next experiment. Subsequently, we investigate the impact of combining original and generated data.

Confidence Distribution

The data generation process aims to produce a wide variety of data samples. This includes data samples that are classified with a high confidence and data samples that are classified with a low confidence to cover the full range of prediction uncertainties. The following analyses are shown exemplarily on the *Soybean* dataset. This dataset has 35 features and 19 classes. First, we analyze the generated data with a fixed number of decision trees, i.e., the number of sampled decision trees in RF_{sub}. The resulting confidence distributions for different numbers of decision trees are shown in the first column of Fig. 5.9. When adopting the data sample to only a few decision trees, the confidence of the generated samples is lower (around 0.2 for 5 samples per class). Using more decision trees for generating data samples increases the confidence on average.

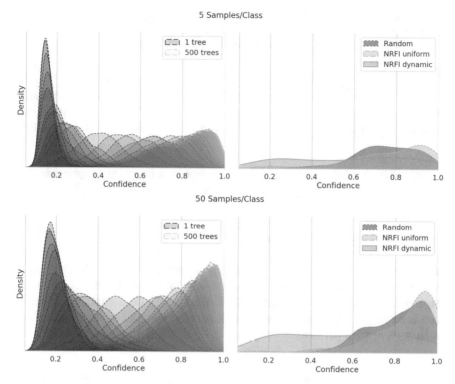

Fig. 5.9 Probability distribution of the predicted confidences for different data generation settings on *Soybean* with 5 (top) and 50 samples per class (bottom). Generating data with different numbers of decision trees is visualized in the left column. Additionally, a comparison between random sampling (red), NRFI uniform (orange), and NRFI dynamic (green) is shown in the right column. By optimizing the decision tree sampling, NRFI dynamic automatically balances the confidences and generates the most diverse and evenly distributed data

NRFI uniform and NRFI dynamic sample the number of decision trees for each data point uniformly, respectively, optimized via automatic confidence distribution (see Sect. 5.4.2.1). The confidence distributions for both sampling modes are visualized in the second column of Fig. 5.9. Additionally, sampling random data points without generating data from the random forest is included as a baseline. The analysis shows that random data samples and uniform sampling have a bias to generate data samples that are classified with high confidence. NRFI dynamic automatically balances the number of decision trees and archives an evenly distributed data distribution, i.e., generates the most diverse data samples.

In the next step, the imitation learning performance of the sampling modes is evaluated. The results are shown in Table 5.3. Random data generation reaches a mean accuracy of 63.80%, while NRFI uniform and NRFI dynamic achieve 87.46% and 88.14%, respectively. This shows that neural random forest imitation is able to generate significantly better data samples based on the knowledge in the random

Table 5.3 Imitation learning performance (in accuracy [%]) of different data sampling modes on *Soybean*. NRFI achieves better results than random data generation. When optimizing the selection of the decision trees, the performance is improved due to more diverse sampling

Method	Samples per class				mean
	5	10	20	50	
Random	58.70 ± 4.15	58.65 ± 1.34	64.61 ± 6.91	73.24 ± 0.79	63.80 ± 3.30
NRFI uniform	84.27 ± 2.57	87.43 ± 1.76	88.63 ± 1.35	89.52 ± 1.03	87.46 ± 1.67
NRFI dynamic	84.82 ± 2.75	88.16 ± 1.64	89.10 ± 1.65	90.49 ± 1.47	88.14 ± 1.88

forest. NRFI dynamic improves the performance by automatically optimizing the decision tree sampling and generating the largest variation in the data.

Original and Generated Data

In the next experiment, we study the effects of training with original data, NRFI data, and combinations of both. For that, the fraction of NRFI data w_{gen} is varied, which weights the loss of the generated data. Accordingly, the weight for the original data is set to $w_{ori} = 1 - w_{gen}$. The average accuracy over all datasets for different number of samples per class is shown in Fig. 5.10. When the fraction of NRFI data

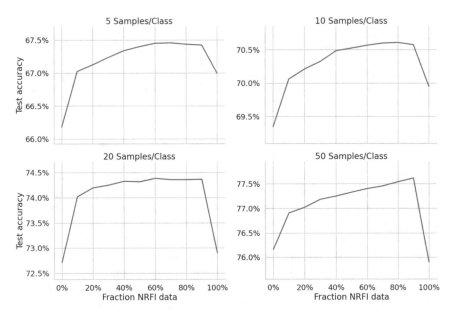

Fig. 5.10 Analyzing the influence of training with original data, NRFI data, and combinations of both for different number of samples per class. Using only NRFI data ($w_{gen} = 100\%$) achieves better results than using only the original data ($w_{gen} = 0\%$) for less than 50 samples per class. Combining the original data and generated data improves the performance

is set to 0%, the network is trained with only the original data. When the fraction is set to 100%, the network is trained completely with the generated data. The study shows that training with NRFI data performs better than training with original data except for 50 samples per class where training with original data is slightly better. Combining original and NRFI data improves the performance. The best result is achieved when using mainly NRFI data with a small fraction of original data.

5.5   Conclusion

In this work, we brought two worlds together by combining neural networks and random forests. First, we presented an object detection framework for analyzing the road and traffic situations for cyclists and pedestrians. The combination of convolutional neural networks and random forests enables the training with very few labeled examples. Both methods are combined in an end-to-end pipeline by transforming the random forest into a neural network and generating a fully convolutional network.

Because existing approaches for mapping random forests into neural networks generate inefficient networks, we presented a novel method for transforming random forests into neural networks. Instead of a direct mapping, we introduced a process for generating data from random forests by analyzing the decision boundaries and guided routing of data samples to selected leaf nodes. Based on the generated data and corresponding labels, a network is trained that imitates the random forest. Experiments on several real-world benchmark datasets demonstrate that NRFI is capable of learning the decision boundaries very efficiently. Compared to state-of-the-art methods, the presented implicit transformation significantly reduces the number of parameters of the networks while achieving the same or even slightly improved accuracy due to better generalization. Our approach has shown that it scales very well and is able to imitate highly complex classifiers.

Acknowledgments This research was supported by the German Research Foundation DFG (COVMAP—RO 2497/12-2) within Priority Research Program 1894 *Volunteered Geographic Information: Interpretation, Visualization, and Social Computing*; the Federal Ministry of Education and Research (BMBF), Germany, under the project LeibnizKILabor (grant no. 01DD20003); the Center for Digital Innovation (ZDIN); and the German Research Foundation (DFG) under Germany's Excellence Strategy within the Cluster of Excellence PhoenixD (EXC 2122).

References

Barz B, Denzler J (2020) Deep learning on small datasets without pre-training using cosine loss. In: IEEE Winter Conference on Applications of Computer Vision (WACV), pp 1360–1369
Biau G, Scornet E, Welbl J (2019) Neural Random Forests. Sankhya A 81:347–386

Bornschein J, Visin F, Osindero S (2020) Small data, big decisions: Model selection in the small-data regime. In: Proceedings of the 37th International Conference on Machine Learning, PMLR, vol 119, pp 1035–1044

Breiman L (2001) Random forests. Mach Learn 45(1):5–32

Breiman L, Friedman JH, Olshen RA, Stone CJ (1984) Classification and Regression Trees. Wadsworth and Brooks, Monterey

Cai Z, Vasconcelos N (2018) Cascade r-cnn: Delving into high quality object detection. In: 2018 IEEE/CVF Conference on Computer Vision and Pattern Recognition, pp 6154–6162

Carion N, Massa F, Synnaeve G, Usunier N, Kirillov A, Zagoruyko S (2020) End-to-end object detection with transformers. In: Vedaldi A, Bischof H, Brox T, Frahm JM (eds) Computer Vision—ECCV 2020. Springer, Cham, pp 213–229

Chang CC, Lin CJ (2011) LIBSVM: a library for support vector machines. ACM Trans Intell Syst Technol 2(3):1–27

Chen Q, Wang Y, Yang T, Zhang X, Cheng J, Sun J (2021) You only look one-level feature. In: IEEE Conference on Computer Vision and Pattern Recognition

Collobert R, Weston J, Bottou L, Karlen M, Kavukcuoglu K, Kuksa P (2011) Natural language processing (almost) from scratch. J Mach Learn Res 12:2493–2537

Dockhorn A, Braune C, Kruse R (2015) An alternating optimization approach based on hierarchical adaptations of dbscan. In: 2015 IEEE Symposium Series on Computational Intelligence (SSCI), 2, pp 749–755

Dockhorn A, Braune C, Kruse R (2016) Variable density based clustering. In: 2016 IEEE Symposium Series on Computational Intelligence (SSCI), pp 1–8

Dockhorn A, Doell C, Hewelt M, Kruse R (2017) A decision heuristic for Monte Carlo tree search doppelkopf agents. In: 2017 IEEE Symposium Series on Computational Intelligence (SSCI), pp 1–8

Dua D, Graff C (2017) UCI machine learning repository. http://archive.ics.uci.edu/ml

Duan K, Bai S, Xie L, Qi H, Huang Q, Tian Q (2019) Centernet: keypoint triplets for object detection. In: Proceedings of the IEEE/CVF International Conference on Computer Vision (ICCV)

Fang T, Lu N, Niu G, Sugiyama M (2020) Rethinking importance weighting for deep learning under distribution shift. In: Advances in Neural Information Processing Systems (NeurIPS)

Fernández-Delgado M, Cernadas E, Barro S, Amorim D (2014) Do we need hundreds of classifiers to solve real world classification problems? J Mach Learn Res 15(1):3133–3181

Fukunaga K, Hostetler L (1975) The estimation of the gradient of a density function, with applications in pattern recognition. IEEE Trans Inform Theory 21(1):32–40

Girshick R (2015) Fast R-CNN. In: 2015 IEEE International Conference on Computer Vision (ICCV), pp 1440–1448

Girshick R, Donahue J, Darrell T, Malik J (2014) Rich feature hierarchies for accurate object detection and semantic segmentation. In: Computer Vision and Pattern Recognition

Graves A, Mohamed A, Hinton G (2013) Speech recognition with deep recurrent neural networks. In: IEEE International Conference on Acoustics, Speech and Signal Processing, pp 6645–6649

Hekler EB, Klasnja PV, Chevance G, Golaszewski NM, Lewis DM, Sim I (2019) Why we need a small data paradigm. BMC Medicine 17:133

Humbird KD, Peterson JL, McClarren RG (2019) Deep neural network initialization with decision trees. IEEE Trans Neural Netw Learn Syst 30(5):1286–1295

Krizhevsky A, Sutskever I, Hinton GE (2012) Imagenet classification with deep convolutional neural networks. In: Advances in Neural Information Processing Systems (NeurIPS), vol 25

Li Y, Wang S (2022) R(det)2: Randomized decision routing for object detection. In: Proceedings of the IEEE/CVF Conference on Computer Vision and Pattern Recognition (CVPR), pp 4825–4834

Lin TY, Goyal P, Girshick R, He K, Dollár P (2017) Focal loss for dense object detection. In: 2017 IEEE International Conference on Computer Vision (ICCV), pp 2999–3007

Liu W, Anguelov D, Erhan D, Szegedy C, Reed S, Fu CY, Berg AC (2016) Ssd: Single shot multibox detector. In: European Conference on Computer Vision. Springer, Berlin, pp 21–37

MacQueen J, et al (1967) Some methods for classification and analysis of multivariate observations. In: Proceedings of the fifth Berkeley Symposium on Mathematical Statistics and Probability, Oakland, CA, USA, vol 1, pp 281–297

Massiceti D, Krull A, Brachmann E, Rother C, Torr PH (2017) Random forests versus neural networks—what's best for camera localization? In: IEEE International Conference on Robotics and Automation (ICRA), pp 5118–5125

Menze BH, Kelm BM, Splitthoff DN, Koethe U, Hamprecht FA (2011) On oblique random forests. In: Machine Learning and Knowledge Discovery in Databases, pp 453–469

Mnih V, Kavukcuoglu K, Silver D, a Rusu A, Veness J, Bellemare MG, Graves A, Riedmiller M, Fidjeland AK, Ostrovski G, Petersen S, Beattie C, Sadik A, Antonoglou I, King H, Kumaran D, Wierstra D, Legg S, Hassabis D (2015) Human-level control through deep reinforcement learning. Nature 518:529–533

Nair V, Hinton GE (2010) Rectified linear units improve restricted boltzmann machines. In: Proceedings of the 27th International Conference on International Conference on Machine Learning, pp 807–814

Olson M, Wyner A, Berk R (2018) Modern neural networks generalize on small data sets. In: Advances in Neural Information Processing Systems (NeurIPS), vol 31

Otter D, Medina JR, Kalita JK (2021) A survey of the usages of deep learning for natural language processing. IEEE Trans Neural Netw Learn Syst 32:604–624

Park DS, Chan W, Zhang Y, Chiu CC, Zoph B, Cubuk ED, Le QV (2019) SpecAugment: a simple data augmentation method for automatic speech recognition. In: Proc. Interspeech 2019, pp 2613–2617

Phoo CP, Hariharan B (2021) Self-training for few-shot transfer across extreme task differences. In: Proceedings of the International Conference on Learning Representations

Qi GJ, Luo J (2020) Small data challenges in big data era: a survey of recent progress on unsupervised and semi-supervised methods. IEEE Trans Pattern Anal Mach Intell 44(4):2168–2187

Qiao S, Chen LC, Yuille AL (2021) Detectors: Detecting objects with recursive feature pyramid and switchable atrous convolution. In: The IEEE Conference on Computer Vision and Pattern Recognition (CVPR)

Quinlan JR (1993) C4.5: programs for machine learning. Morgan Kaufmann Publishers, San Francisco

Redmon J, Farhadi A (2016) Yolo9000: better, faster, stronger. arXiv preprint arXiv:161208242

Reinders C, Ackermann H, Yang MY, Rosenhahn B (2018) Object recognition from very few training examples for enhancing bicycle maps. In: IEEE Intelligent Vehicles Symposium (IV)

Reinders C, Ackermann H, Yang MY, Rosenhahn B (2019) Learning convolutional neural networks for object detection with very little training data. Multimodal Scene Understanding

Reinders C, Schubert F, Rosenhahn B (2022) ChimeraMix: Image classification on small datasets via masked feature mixing. In: Proceedings of the Thirty-First International Joint Conference on Artificial Intelligence, IJCAI, pp 1298–1305

Ren S, He K, Girshick R, Sun J (2015) Faster R-CNN: towards real-time object detection with region proposal networks. In: NIPS

Richmond D, Kainmueller D, Yang M, Myers E, Rother C (2016) Mapping auto-context decision forests to deep convnets for semantic segmentation. In: Proceedings of the British Machine Vision Conference (BMVC)

Rudolph M, Wehrbein T, Rosenhahn B, Wandt B (2022) Fully convolutional cross-scale-flows for image-based defect detection. In: Winter Conference on Applications of Computer Vision (WACV)

Rumelhart DE, Hinton GE, Williams RJ (1988) Learning representations by back-propagating errors. MIT Press, Cambridge, pp 696–699

Schier M, Reinders C, Rosenhahn B (2022) Constrained mean shift clustering. In: Proceedings of the 2022 SIAM International Conference on Data Mining (SDM)

Sethi IK (1990) Entropy nets: from decision trees to neural networks. Proc IEEE 78(10):1605–1613

Shelhamer E, Long J, Darrell T (2017) Fully convolutional networks for semantic segmentation. IEEE Trans Pattern Anal Mach Intell 39(4):640–651

Simonyan K, Zisserman A (2015) Very deep convolutional networks for large-scale image recognition. In: International Conference on Learning Representations

Springenberg JT, Dosovitskiy A, Brox T, Riedmiller M (2015) Striving for simplicity: the all convolutional net. In: ICLR, pp 1–14, 1412.6806

Stallkamp J, Schlipsing M, Salmen J, Igel C (2012) Man vs. computer: benchmarking machine learning algorithms for traffic sign recognition. Neural Netw 32:323–332

Sun B, Li B, Cai S, Yuan Y, Zhang C (2021a) Fsce: Few-shot object detection via contrastive proposal encoding. In: The IEEE Conference on Computer Vision and Pattern Recognition (CVPR)

Sun P, Zhang R, Jiang Y, Kong T, Xu C, Zhan W, Tomizuka M, Li L, Yuan Z, Wang C, Luo P (2021b) Sparse R-CNN: End-to-end object detection with learnable proposals. 2021 IEEE/CVF Conference on Computer Vision and Pattern Recognition (CVPR), pp 14449–14458

Sutskever I, Vinyals O, Le QV (2014) Sequence to sequence learning with neural networks. In: Advances in Neural Information Processing Systems (NeurIPS), vol 27

Tian Z, Shen C, Chen H, He T (2019) Fcos: Fully convolutional one-stage object detection. In: Proceedings of the IEEE/CVF International Conference on Computer Vision (ICCV)

Wang P, Fan E, Wang P (2021) Comparative analysis of image classification algorithms based on traditional machine learning and deep learning. Pattern Recogn Lett 141:61–67

Welbl J (2014) Casting random forests as artificial neural networks (and profiting from it). In: German Conference on Pattern Recognition

Zhang C, Bengio S, Hardt M, Recht B, Vinyals O (2017) Understanding deep learning requires rethinking generalization. In: 5th International Conference on Learning Representations (ICLR)

Zhao H, Shi J, Qi X, Wang X, Jia J (2017) Pyramid scene parsing network. In: 2017 IEEE Conference on Computer Vision and Pattern Recognition (CVPR), pp 6230–6239

Part II
Geovisualization and User Interactions Related to VGI

VGI has been employed to support a variety of applications that benefit from the georeferencing of data, from navigation to exploratory data analysis. VGI data is large in quantity, dynamic, and spread across many different types of modalities, such as text, images, or positional information. To make full use of this data and to make it understandable for non-experts, the effective use of visualizations is a primary tool. For expert users, interactive visualizations enable varied user interactions to explore and interpret large, dynamic, and heterogeneous datasets and enable extraction of knowledge and patterns. For non-expert users, geovisualizations can help to communicate relevant patterns and properties like uncertainty of locations.

Part II of the book explores geovisualizations and user interactions supporting the analysis and presentation of VGI data. When designing these visualizations and user interactions, we need to consider the specific properties of VGI data, the knowledge and abilities of different target users, as well as the technical viability of solutions. It is crucial to find the right trade-off between information density and complexity, on the one hand, and interpretability or ease of use, on the other hand. Algorithmic advances for the optimization of visualizations can support the process of revealing relevant information in visualizations; however, we should be aware to keep faithful representations of data. User interactions need to be efficient and powerful for analysts while not overwhelming possible users. The chapters in this part tackle these challenges in the space of VGI.

We start by discussing strategies for visually analyzing dynamic social messages and news articles containing geo-referenced information based on visual analytics. We then discuss the use of interactive visual reporting solutions for leveraging natural language and visualizations for geo-related data. Following this, we examine the effects of landmark position and design in VGI-based maps on visual attention and cognitive processing. Related to this, we discuss landmark uncertainty in VGI-based maps and its effects on orientation and navigation performance. Finally, we present a study to analyze the user behavior while solving interpretation tasks with VGI data and use the findings to improve task-oriented visual interpretation of VGI point data.

We hope that this research strengthens the extraction of usable knowledge from VGI data for both experts and non-experts, integrating different data types such as text, and guiding further research into visual saliency for maps and task-specific representations of VGI data.

Chapter 6
Toward Visually Analyzing Dynamic Social Messages and News Articles Containing Geo-Referenced Information

Johannes Knittel, Franziska Huth, Steffen Koch, and Thomas Ertl

Abstract The number of social media posts and news articles that are being published every day is high. This makes them an attractive source of human-generated information for different domain experts such as journalists and business analysts but also emergency responders, particularly if posts contain references to geolocations. Visual analytics approaches can help to gain insights into such datasets and inform decision-makers. However, the high volume and the veracity of the data, as well as the velocity in the case of streaming data, pose challenges when supporting explorative analysis with interactive visualization. Based on four exemplary approaches, we outline recently proposed strategies to tackle these challenges. We describe how geo-aware filtering and anomaly detection methods can help to inform stakeholders based on geolocated tweets. We show that data-aware tag maps can provide analysts with an overview-first, details-on-demand visual summary of large amounts of text content over time. With space-filling curves, we can visualize the temporal evolution of geolocations in a two-dimensional plot without relying on animations that would impede comparative analyses. Additionally, we discuss the use of an efficient dynamic clustering algorithm for enabling large-scale visual analyses of streaming posts.

Keywords Social media · Documents · Streaming data · Visualization · Visual analytics

6.1 Introduction

Unstructured or semi-structured data such as news articles and social media posts contain a significant amount of human knowledge. Analyzing such vast amounts of data enables several stakeholders from different domains to obtain insights and

J. Knittel (✉) · F. Huth · S. Koch · T. Ertl
University of Stuttgart, Stuttgart, Germany
e-mail: johannes.knittel@vis.uni-stuttgart.de; franziska.huth@vis.uni-stuttgart.de; steffen.koch@vis.uni-stuttgart.de; thomas.ertl@vis.uni-stuttgart.de

© The Author(s) 2024
D. Burghardt et al. (eds.), *Volunteered Geographic Information*,
https://doi.org/10.1007/978-3-031-35374-1_6

inform their decision-making, for instance, business traders that need up-to-date information about new developments and first responders that benefit from timely witness reports on social media. In some cases, we can leverage the structured metadata associated with some of these documents such as the geolocation of posts or the timestamp of news articles, but it is generally challenging to gain insights into the actual content comprising text data.

The field of visual analytics (Thomas and Cook 2005; Keim et al. 2010) particularly aims at solving such complex problems of analyzing and exploring large amounts of data with often open and ill-defined goals. It combines the domain knowledge and intelligence of human experts with interactive visualizations that are sourced from advanced automated data analytics and machine learning models. If we want to harvest news reports and social media posts for timely insights using visual analytics, we need efficient algorithms to deal with the volume of the data, we need to extract information from unstructured data such as text, we need to integrate and combine this information with additional metadata such as geolocations into interactive visualizations, and we need to develop adaptive visualizations and streaming-aware algorithms that can deal with dynamic data sources. This chapter outlines recently proposed visual analytics approaches to tackle the said challenges.

6.2 Analyzing the Temporal Evolution of Text Data with PyramidTags

Making sense of large document corpora is a challenging endeavor, since it is inherently difficult for machines to grasp the meaning of natural language. Visual analytics approaches that combine methods of automatic information retrieval and data analytics with interactive visualizations help to tackle such challenges by incorporating human expertise and human interpretability. However, it remains challenging to provide a comprehensive overview of large amounts of text data such as news articles or social media posts due to the unstructured nature of the data, the variety of how people express similar things, and the inherent ambiguity of natural languages.

PyramidTags (Knittel et al. 2021c) proposes a novel tag layout for exploring large document collections such as tweets that aims at providing analysts with an overview of the content at hand and the temporal evolution of its themes without introducing hard clusters or topics. The approach utilizes an optimization process to place extracted relevant keywords and keyphrases from articles or posts onto a two-dimensional plot such that related tags ideally appear close to each other, while it is also possible to infer in which date ranges tags mostly appear in the dataset based on their position on the map (Fig. 6.1).

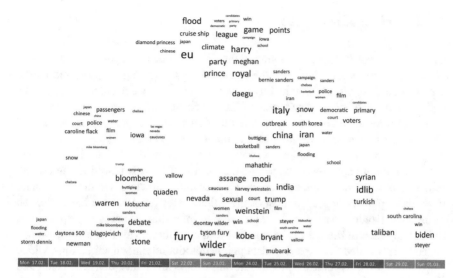

Fig. 6.1 PyramidTags aims at providing an interactive overview of large document collections. This is an example of the main visualization based on 70,000 news articles published in late January 2020

6.2.1 Processing and Objectives

PyramidTags first extracts the top k relevant keywords and keyphrases from the document collection using ELSKE (Knittel et al. 2021b), a fast keyphrase extraction library specialized in summarizing text collections. These tags serve as a summary of the content, and the way how they are placed on the two-dimensional tag map should support analysts making sense of the data with a date-aware, context-aware, and word-order-aware layout. In a second step, we process the dataset again to infer which tags are related based on how often and how close they appear in the same paragraph, whether there seems to be a preferred reading order of tag pairs (e.g., *John | Doe* vs. *Doe | John*), as well as in which date ranges tags and tag pairs mostly appear.

The resulting data structure informs the subsequent layouting process, which optimizes an objective function using particle swarm optimization (Kennedy and Eberhart 1995). Minimizing this function therefore corresponds to finding a balanced trade-off between different objectives such as that (1) related tags are placed nearby, (2) the position of the tag conveys the associated date range, (3) the preferred reading order is preserved for important pairs (if applicable), and (4) tags should not overlap.

6.2.2   Triangular Layout

One of the defining aspects of PyramidTags is its triangular layout, which aims to convey the temporal evolution of the extracted tags. Each tag is associated with a specific date range in which it mainly appears in the data (we may have several tags with the same text in case they appear in distinct clusters of date ranges). The vertical position on the map corresponds to the duration of this range, and the horizontal position to the mid-point of the said date range. At the bottom of the visualization, we place a timeline that depicts the entire date range of the dataset. For instance, if a word mainly appears on a specific date in the data, it is placed at the bottom of the map, right above the corresponding date in the timeline. On the other hand, if words or phrases appear in most of the articles, they are placed at the center-top of the visualization. Analysts should be able to infer this date range by spanning a right triangle from the tag to the timeline at the bottom. With this layout, we can visualize associated time spans of data points without relying on animations, which helps analysts to hypothesize about relevant events since tags that are mentioned during similar date ranges are also placed in the same neighborhood.

For instance, in Fig. 6.1, the tags *diamond princess* and *cruise ship* appear at the top of the map, indicating that the discussion about the Covid-19 cases on the said cruise ship was in the news during most of the depicted date range, whereas *storm dennis* and *flooding* are placed at the bottom-left around the first day of the 2-week date range.

6.2.3   Interactions and Document Retrieval

If users hover over certain tags, a lightly colored trapezoid visualizes the associated date range of the respective tag. While the optimization process tries to place related tags nearby, due to the inherent information loss of projecting data to two-dimensional spaces, not all tags that are close to each other are necessarily related, and there might be tags placed further away that are nevertheless related. The system therefore shades all other tags on the map depending on how related they are to the currently hovered tag (the more opaque, the less related). Users can also select one or several keywords by clicking on them. For instance, Fig. 6.2 shows an example in which the analyst has selected three tags (A). PyramidTags will then list the most related documents that contain the chosen selection of words or phrases, ranking the results based on the number of occurrences and the relative position of keywords to each other in the document (B). Users may also retrieve individual documents (C).

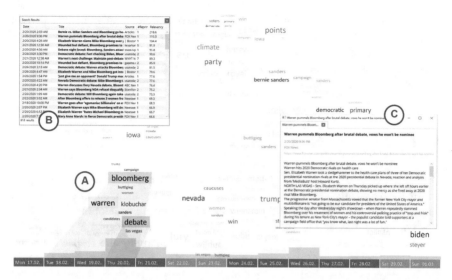

Fig. 6.2 Once one or several tags are selected (A), the PyramidTags system shades terms based on how related they are to the selected keywords. For each tag, a colored trapezoid visualizes the associated date range in which the respective term mostly appears in. The system also provides a list of related documents (B) from which individual documents can be retrieved (C)

6.3 Leveraging Geodata to Scale the Visual Analysis of Posts

When dealing with large amounts of streaming text data, we can leverage geographical references to scale the analysis (e.g., the current position of the person that has just posted the respective tweet). Such geo-tags not only provide additional context to the textual content; we can also use them to cluster items and their content in a geospatial way, providing several important benefits. We can visualize content on top of a geographical map to help analysts focusing on specific regions of instance, drastically reducing the actual amount of data analysts have to cope with. Grouping content by geographic region may also help with providing thematic aggregations, as people in the same region within a certain time span may also have a higher chance of posting content with similar topics (e.g., football match in a city). Another advantage is that we can compare metadata and extracted aggregated information of documents with a spatial-aware baseline (e.g., typical occurrences of tags within a region). This also helps to develop anomaly detection algorithms, for instance, to notify first responders in a very timely manner about evolving situations.

ScatterBlogs (Thom et al. 2012, 2015; Bosch et al. 2013) is a visual analytics approach that proposed to leverage the geographical annotations of tweets in these ways to scale the visual analysis of streaming posts. Case studies with domain experts from crisis management groups and critical infrastructure companies underlined the need for such systems so that analysts can obtain important additional information in real time regarding critical situations, despite the apparent learning curve and the need for specialized human labor to monitor this channel (Thom et al. 2015). However, they also showed that the velocity of newly published posts, even regarding specific events, and the dynamically evolving nature of the content itself (e.g., novel hashtags) still pose significant challenges for analysts and the development of such interactive monitoring systems (Fathi et al. 2020).

6.3.1 Geospatial Clustering of Terms

One of the core ideas in ScatterBlogs is to continuously extract terms that appear unusually frequently in certain geographic regions and visualize them on top of a map such that analysts can get an overview of interesting developments that take place at specific locations. In the beginning, every term (except for stop words) defines a cluster comprising all geo-tagged posts containing the respective term; new received posts are added to these clusters based on the terms they contain. Once the distortion of any such cluster is too high (i.e., the geographic positions of related messages are too widely scattered), we split the cluster using the k-means algorithm (with $k = 2$). The system visualizes such dynamic *term clusters* by placing the respective tags (or representative dots) on a map based on the average geographic position of corresponding posts. The decision which terms to display in which size also incorporates how anomalous the usage of this term is. This score depends on the number of unique users and the geographic density of the corresponding posts, that is, the importance is high if many different users post messages at a specific location. Figure 6.3 depicts the main user interface of the system, visualizing anomalous terms in green on top of a map.

6.3.2 Keyword Lens and Topic Modeling

ScatterBlogs provides additional views to support explorative tasks. Analysts can move a lens across the map, which will highlight the most important keywords of posts that were sent in the corresponding geographic regions under the lens. When selecting specific term clusters, a histogram depicts the number of posts over time. Text-based and date/time-based filters can be applied to select a subset of tweets. For such a selection of posts, the system can provide a thematic overview of the content using LDA topic modeling (Blei et al. 2003).

6.3.3 Interactive Classifier

In addition to keyword-based filtering, ScatterBlogs also offers means to train and apply SVM-based classifiers interactively, which can be mapped to a color and icon

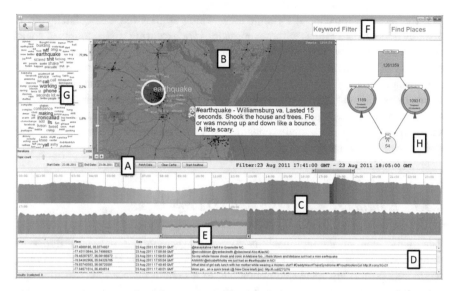

Fig. 6.3 Main interface of the ScatterBlogs system (Thom 2015) for visually analyzing geolocated tweets in real time or retrospectively. Terms which exhibit anomalous usage in specific geographic regions are placed on top of a dynamic map visualization to guide analysts to potentially developing situations (B). Analysts can select a subset of tweets (D) based on date ranges (A, C, E), which will also trigger the extraction of topics using LDA (G). The right side allows defining and combines classifiers interactively for filtering messages (F, H)

to support the visual indication of classified posts. An initial training set can be labeled greedily based on keyword searches and geographic filters. The system then provides visual feedback of the classifier in its current state (e.g., visualization of affected posts on the map) so that analysts can refine them iteratively. It is also possible to combine several such trained classifiers with a visual graph structure (right side of Fig. 6.3) that helps to define Boolean chains.

6.4   Space-Filling Curves for Visualizing the Spatiotemporal Evolution of Data

In addition to the publishing date, a subset of posts and articles also contain a geographic reference (e.g., location of the tweeter). As outlined in the previous section, such geo-references enable analysts to filter data based on relevant regions, but evolving geographic patterns and anomalies can also hint at interesting developments and inform the decision-making process. The ScatterBlogs system focuses on the real-time analysis of streaming posts, and thus, the temporal aspect is mostly implicitly encoded by the dynamic nature of the visualizations. However, for certain analytical tasks, it might be important to analyze larger time ranges in retrospect.

The first section introduced PyramidTags, which applies the triangular layout to visually encode date ranges without animations for exploring vast amounts of social media posts and news articles. However, it is challenging to visualize the temporal evolution of *geolocated* data without animations, since the two main dimensions are typically already reserved to visually encode the geographic location on a map. Animations, though, need to capture the attention of the analyst over a longer time span and impede comparative analyses. Franke et al. (2021) proposed the use of space-filling curves for visually encoding spatial data into just one dimension so that we can depict the evolution alongside the y-axis.

6.4.1 Neighborhood-Preserving 1D Projections

The main idea of the approach is to project geographic positions into one-dimensional positions using space-filling curves, such as Hilbert and Morton, while still preserving local neighborhoods to a certain degree. We can then plot a representative scalar value of geo-referenced data points across time in a two-dimensional plot so that analysts can better assess the temporal evolution of geographic neighborhoods, as well as spot spreading patterns, geographic hotspots, similar patterns across different regions, and trendsetters. In a preprocessing step, the system clusters the data points hierarchically based on their spatial position (if the spatial hierarchy is not already given). This clustering allows us to aggregate larger datasets at different levels of spatial granularity and enables analysts to focus their analysis on specific geographic regions, which also aids in the interpretation of the resulting geo-projections.

6.4.2 Main Interface

For each aggregation level in the spatial hierarchy, the timeline view (Fig. 6.4 underneath the map) visualizes the temporal evolution of each *entity* (e.g., aggregated cluster or single data point) alongside the y-axis. The bars correspond to geographic entities in the clustering and are ordered based on their calculated position in the respective space-filling curve. Analysts can select a specific entity to focus on (highlighted by a red border), which will trigger the system to re-order close entities in the detail view at the bottom. The map at the top serves as an aggregated overview of the different geospatial entities based on a specific point in time that users can specify with the slider at the top-right of the interface.

The system supports several methods for computing space-filling curves (top-left panel). Upon hovering over a specific method, the respective curve is plotted on top of the map, and the differences in the ordering of the elements to the currently selected curve are visualized. Several computed metrics help analysts better assess the quality of the projections.

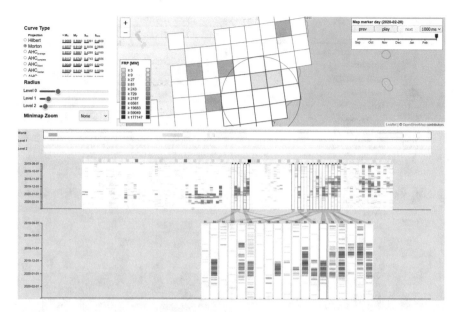

Fig. 6.4 Space-filling curves help to visualize the temporal evolution of geospatial patterns in local neighborhoods (Franke et al. 2021). The map at the top shows an aggregated view at a specific time, whereas the timeline view underneath depicts the value of each geographic entity in the respective hierarchy layer across time. Analysts can select one of the bars, representing entities in the hierarchical clustering, which will highlight close entities in dark orange

6.5 Clustering Posts Dynamically to Analyze Posts in Real Time

Leveraging geo-annotations helps to scale the visual analysis of streaming data and enable geo-specific baselines as well as anomaly detection methods. However, while people still post textual geo-references (e.g., names of cities), the percentage of geolocated social media posts has steadily decreased in recent years. Thus, we need different strategies to enable the real-time analysis of social media posts, and we need to facilitate more context-rich analyses of the actual textual content.

To achieve this, Knittel et al. (2022) have proposed a visual analytics system that employs an efficient dynamic clustering algorithm, providing analysts a continuous overview of what people talk about on Twitter. A dynamic visualization of frequently used phrases and a stream of representative posts help analysts to monitor topics they are interested in, and they can also dive deeper into such topics while increasing the resolution of the analysis. Figure 6.5 depicts the main interface of the approach.

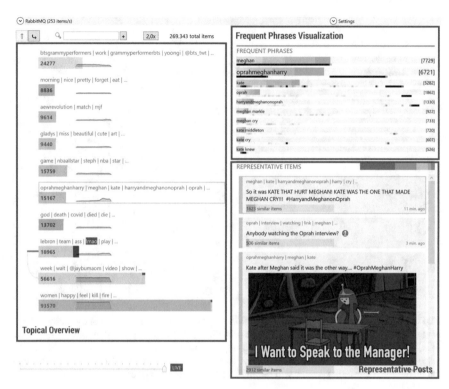

Fig. 6.5 The approach of Knittel et al. (2022) enables large-scale real-time analyses of streaming posts using an efficient dynamic clustering algorithm. The interface provides a thematic overview of what people are currently posting on Twitter (left side). Each row corresponds to one such topic, conveying its size with the length of the bar and the number of posts over time in a small line chart. Once analysts have selected one or several such topics, frequent phrases are continuously extracted and visualized on the right side at the top. In addition to a manageable stream of representative posts underneath, they help to provide a more context-rich visual summary of the streamed content. It is also possible to create filters based on such a topic selection so that the clustering processes and visualizations operate on this filtered stream to enable more fine-grained analyses

6.5.1 Dynamic Clustering

The system stores each new post in a sliding window of configurable size (e.g., the last 20 minutes) and computes corresponding bag-of-words vector representations (Salton and Buckley 1988). For the dynamic clustering, the approach adapts the k-means clustering algorithm (Lloyd 1982) based on a more efficient implementation for sparse vectors (Knittel et al. 2021a). The idea is to regularly cluster the documents in the sliding window with different cluster sizes while using the centroids from the previous clustering run (if available) to obtain more coherent clusters between runs. The Davies-Bouldin Index (DBI) (Davies and Bouldin 1979) determines which of the different clusterings is deemed best. The algorithm then tries to map the final cluster centroids to the ones from the previous run to determine which clusters are actually new, are deprecated, or have just been updated. The system runs two independent clustering processes with different levels of granularity. The more coarse-grained clustering provides a topical overview of the tweets in the window; the more fine-grained clustering facilitates a stream of representative posts.

6.5.2 Topical Overview

The left side of the main user interface (Fig. 6.5) visualizes the topics as computed by the coarse-grained dynamic clustering process. Each row represents one topic, conveying its size with the length of the bar and the number of posts over time in a small line chart, as well as providing a short topic summary with the most important words that define that cluster. Once the clustering has been updated with new posts, the visualization changes dynamically to represent the new state, but in a staggered way so that the mental map is preserved. Each row gets updated step by step (e.g., the *lebron* topic in Fig. 6.5), visualizing the number of new posts in dark green, the number of removed posts in red, and the number of posts that were moved to a different topic in magenta (while also depicting this flow with curves to the left of the bars). New terms are highlighted in dark green. The speed of this dynamic rollout is adjustable.

6.5.3 Frequent Phrases and Stream of Representative Posts

Users can select one or several of the main topics on the left, which will update the right side of the interface with more detailed views for summarizing the content of these topics. The system continuously extracts unusually frequent words and phrases in the selection with ELSKE (Knittel et al. 2021b) and lists them on the right side, highlighting new entries in dark green. Below each such phrase, a small barcode-like visualization allows analysts to infer which of these text parts co-occur by mentally overlapping their respective barcodes. The system can also emphasize this overlap in orange if analysts select several such phrases.

Below this view, a stream of posts appears that resembles a typical feed users would also see on Twitter. One of the challenges the case study of Fathi et al. (2020) with emergency managers identified is the dynamic nature of quickly evolving situations, which can easily lead to situations in which the sheer number of published posts related to a specific event or topic can overwhelm analysts. Hence, the idea of the feed below the frequent phrases is that the appearing posts should cover the thematic variety of the selected topics while keeping the number of new posts in a given time span low, ensuring that analysts can focus on a digestible set of posts.

The system selects for each cluster in the fine-grained dynamic clustering process one post that should represent this fine-grained theme by calculating distances between the vector representations with the respective cluster centroid. Due to their similarity with the respective centroids, these *representative posts* ideally cover a large proportion of tweets in the same cluster, but they should also provide a diverse summary as they were sourced from different clusters in the fine-grained clustering process. If such a representative post has not been added yet, it will be queued up, and a small badge in light blue appears that notifies the user about new posts. For each such post, a small bar chart depicts the number of similar posts, which can also be retrieved if analysts click on the bar.

6.5.4   Diving into Topics

While the frequent phrases and representative posts already provide more context-rich summaries of the selected topics, the general thematic variety on social media platforms is typically high, so it is challenging to group all posts into just 10 to 20 overview clusters. To alleviate this, the proposed system allows users to gradually dive into topics. After selecting one or several such coarse topics on the left side, analysts can click on the fork button at the top of the interface. This will define a new filter layer, in which only posts that fit the selection will be processed and visualized (it is also possible to create keyword-based filters). As a result, the topical overview on the left and the fine-grained clustering process now only operate on this filtered stream of data, increasing the resolution of the topics and aggregations and, thus, increasing the resolution and specificity of the analysis.

6.6   Conclusion

Geo-referenced social media posts and news articles are a rich source for harvesting information and knowledge, but the unstructured nature of the main content and the volume, veracity, and velocity of the data pose significant challenges for developing such visual analysis systems. This chapter outlined several recent approaches for tackling these challenges. The ScatterBlogs system scales the visual analysis of streaming geo-referenced posts by continuously extracting terms that exhibit spatiotemporal anomalies. Our proposed dynamic clustering algorithm enables the continuous monitoring of posts even if they are not geo-tagged. PyramidTags is a novel tag map layout for exploring large time-stamped text collections. We further outlined how we can utilize one-dimensional projection methods to visualize geo-referenced time series data such that we can still observe important spatial trends and patterns.

There are several benefits if we can incorporate geographic locations into our analysis, since this allows us to better detect interesting events and helps to filter the content that has to be processed. However, due to the decrease of geolocated documents, we need to develop new strategies and techniques for leveraging geographic references in social media posts and news articles.

Acknowledgments This research was funded by the Deutsche Forschungsgemeinschaft (DFG, German Research Foundation)—VA4VGI, 314647693.

References

Blei DM, Ng AY, Jordan MI (2003) Latent Dirichlet allocation. J Mach Learn Res 3:993–1022. https://doi.org/10.1016/b978-0-12-411519-4.00006-9

Bosch H, Thom D, Heimerl F, Puttmann E, Koch S, Kruger R, Worner M, Ertl T (2013) ScatterBlogs2: Real-time monitoring of microblog messages through user-guided filtering. IEEE Trans Vis Comput Graph 19(12):2022–2031. https://doi.org/10.1109/TVCG.2013.186

Davies DL, Bouldin DW (1979) A cluster separation measure. IEEE Trans Pattern Anal Mach Intell PAMI-1(2):224–227. https://doi.org/10.1109/TPAMI.1979.4766909

Fathi R, Thom D, Koch S, Ertl T, Fiedrich F (2020) VOST: A case study in voluntary digital participation for collaborative emergency management. Inf Process Manag 57(4):102174. https://doi.org/https://doi.org/10.1016/j.ipm.2019.102174. https://www.sciencedirect.com/science/article/pii/S0306457319302316

Franke M, Martin H, Koch S, Kurzhals K (2021) Visual analysis of spatio-temporal phenomena with 1D projections. Comput Graph Forum 40(3):335–347. https://doi.org/https://doi.org/10.1111/cgf.14311. https://onlinelibrary.wiley.com/doi/abs/10.1111/cgf.14311

Keim D, Kohlhammer J, Ellis G, Mansmann F (2010) Mastering the information age: solving problems with visual analytics. Goslar: Eurographics Association. https://diglib.eg.org/handle/10.2312/14803

Kennedy J, Eberhart R (1995) Particle swarm optimization. In: Proceedings of the International Conference on Neural Networks, ICNN 1995, vol 4, pp 1942–1948. https://doi.org/10.1109/ICNN.1995.488968

Knittel J, Koch S, Ertl T (2021a) Efficient sparse spherical K-means for document clustering. In: Proceedings of the 21st ACM Symposium on Document Engineering, DocEng 2021. Association for Computing Machinery, New York, NY, USA. https://doi.org/10.1145/3469096.3474937

Knittel J, Koch S, Ertl T (2021b) ELSKE: efficient large-scale keyphrase extraction. In: Proceedings of the 21st ACM Symposium on Document Engineering, DocEng 2021. Association for Computing Machinery, New York, NY, USA. https://doi.org/10.1145/3469096.3474930. https://dl.acm.org/doi/10.1145/3469096.3474930

Knittel J, Koch S, Ertl T (2021c) PyramidTags: context-, time- and word order-aware tag maps to explore large document collections. IEEE Trans Vis Comput Graph 27(12):4455–4468. https://doi.org/10.1109/TVCG.2020.3010095

Knittel J, Koch S, Tang T, Chen W, Wu Y, Liu S, Ertl T (2022) Real-time visual analysis of high-volume social media posts. IEEE Trans Vis Comput Graph 28(1):879–889. https://doi.org/10.1109/TVCG.2021.3114800

Lloyd SP (1982) Least squares quantization in PCM. IEEE Trans Inf Theory 28(2):129–137. https://doi.org/10.1109/TIT.1982.1056489

Salton G, Buckley C (1988) Term-weighting approaches in automatic text retrieval. Inf Process Manag 24(5):513–523. https://doi.org/10.1016/0306-4573(88)90021-0

Thom D (2015) Visual analytics of social media for situation awareness. PhD thesis, University of Stuttgart. https://doi.org/10.18419/opus-3540

Thom D, Bosch H, Koch S, Worner M, Ertl T (2012) Spatiotemporal anomaly detection through visual analysis of geolocated Twitter messages. In: Proceedings of the 2012 IEEE Pacific Visualization Symposium, PacificVis 2012, pp 41–48. https://doi.org/10.1109/PacificVis.2012.6183572

Thom D, Kruger R, Ertl T, Bechstedt U, Platz A, Zisgen J, Volland B (2015) Can twitter really save your life? A case study of visual social media analytics for situation awareness. In: Proceedings of the 2015 IEEE Pacific Visualization Symposium, PacificVis 2015, pp 183–190. https://doi.org/10.1109/PACIFICVIS.2015.7156376

Thomas JJ, Cook KA (2005) Illuminating the path: The research and development agenda for visual analytics. Pacific Northwest National Laboratory (PNNL), Richland, WA

Chapter 7
Visually Reporting Geographic Data Insights as Integrated Visual and Textual Representations

Fabian Beck and Shahid Latif

Abstract Geographic information volunteered by the public is usually also of public interest. However, just publishing the data is not enough to make the data accessible and usable for the public. The raw data might need to be abstracted and interpreted, as well as visually presented to be understandable to non-experts. To address this, we propose interactive visual reporting solutions that leverage natural language and visualizations for geo-related data. We present these reports as interactive documents, but also in other media such as virtual reality environments. First, we have studied the interplay of textual and visual content in such reports. To ease the creation of content, we have developed solutions for authoring interactive documents with a close linking of textual contents and visually presented data. Moreover, we propose automatic report generation approaches that specifically support the exploration of the geo-related data starting from an explanatory summary.

Keywords Geographic visualization · Data-driven storytelling · Interactive documents · Authoring interfaces · Text generation

7.1 Introduction

Data-driven storytelling embeds data into a narration and usually combines a textual representation with visualizations (Segel and Heer 2010). While magazine-style stories might be most common, there exist other presentation genres like animations and slide shows, comics, or annotated charts (Segel and Heer 2010). The power of data-driven storytelling lies in making data accessible to a wider audience, by guiding through the analysis insights while inviting users to engage through simple interactions. Textual and visual data descriptions complement each other and, together, form an integrated representation that is both easy to follow and rich

F. Beck (✉) · S. Latif
University of Bamberg, Bamberg, Germany
e-mail: fabian.beck@uni-bamberg.de

© The Author(s) 2024
D. Burghardt et al. (eds.), *Volunteered Geographic Information*,
https://doi.org/10.1007/978-3-031-35374-1_7

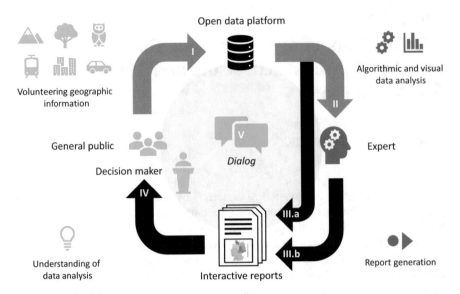

Fig. 7.1 Conceptual diagram of the applied analysis cycle; steps **I–II** in gray mark previous research and results; steps **III–IV** as black arrows represent the focus of our research, which is intended to facilitate a joint dialog between stakeholders (**V**)

of information. For instance, a story on flight data might link a globe or map that shows flight trajectories with explanations on busy routes and airports but also could provide insights into specific examples like the longest flights.

As an expressive and interactive medium, data-driven stories fit volunteered geographic information. For data that is volunteered by the public, it is a natural choice to make that data also available and understandable for a broad audience. Figure 7.1 illustrates this concept as *closing a circle*. Public groups or individuals volunteer geographic information that is then made available on open data platforms (**I**). While experts have already analyzed such data in various application scenarios (**II**), we contribute methods for bringing the derived insights back to the general public and decision-makers. From the open data itself (**III.a**) and insights of the experts (**III.b**), we support authoring and automatically generating reports that can be understood by this broad group of users (**IV**). Computer support and automation are necessary to efficiently create summaries of varying data and to allow personalized reporting. With these reporting solutions, we intend to facilitate non-experts users and foster a dialog between the public, decision-makers, and experts (**V**). This cycle aligns with endeavors of others to make volunteered geographic data directly usable to the public, for instance, within project *IDEAL-VGI* (Chap. 2).

The challenges of this research are in the identification and selection of relevant insights, as well as in their reporting as integrated visual and textual representations. The produced reports should furthermore invite the readers and users to explore the data. We have approached these challenges by first studying the interplay of text and visualization in existing examples of geographic data-driven stories

(Sect. 7.2). Then, we investigated solutions that help author reports with close linking between the two representations (Sect. 7.3). Finally, designing automatically generated reports allowed us to provide, aside to a data-driven story, solutions with certain support for exploration (Sect. 7.4). While geodata plays a role in all presented research, we also consider the visualization of additional, non-geographic data.

This chapter describes the results of the project *vgiReports* and summarizes as well as connects an excerpt of project-related publications (Latif et al. 2021a,b, 2022a,b) and a preliminary work (Latif and Beck 2019a). Two of these works (Latif et al. 2021a,b) report results from collaborations with other projects of the priority program.

7.2 The Interplay of Text and Visualization

The way textual and visual descriptions are combined is crucial in data-driven stories. If integrated well, it could avoid a *split attention effect* between the two media (Ayres and Sweller 2005) and might even help identify misaligning information (Zheng and Ma 2022). Interactive linking of text and visualization can increase user engagement (Zhi et al. 2019) and guide user attention while supporting specifically less experienced users in correctly mapping the text and data (Barral et al. 2021).

Journalistic outlets provide many high-quality, manually crafted examples of data-driven stories. For instance, *The New York Times* has published in 2021 more than a hundred carefully designed visual stories and interactive graphics (The New York Times 2021). As some of such stories cover geographic aspects, we can leverage them to study how geographic data is successfully reported to a wide audience. Previous research has already studied in existing stories structure and sequence (Hullman et al. 2013), patterns of visual narrative flow (McKenna et al. 2017), and narrative order in time-oriented stories (Lan et al. 2021). Text in such stories can have different roles, ranging from introductory texts to detailed annotations of the visualization (Segel and Heer 2010). We have focused on a fine-grained analysis of such categories and the explicit and implicit interplay of text and visualization in stories, with a certain focus on geographic aspects. In a first study, we analyzed 22 full stories from a variety of news media (Latif et al. 2021b). A second study looked at a set of 110 paragraph-chart pairs stemming from 77 articles of different news media (Latif et al. 2022b). Using a qualitative methodology in both studies, we investigated the text on sentence and word level and classified the cases into different categories. Specifically, we have addressed the following research questions.

***What Are the Reported Analysis Insights, and How Is the Related Data Visually Communicated?* (Latif et al. 2021b)** We observed two categories of textual narrative: data-driven text and contextual embedding text. The former directly relates to the data and describes analysis insights. These insights link to the analysis tasks, namely, *identify*, *summarize*, and *compare*. In stories with geographic focus, location and time are generally key concepts. In particular, locations associated with extreme values as well as those depicting very dissimilar behavior (outliers) are identified and explained in the narrative. Likewise, clusters of locations are discussed together to highlight their similarities. Other reported insights include geographic and temporal variations of variable values across a geographic region or time span. The narrative either uses measures of central tendency like mean, median, or mode to summarize them or describes these variations in plain words. Lastly, the narrative compares locations or other data items using part-to-whole contrasts, correlation, or statistical ranking. Apart from these insights, as contextual embedding, a comparable proportion of the narrative blends in the background of the story and data, necessary domain knowledge, and quotes from external sources or people to make these stories self-sufficient units of information. Moreover, authors of the stories interpret analysis insights, relate to other data and information sources, and attach judgment. It is important to note that the data-driven text and the contextual embedding text are often intermingled and cannot be always unambiguously separated. As the textual narrative explains the analysis insights, visualizations act as a complement to show the relevant data. Visualizations can serve a specific purpose, for instance, to provide an overview of the data, to support comparisons, or to highlight details. The use of simple visualizations like maps, visually enriched tables, bar charts, and line plots is more common compared to slightly more advanced ones like distribution plots and scatter plots.

***How Do Textual Narration and Visualization Interplay?* (Latif et al. 2021b)** We discovered different kinds of linking strategies to convert visualizations and an associated textual narrative into a single engaging story. First, the visualizations are almost always placed close to the text that describes them. Likewise, the sequence of visualizations in a story is important: The overview visualizations often appear first and are followed by detailed visualizations. Second, to strengthen the linking further, textual elements like captions, annotations, or tooltips are employed inside or next to visualizations. These textual elements often explain the key insight of a visual or help users in better interpreting the visualization. We observed that the use of descriptive annotations even enabled authors to include comparatively complex and non-standard visualizations into their stories. Third, visualizations as a whole or parts of them are explicitly referenced from the textual narrative. Authors sometimes also use the same colors in text and visualization to show connections between the two media.

***What Implicit References Exist Between Text and Visualization, and How Do They Relate to the Data?* (Latif et al. 2022b)** Implicit references can be defined as connections between a textual narrative and a visualization if both refer to the same data items. For instance, the mentions of countries aside showing a world map make country names implicit references. However, such connections are not limited to just single entities or values but also include group references (referring to many data points, e.g., *EU*) and interval references (referring to numerical ranges).

Furthermore, individual references can be grouped together to form higher-order references. We found that these implicit references can correspond to analysis tasks such as *identification, summarization,* and *comparison.* Almost half of the implicit references directly matched a chart feature (e.g., axis label, legend, annotation, caption). However, the other half contained linguistic variations (e.g., inferences, synonyms, abbreviations, stems, or lemmas) or numerical variations (e.g., rounded off numbers, approximations, computed measures) and are harder to map to the visualized data.

7.3 Authoring Interactive Reports

Creating a data-driven story requires effort: Aside from writing the text, data needs to be analyzed and visually presented. Web technologies provide a good basis for making the content available. However, whereas many content management systems allow placing textual and visual content side by side, they do not support integrating both representations closer and, through this, creating interactive documents. Filling this gap, various authoring tools and supporting approaches have already been suggested for data-driven storytelling (Tong et al. 2018, Section 3). For instance, Chen et al. (2020) developed a framework to synthesize stories from insights identified using a visual analytics systems. It allows an author to arrange insights in different simplified visualizations, annotating and connecting them to tell a story.

Whereas most of these approaches support efficiently generating stories of different kinds, they do not directly address creating explicit and interactive links between the text and visualization. In contrast, *VizFlow* (Sultanum et al. 2021) focuses on links between text and visualization for authoring; while their links are limited to manually created links to image-based features of the visualization, they investigate in more detail how to leverage such links for document layout. *Ellipsis* (Satyanarayan and Heer 2014) allows authoring staged slide show stories with annotations that can be bound to data values and adapt with it. *Elastic Documents* (Badam et al. 2019) does not allow creating links directly but extracts text and tables that relate and connects them using new visualizations. Related are, furthermore, general approaches for annotating charts by, among other visual marks, textual content (Ren et al. 2017).

Focusing on an easy and efficient creation of valuable links between the text and visualizations, we have developed *Kori* (Latif et al. 2022b). The system, as demonstrated in Fig. 7.2, supports both manual creation of links and automatic suggestions for links. The computed links are based on processing the text and consider the hierarchical structure of references discussed in Sect. 7.2. While an author is composing an interactive story, the system offers unobtrusive suggestions, which can then be inspected and accepted or discarded. For the reader, the links finally act as interactive references and, when triggered, take the users' attention to the respective portion of the visualization. Not only do they reduce a split attention effect but could be starting points from where to explore the data further. For the

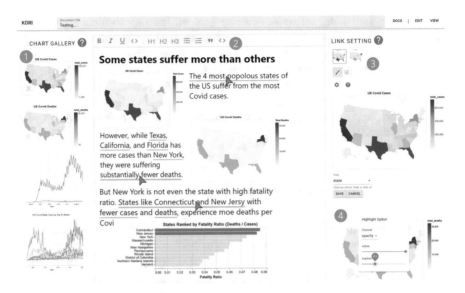

Fig. 7.2 The user interface of *Kori*. It consists of a chart gallery (1) and an editing interface (2). It supports manual creation of links through simple interactions (3). Users can choose highlighting options and their properties (4)

manual creation and adaption of links, the system offers an interface that requires only a few interactions to define the references. There are two modes of manual construction: First, the authors can directly select visual marks in the chart using a direct manipulation mode (e.g., rectangular brush selection). Alternatively, authors can apply a series of filters to select visualized data points. Through these means, authors can effortlessly create references and focus on composing their story.

In a study, we asked 11 participants having diverse background and experience to create references to link text and visualizations in three examples. In the first and second task, they reproduced given links of various kinds with the tool. The third task was more open-ended; the participants were only given a set of visualizations and had to also textually summarize some findings as a short story while linking the text to the visualizations. The results indicated that participants did not have difficulties using the interface and were able to construct meaningful references in all three tasks. Among 64 automatic suggestions that occurred in the sessions, 48 were correct and 16 incorrect. Participants also used the manual construction mode and rated it comparable to the automatic suggestion feature, both with a median of 4 on a scale from 1 (worst) to 5 (best). The feedback for the automatic suggestions was mostly positive and confirmed that many recommendations were helpful and did not disturb the users' workflow. "Smarter" reference detection methods, however, could still improve the experience.

7.4 Explorative Reporting

Data-driven stories and visual reports of data might be presented as interactive documents but often remain rather static. Users can interactively navigate through the story and retrieve some details on demand, but the documents mostly lack support for starting an explorative analysis that goes beyond the original story. Moreover, stories do not adapt to personal interests or current data. The explanation for these restrictions is simple: the stories are manually written as static texts. However, if partly automizing the generation of the natural-language content, we could provide extended options for explorative data analysis and personalization.

Various techniques exist for natural-language generation (Gatt and Krahmer 2018). Whereas the use of advanced generation methods for automatic reporting of data is not yet common in journalistic and industrial practices, some research prototypes have already investigated its potential for more adaptive reporting. For instance, such generation techniques have been used to provide guidance in the data exploration process by reporting automatically derived data facts (Srinivasan et al. 2019). But also whole stories can be generated. Unlike approaches that rather target at fully automatic generation (Shi et al. 2021), we are interested in still human-authored reports, however, which can adapt to different data automatically. Used in interactive documents, the generated text blends with visualizations in a data-driven story. In earlier works, we have explored such representations, for instance, to generate profiles of scientific authors (Latif and Beck 2019b) or, in software engineering use cases, to summarize program executions (Beck et al. 2017) and code quality (Mumtaz et al. 2019). More and more, the generated reports allow greater flexibility regarding the interactive exploration of the data, which complements explanatory texts and guided data analysis. The general idea can be described as *exploranation*, mixing *exploration* with *explanation* (Ynnerman et al. 2018).

Now, we have investigated such approaches to apply them to geographic data in the context of different media and usage modalities. These cover novel aspects such as comparative descriptions of selected entities, novel forms of presentation such as adaptive audio guides, and novel blends of interaction forms and presentations such as chatbots. This set of diverse examples shows early prototypes that demonstrate promising directions of visual reporting; we have not evaluated them yet in detail or connected them into a more comprehensive framework.

7.4.1 Maps with Data-Driven Explanations

Maps that show statistical information are widely used in data-driven storytelling. *Choropleth maps* visualize variation in one variable for a set of regions (e.g., countries). However, oftentimes, it is desirable to describe the relationship between two variables, which would require simultaneous visualization of two values per region. For instance, *per capita spending on education* could be compared to *per*

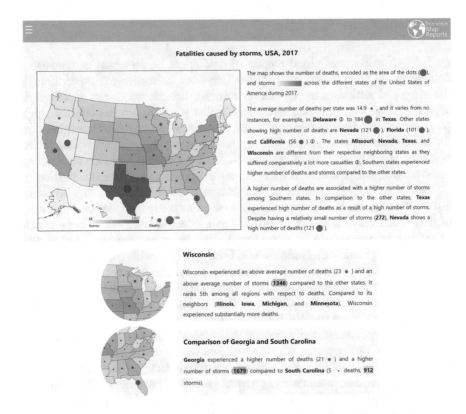

Fig. 7.3 A report describing casualties due to storms in the US as a bivariate map and textual summary (top). Details on a selected region and a comparison of two selected regions are available on demand (cutouts at the bottom)

capita spending on defense to understand different geopolitical roles of countries. An established way of visualizing such bivariate data is to employ graduated symbols overlaid on a choropleth map (Elmer 2012). However, by construction, these bivariate map visualizations are more complex to interpret, and it gets harder to spot visual patterns. Additional textual explanations might counterbalance and could hint at interaction effects of the variables that would go unnoticed otherwise. Encoding more variables per region in a more complex visual glyph is doable but would render the communication to a wider audience even more challenging.

To visually and textually report at least bivariate geographic data in a more accessible way, we developed *Interactive Map Reports* (Latif and Beck 2019a). It employs well-established statistical methods to detect notable relationships, geographic patterns, and outliers in given bivariate data. These insights are automatically transformed into a natural-language narrative that is, then, presented alongside a bivariate map visualization as shown in Fig. 7.3. The given example relates, for states of the USA, the *number of fatalities caused by storms* to the *number of storms* to reflect if the quantity of storms is directly related to the death toll. The textual narrative serves as a guide and explains findings. Small graphics in the text help establish linking between the two representations. Users can explore the map visualization as they read through the narrative by activating interactive links (printed in boldface). Likewise, while exploring the map, users can either get additional details on a selected geographic region or a comparative text for two selected regions.

The system is capable of generating interactive reports for different bivariate geographic datasets. Through a small set of parameters that the user provides about variables, geographic region and granularity, and general terminology, it adapts the generated narrative and visualization.

7.4.2 Interactive Audio Guides in Virtual Reality

Virtual reality is emerging to be an engaging medium for interactive data visualization, and it has just been started to be explored for data-driven storytelling (Isenberg et al. 2018). The idea of *exploranation* is also applicable to virtual reality visualizations. However, as used previously in documents, longer textual narrative will not be suitable as reading would counteract immersion. As an alternative, audio can be used in virtual reality applications such as games, movies, or virtual museums. Prerecorded audio narrative might be played at various stages of the story (e.g., in a game) or activated by a user interaction (e.g., in a virtual museum). The prerecording aspect limits the flexibility, and such approaches cannot adapt to changes in the data as a result of interactions.

To support *exploranation*, our approach *Talking Realities* (Latif et al. 2022a) combines a data-driven audio narrative with an immersive virtual-reality visualization. The audio narrative is based on automatic identification of interesting analysis insights. Using speech synthesis services, it is rendered on the fly from generated text and, therefore, adapts to data selections and user interactions. To provide a smooth exploration experience, the narrative should be synchronized with visual animations. To cater to the needs of a larger target user group, *Talking Realities* advocates three modes with varying levels of guidance. On the one hand, fully *guided tours* walk users through a pre-defined sequence of findings with the least freedom to explore. *Free exploration*, on the other hand, lets users investigate the data visualization without any intervention. In the middle lies the *guided exploration* that provides hints at potential perspectives that are worthy of exploration. We have tested the approach with different immersive visualizations, ranging from multivariate statistical data to astronomic data. Figure 7.4 shows an example of intercontinental air traffic data projected onto a globe.

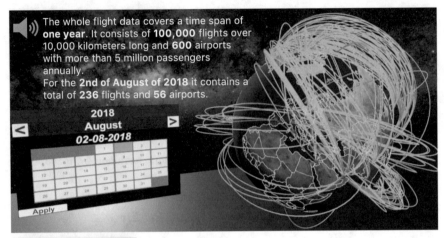

The whole flight data covers a time span of **one year**. It consists of **100,000** flights over 10,000 kilometers long and **600** airports with more than 5 million passengers annually.
For the **2nd of August of 2018** it contains a total of **236** flights and **56** airports.

The longest inter-continental flight is to **Dubai International Airport**, which is **14** hours **46** minutes. An average flight length from **San Francisco International Airport** for a long-distance flight is **12** hours and **49** minutes.

Most long-distance flights are going to **Los Angeles International Airport** which is in the **United States**.

Fig. 7.4 Scenes and audio explanations (here, transcribed) from our prototype implementing the *Talking Realities* approach for air traffic data. (Top) A description of the aggregated intercontinental flights for one day. (Bottom) Scenes reporting the longest flight from an airport and most flights to any other airport

7.4.3 A Chatbot Interface Providing Visual and Textual Answers

Using natural language can make interactions with a machine effortless. Chatbots that reply to textual messages are an example of that. Instead of going through context menus and then choosing the relevant option, chatbots let us verbalize our requests as we would to another human being. However, research on supporting chatbot interfaces for data analysis and visualization is still in its infancy. Although some systems are already powerful, the use of chatbots can still lead to false expectations, misunderstood questions, and unexpected replies (Tory and Setlur

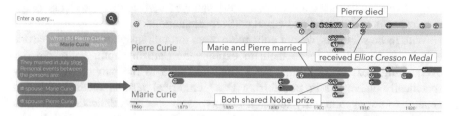

Fig. 7.5 *VisKonnect* answers user questions with a mix of textual reply (left) and explorable visualizations (right). The cutout of the visualization shows a timeline for the two identified scientists; annotations are placed manually for highlighting events that the users might further explore

2019). We believe that a chatbot interface could be a good starting point to make the first contact with the data. In response to the user query, a resulting *exploranative* representation of data should then enable users to verify and validate presented facts and to explore related ones.

For a specialized use case of exploring relationships among historical public figures, we have developed *VisKonnect* (Latif et al. 2021a) together with project *WorldKG* (Chap. 1). The approach offers a chatbot interface to ask questions about said historical figures. Given a question, it uses a rule-based approach to understand the intent of the question and extract meaningful entities (e.g., people, places). Based on this information, it formulates a SPARQL query to pull the relevant data from an event knowledge graph (Gottschalk and Demidova 2019). This data is then visualized in multiple linked visualizations, highlighting the timelines of individual and shared events, as well as where these events take place. These visualizations are augmented with a textual explanation that aims at answering the user question (either through simple text templates or the GPT-3 language model). Additionally, related events are listed and serve as interactive links to explore them in the associated visualization. Figure 7.5 demonstrates a query about two well-known scientists and the response generated by *VisKonnect*.

7.5   Conclusion and Future Work

Within the presented research, we have empirically investigated in-depth how geographic data and related information can be jointly described and linked in textual and visual representations. For creating data-driven stories as integrated reports, we provide authoring support of such links that better connect the two representations. While links can be manually added in a flexible and easy-to-use way, our solution also automatically recommends specific linking through analyzing the data-driven text. In different reporting solutions, we were able to demonstrate the flexibility and broad applicability of our reporting solutions as automatically generated descriptions of statistical maps, as audio guides in virtual reality for immersive visualizations, and as a natural-language interface to a knowledge graph that responds with textual and visual data representations.

With these solutions, we not only guide users through insights of a data analysis but at the same time invite them to explore the data in depth. Following our overarching goal to loop back the volunteered data to the public, we are now specifically interested in transferring these empirical results, methods-oriented general solutions, and early research prototypes to specific application examples and inviting a broader audience to use them. Ongoing work targets at this already, for instance, investigating a visual reporting solution for personalized, comparative summarizations of hotel reviews.

Our research generally emphasizes that citizen participation in research is not one directional. Reflecting back results and providing options to explore the data support an even higher level of participation and should be considered in all citizen science projects and data volunteering platforms. Our ideas can be brought together with analysis solutions for volunteered geographic information, and we invite researchers developing such solutions to also investigate this perspective. Still, empirical studies are necessary that explore the effect of providing visual reporting solutions on the engagement of volunteers and the influence on decision processes.

Acknowledgments The work is funded by the *Deutsche Forschungsgemeinschaft* (DFG, German Research Foundation)—424960846.

References

Ayres P, Sweller J (2005) The split-attention principle in multimedia learning. In: Mayer RE (ed) The Cambridge handbook of multimedia learning. Cambridge University Press, Cambridge, pp 135–146

Badam SK, Liu Z, Elmqvist N (2019) Elastic documents: coupling text and tables through contextual visualizations for enhanced document reading. IEEE Trans Vis Comput Graph 25(1):661–671. https://doi.org/10.1109/tvcg.2018.2865119

Barral O, Lallé S, Iranpour A, Conati C (2021) Effect of adaptive guidance and visualization literacy on gaze attentive behaviors and sequential patterns on magazine-style narrative visualizations. ACM Trans Interact Intell Syst 11(3-4):1–46. https://doi.org/10.1145/3447992

Beck F, Siddiqui HA, Bergel A, Weiskopf D (2017) Method execution reports: generating text and visualization to describe program behavior. In: Proceedings of the 2017 IEEE Working Conference on Software Visualization. IEEE, pp 1–10. https://doi.org/10.1109/vissoft.2017.11

Chen S, Li J, Andrienko G, Andrienko N, Wang Y, Nguyen PH, Turkay C (2020) Supporting story synthesis: bridging the gap between visual analytics and storytelling. IEEE Trans Vis Comput Graph 26(7):2499–2516. https://doi.org/10.1109/tvcg.2018.2889054

Elmer ME (2012) Symbol considerations for bivariate thematic mapping. PhD thesis, University of Wisconsin–Madison

Gatt A, Krahmer E (2018) Survey of the state of the art in natural language generation: core tasks, applications and evaluation. J Artif Intell Res 61:65–170. https://doi.org/10.1613/jair.5477

Gottschalk S, Demidova E (2019) EventKG—the hub of event knowledge on the web—and biographical timeline generation. Semant Web 10(6):1039–1070

Hullman J, Drucker S, Henry Riche N, Lee B, Fisher D, Adar E (2013) A deeper understanding of sequence in narrative visualization. IEEE Trans Vis Comput Graph 19(12):2406–2415. https://doi.org/10.1109/tvcg.2013.119

Isenberg P, Lee B, Qu H, Cordeil M (2018) Immersive visual data stories. In: Immersive analytics. Springer, New York, pp 165–184. https://doi.org/10.1007/978-3-030-01388-2_6

Lan X, Xu X, Cao N (2021) Understanding narrative linearity for telling expressive time-oriented stories. In: Proceedings of the 2021 CHI Conference on Human Factors in Computing Systems. ACM, pp 1–13. https://doi.org/10.1145/3411764.3445344

Latif S, Beck F (2019a) Interactive map reports summarizing bivariate geographic data. Vis Inf 3(1):27–37. https://doi.org/10.1016/j.visinf.2019.03.004

Latif S, Beck F (2019b) VIS author profiles: interactive descriptions of publication records combining text and visualization. IEEE Trans Vis Comput Graph 25(1):152–161. https://doi.org/10.1109/tvcg.2018.2865022

Latif S, Agarwal S, Gottschalk S, Chrosch C, Feit F, Jahn J, Braun T, Tchenko YC, Demidova E, Beck F (2021a) Visually connecting historical figures through event knowledge graphs. In: Proceedings of the 2021 IEEE Visualization Conference. IEEE, pp 156–160. https://doi.org/10.1109/vis49827.2021.9623313

Latif S, Chen S, Beck F (2021b) A deeper understanding of visualization-text interplay in geographic data-driven stories. Comput Graph Forum 40(3):311–322. https://doi.org/10.1111/cgf.14309

Latif S, Tarner H, Beck F (2022a) Talking realities: audio guides in virtual reality visualizations. IEEE Comput Graph Appl 42(1):73–83. https://doi.org/10.1109/mcg.2021.3058129

Latif S, Zhou Z, Kim Y, Beck F, Kim NW (2022b) Kori: interactive synthesis of text and charts in data documents. IEEE Trans Vis Comput Graph 28(1):184–194. https://doi.org/10.1109/tvcg.2021.3114802

McKenna S, Henry Riche N, Lee B, Boy J, Meyer M (2017) Visual narrative flow: exploring factors shaping data visualization story reading experiences. Comput Graph Forum 36(3):377–387. https://doi.org/10.1111/cgf.13195

Mumtaz H, Latif S, Beck F, Weiskopf D (2019) Exploranative code quality documents. IEEE Trans Vis Comput Graph 26(1):1129–1139. https://doi.org/10.1109/tvcg.2019.2934669

Ren D, Brehmer M, Lee B, Hollerer T, Choe EK (2017) ChartAccent: annotation for data-driven storytelling. In: Proceedings of the 2017 IEEE Pacific Visualization Symposium. IEEE, pp 230–239. https://doi.org/10.1109/pacificvis.2017.8031599

Satyanarayan A, Heer J (2014) Authoring narrative visualizations with Ellipsis. Comput Graph Forum 33(3):361–370. https://doi.org/10.1111/cgf.12392

Segel E, Heer J (2010) Narrative visualization: telling stories with data. IEEE Trans Vis Comput Graph 16(6):1139–1148. https://doi.org/10.1109/tvcg.2010.179

Shi D, Xu X, Sun F, Shi Y, Cao N (2021) Calliope: automatic visual data story generation from a spreadsheet. IEEE Trans Vis Comput Graph 27(2):453–463. https://doi.org/10.1109/tvcg.2020.3030403

Srinivasan A, Drucker SM, Endert A, Stasko J (2019) Augmenting visualizations with interactive data facts to facilitate interpretation and communication. IEEE Trans Vis Comput Graph 25(1):672–681. https://doi.org/10.1109/TVCG.2018.2865145

Sultanum N, Chevalier F, Bylinskii Z, Liu Z (2021) Leveraging text-chart links to support authoring of data-driven articles with VizFlow. In: Proceedings of the 2021 CHI Conference on Human Factors in Computing Systems. ACM. https://doi.org/10.1145/3411764.3445354

The New York Times (2021) 2021: The year in visual stories and graphics. https://www.nytimes.com/interactive/2021/12/29/us/2021-year-in-graphics.html

Tong C, Roberts R, Borgo R, Walton S, Laramee R, Wegba K, Lu A, Wang Y, Qu H, Luo Q, Ma X (2018) Storytelling and visualization: an extended survey. Information 9(3):65. https://doi.org/10.3390/info9030065

Tory M, Setlur V (2019) Do what I mean, not what I say! Design considerations for supporting intent and context in analytical conversation. In: Proceedings of the 2019 IEEE Conference on Visual Analytics Science and Technology. IEEE, pp 93–103. https://doi.org/10.1109/vast47406.2019.8986918

Ynnerman A, Löwgren J, Tibell LAE (2018) Exploranation: a new science communication paradigm. IEEE Comput Graph Appl 38(3):13–20. https://doi.org/10.1109/MCG.2018.032421649

Zheng C, Ma X (2022) Evaluating the effect of enhanced text-visualization integration on combating misinformation in data story. In: Proceedings of the 15th IEEE Pacific Visualization Symposium. IEEE, pp 141–150. https://doi.org/10.1109/pacificvis53943.2022.00023

Zhi Q, Ottley A, Metoyer R (2019) Linking and layout: exploring the integration of text and visualization in storytelling. Comput Graph Forum 38(3):675–685. https://doi.org/10.1111/cgf.13719

Chapter 8
Effects of Landmark Position and Design in VGI-Based Maps on Visual Attention and Cognitive Processing

Julian Keil, Frank Dickmann, and Lars Kuchinke

Abstract Landmarks play a crucial role in map reading and in the formation of mental spatial models. Especially when following a route to get to a fixed destination, landmarks are crucial orientation aids. Which objects from the multitude of spatial objects in an environment are suitable as landmarks and, for example, can be automatically displayed in navigation systems has hardly been clarified. The analysis of Volunteered Geographic Information (VGI) offers the possibility of no longer having to separate methodologically between active and passive salience of landmarks in order to gain insights into the effect of landmarks on orientation ability or memory performance. Since the users (groups) involved are map producers and map users at the same time, an analysis of the user behavior of user-generated maps provides in-depth insights into cognitive processes and enables the direct derivation of basic methodological principles for map design. The landmarks determined on the basis of the VGI and entered as signs in maps can provide indications of the required choice, number, and position of landmarks that users need in order to orientate themselves in space with the help of maps. The results of several empirical studies show which landmark pictograms from OpenStreetMap (OSM) maps are cognitively processed quickly by users and which spatial position they must have in order to be able to increase memory performance, for example, during route learning.

Keywords Landmarks · Salience · Pictograms · Spatial memory

J. Keil (✉) · F. Dickmann
Ruhr-Universität Bochum, Bochum, Germany
e-mail: julian.keil@rub.de; frank.dickmann@rub.de

L. Kuchinke
International Psychoanalytic University Berlin, Berlin, Germany
e-mail: lars.kuchinke@ipu-berlin.de

© The Author(s) 2024
D. Burghardt et al. (eds.), *Volunteered Geographic Information*,
https://doi.org/10.1007/978-3-031-35374-1_8

8.1   Introduction

The increasing relevance of mobility is linked to an increasing extent of using mobile navigation technology. Modern navigation and visualization technology depend on the processing of extensive spatial data, which must be provided and maintained in real time. A sub-project of the SPP 1894 "Volunteered Geographic Information (VGI): Interpretation, Visualization and Social Computing" of the Ruhr University Bochum and the International Psychoanalytical University Berlin investigates how wayfinding and navigation can be supported by user-generated geodata. The project addresses the growing interest in reliable spatial data, which extends far beyond the capacities of official geodata sets. The aim is to obtain quantitative data on the use of landmarks in maps in the formation of mental spatial models by means of empirical studies and to transfer these to cartographic applications. The focus of this research is thus on the representation, use, and interpretation of landmarks as they appear in map-based Web 2.0 services such as OpenStreetMap (OSM). The results are expected to further contribute to the development of automatic extraction processes in mobile map design (cf. Klippel and Winter 2005; Elias and Paelke 2008).

8.2   The Role of Landmarks in Self-Localization, Navigation, and the Formation of Mental Spatial Models

Effective navigation in space depends on continuous self-localization (Loomis et al. 1999). By using topographic objects in the environment as spatial reference points, self-localization can be achieved (Meilinger et al. 2006). In this context, landmarks, salient spatial objects (Sorrows and Hirtle 1999; Bestgen et al. 2017), play an important role in real-world and map-based navigation. From the perspective of their perceiver, landmarks pop out of their surrounding objects (Bestgen et al. 2017; Röser 2017). This increases their likelihood to be perceived and processed. Therefore, landmarks are more likely used as the spatial reference points required for self-localization (Sorrows and Hirtle 1999; Bestgen et al. 2017; Elias and Paelke 2008; Millonig and Schechtner 2007).

Due to their relevance in the context of self-localization, landmarks are also known to play an important role for orienting. Turn-off points of a route can be identified by recognizing landmarks located close to these turn-off points (Millonig and Schechtner 2007). Furthermore, landmarks along the route can help ascertain that one is still following a specific route correctly, even if these landmarks are not located close to a location where the travel direction needs to be adjusted (Anacta et al. 2017). The relevance of landmarks for navigation is also reflected in findings demonstrating that the availability of landmarks during navigation increases the accuracy of route finding (Ruddle et al. 1997) and user confidence (Ross et al. 2004).

When people perceive space, either directly in an environment or as a map representation, they gradually integrate the perceived spatial information into a mental spatial model (Millonig and Schechtner 2007). According to Siegel and White (1975), landmark-based spatial knowledge is the first building block for the development of a mental spatial model. By focusing on salient landmarks, people make sense of otherwise (too) complex environments. Thus, landmarks provide an abstraction layer that is easier to memorize than the unfiltered environment (Millonig and Schechtner 2007; Presson and Montello 1988). In this abstraction layer, spatial elements can be memorized based on their relative location of the reference points provided by landmarks (Golledge 1999; Richter and Winter 2014). In other words, landmarks form a framework with the help of which other topographic objects in space are remembered. However, memorizing only landmarks as spatial reference points does not result in a usable mental spatial model in the context of spatial navigation. People also need to identify and memorize route segments connecting the spatial reference points. Werner et al. (2000) describe such a network as consisting of "nodes" and "edges" (see Fig. 8.1).

If acquired within real-world space, spatial knowledge in the form of landmarks and routes is incipiently egocentric (Millonig and Schechtner 2007). However, with the integration of different viewing perspectives of space, an allocentric survey model is likely also being developed (Werner et al. 1997). The map-like structure of such an allocentric mental spatial model allows more complex and flexible spatial evaluations. Even landmark pairs, for example, that are not directly connected with a route segment can be set into spatial relation to each other within such an allocentric model. Additionally, each added route segment increases the number of potential routes (see Fig. 8.1). This allows, for example, estimating distances or planning alternative routes between specific landmark pairs.

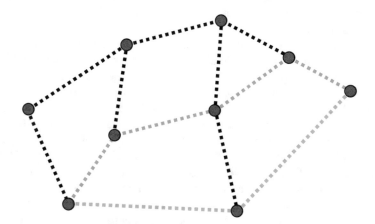

Fig. 8.1 Landmark-based mental spatial model. Memorized landmark locations are represented by red dots (nodes). The dotted line connections (edges) between the landmarks each represent a memorized route between two nodes of the mental spatial model. The blue and green lines demonstrate that different routes between a start and target location can be selected if enough edges between node pairs have been memorized. Figure adapted from Keil (2021)

8.3   Identifying Landmarks

As demonstrated above, landmarks play an important role in spatial cognition, spatial behavior, and spatial memory. But how can we distinguish landmarks as relevant spatial reference points from less relevant topographic objects? So far, several attempts have been made to define landmarks from different perspectives. (Lynch 1960) focused on the inherent physical characteristics of urban objects and their visual contrast to features in the environment. Caduff and Timpf (2008) suggested a trilateral approach to identifying landmarks. The trilateral approach assumes that landmarks are created through an interaction between the observer, the spatial object, and its environment. According to Anacta et al. (2017), landmarks are objects with a fixed geographic location that are easy to perceive and recognize.

All of the aforementioned approaches to defining landmarks, either implicitly or explicitly, suggest a common characteristic of landmarks: they need to be salient. The salience of a topographic object describes and comprises characteristics that increase the likelihood of this object to attract the attention of (potential) observers in the environment (Itti 2005, 2007). Conversely, topographic objects that attract more attention than surrounding objects can be interpreted as being more salient (Caduff and Timpf 2008) and thus more likely to be used as a landmark (Röser 2017).

The trilateral approach of Caduff and Timpf (2008) already suggests that salience is not based on one single object characteristic that can be easily measured. Instead, to identify potential landmarks in the environment based on the salience of topographic objects, numerous characteristics need to be considered. To address this complexity of landmark salience, Sorrows and Hirtle (1999) take an approach that does not only refer to the appearance and position of a spatial element in an environment: They propose three categories of landmarks, visual (visual contrast), structural (prominent location), and cognitive (use, meaning) landmarks. Thus, it is proposed that the identification of landmarks also takes the semantics of the spatial object into account. To better distinguish the semantics from cognitive effects that could also affect the visual salience, the term semantic salience has also been established (Klippel and Winter 2005; Quesnot and Roche 2015). Due to the individual experiences and previous knowledge of a viewer, an exact determination of the meaning that a landmark (e.g., a building) has for an individual viewer can only hardly be defined and measured (Golledge 1991; Nuhn and Timpf 2017). One way out is to fall back on generally accepted classifications, e.g., road classifications (federal road, rural road, etc.) or monument classifications in the case of historic buildings, etc. (e.g. Raubal and Winter 2002). However, a study by Nuhn and Timpf (2019) indicates that personal attributions of meaning might be less important for the selection of landmarks than previously assumed. A landmark selection model containing personal information did not perform better than a model without this information. Although these findings do not solve the fundamental problem of the individual assignment of meanings to landmarks, they provide a perspective on the problem, at least in connection with visual and structural salience characteristics.

8.4 Landmark Representations in VGI-Based Maps

The main focus of previous research on landmark salience was directed at landmarks within real-world space (e.g., Anacta et al. 2017; Röser 2017; Sorrows and Hirtle 1999). However, psychological studies demonstrate that salience also affects visual attention in the perception of 2D information, e.g., on computer interfaces (Buscher et al. 2009) or in maps (Fine and Minnery 2009). This raises the question of how salience effects influence the perception and use of landmark representations in maps.

In most maps, real-world landmarks are represented by pictograms. However, the selection of landmarks to be represented in the maps occurs a priori. Thus, the selection of landmarks to be represented in maps is usually not based on direct interaction with the environment and user choices but rather based on established cartographic principles. As stated above, user characteristics (semantic salience) affect the selection of landmarks in real-world space. Additionally, visual salience characteristics (e.g., visibility from a specific viewpoint) can affect the selection of landmarks. In this context, an additional categorization of visual, semantic, and structural salience characteristics into active and passive salience can be used to illustrate the issue of landmark selection in maps. Passive salience depends on physical attributes of landmarks, such as size, color, or their function (Bestgen et al. 2017). It describes the potential of an object to attract attention based on bottom-up processes independent of individual characteristics of a potential perceiver (Caduff and Timpf 2008). The methods for identifying the passive salience of landmarks are explained in detail by Duckham et al. (2010), Klippel and Winter (2005), Nothegger et al. (2004), or Elias (2003). These approaches use, among other things, the results of statistical and data mining methods (Peters et al. 2010; Sadeghian and Kantardzic 2008). The concept of active salience, on the other hand, is based on the perceived, cognitive, and contextual appearance of a landmark, i.e., viewed from the perspective of the traveler (user), including his or her experiences, age, prior (cultural) knowledge, and the way he or she moves (Caduff and Timpf 2008; Millonig and Schechtner 2007; Zhu 2012). Taken together, exogenous (passive) salience patterns are contrasted with endogenous (active) ones. Both concepts influence the processing of map information and need to be taken into account for the identification and representation of landmarks. If the active user- and context-dependent salience characteristics are not considered, the set of landmarks selected to be represented in a map will not match the topographic objects selected as landmarks by the map users.

A special case in the context of landmark selection are maps based on Volunteered Geographic Information (VGI), such as OpenStreetMap (OSM). Producers of VGI map data (influenced by endogenous salience characteristics) simultaneously assume the role of the consumer (influenced by exogenous salience characteristics), i.e., two perspectives on landmark saliency are merged. There is no external decision-making authority, as is usually the case in map production. The resulting landmarks come from the actual spatial experiences of the volunteers (producers = consumers) themselves. Therefore, the study of map elements that are considered landmarks by such volunteers can provide valuable information on how landmarks are identified in maps in terms of their density and spatial arrangement and how user-generated landmark representations influence the formation and accuracy of cognitive representations (mental models) of the mapped space.

The VGI-based selection of spatial objects represented as landmarks in VGI-based maps is assumed to reflect the specific distribution of landmarks to each other (i.e., patterns) but also the relations to other map elements (non-landmarks) in maps that are required to solve spatial tasks. Since the mapping process of volunteers is affected by direct interaction with the mapped environments as well as their mental spatial representation of these environments, the landmarks represented in VGI-based maps are expected to reflect the patterns of landmarks required to represent space in a mental spatial model. This raises the question of how landmarks distributed in maps on the basis of empirically obtained rules (i.e., on the evaluation of landmark patterns created by volunteers) are used to create mental spatial models.

In real-world space, it is not only a single landmark in the landscape that is important for building survey knowledge of an environment. Rather, networks of several landmarks and the relations (e.g., distances or routes) between them must be considered (Herman et al. 1979; Werner et al. 2000). In a map, an interaction or pattern of several landmarks could therefore support the function of structuring (map) spaces and thus possesses a function for the construction of more accurate mental models in the same way (cf. Golledge 1993). Again, if the active user- and context-dependent salience characteristics are not considered, the set of landmarks selected to be represented in a map will not match the topographic objects selected as landmarks by the map users. In addition, it can be shown that landmarks are used in spatial learning to relate spatial objects to them (Ferguson and Hegarty 1994; Golledge 1999). Thus, visual attention directed toward specific landmarks due to their salience may not only affect spatial memory of the landmarks themselves. The availability of salient landmarks may also affect spatial memory of surrounding spatial objects or routes.

However, it needs to be considered that the salience and perception of landmarks in real-world space differs from the perception of landmark representations in maps. In real-world space, landmarks are only potentially salient if they are within the line of sight of the observer. This excludes many proximate landmarks that are not visible, for example, because they are hidden behind other spatial objects (cf. Lynch 1960). In maps on the other hand, each represented landmark is potentially visible. In this case, which landmark representations are perceived depends on the salience characteristics and/or task requirements. Therefore, how salience or spatial tasks affect the use of landmarks in maps (opposed to real-world landmarks) could affect performance in spatial tasks such as self-localization, orientation, and navigation, as well as the formation of mental spatial models based on map perception.

Not all graphic elements of a map share the same relevance for the formation of cognitive map representations as the selected (higher-level) landmark structures. The analysis of task-dependent salience of VGI-based map elements (here OSM maps) is therefore expected to support the identification of a working definition of the required spatial reference points (specifically landmark representations) that need to be visualized to ensure effective spatial information transfer. This might lead to new content structures that make it easier for map users to associate objects to be learned in a map with a higher-level frame of reference. The results of several studies contained in our project on the semantic, visual, and structural salience of VGI-based landmarks provide insights into the selection of spatial reference points in maps during the acquisition of spatial memory. Thus, these findings contribute to the development of guidelines for task-oriented map design that support the formation of mental spatial models.

8.4.1 Semantic Salience

Semantic salience refers to semantic properties that affect the likelihood of an object to draw attention. As demonstrated by Pilarczyk and Kuniecki (2014), semantically salient stimuli attract visual attention, and the semantic features of a stimulus can in some cases even have a stronger effect on the direction of visual attention than the visual features. Different characteristics that affect the semantic salience of topographic objects have been suggested over the years. Some suggestions focus primarily on generalizable characteristics like cultural and historical significance or the purpose or function of an object (Claramunt and Winter 2007; Nothegger et al. 2004; Raubal and Winter 2002; Röser et al. 2011). These characteristics suggest that semantic salience is an intrinsic bottom-up property of an object that is not affected by the observer. Based on these suggestions, a church or a police station should be semantically more salient than a common residential building. Other approaches to assess semantic salience argue that top-down processes based on knowledge and preferences of an observer affect the semantic salience of topographic objects (e.g. Golledge 1991; Nuhn and Timpf 2017; Quesnot and Roche 2015). For example, a residential building can—in some cases—even be semantically more salient than a church or a police station if the observer lives in it.

In the context of map elements like landmark representations, assessing semantic salience faces issues, which are unique to the medium. First, landmarks are usually represented in maps as pictograms. In most cases, these pictograms do not reflect the visual characteristics of the represented landmarks but are designed to reflect the purpose or function of the landmark. This abstract representation of landmarks can override the semantic associations with a landmark. For example, your favorite restaurant located in a beautiful old building could be represented by the same pictogram as each other restaurant. On the other hand, a map pictogram can also communicate semantic information that is not effectively communicated by the visual appearance of the real-world object represented by the pictogram. It is also important to consider how the pictogram design affects to what extent map users are able to interpret the purpose or function of the represented landmark. For example, some pictograms might only be used in specific cultures or might have different meanings in different cultures (Spinillo 2012). Thus, pictogram designs need to be selected based on the intended user group of a map.

In a first study of our project, we investigated to what extent people understand the meaning of landmark pictograms (meaningfulness) and how the ability of a pictogram to communicate its semantics affects the attraction of visual attention (semantic salience) and the ability to memorize the pictogram (see Keil et al. 2019). We chose to investigate a set of 153 pictograms obtained from OSM. The map content and design of OSM are provided and influenced by a large worldwide community of volunteers. This is assumed to be reflected in the pictogram design and the accessibility of pictogram semantics within different cultural groups.

In a recognition design, sets of 12 pictograms (see example in Fig. 8.2) were shown to the participants. After a distractor task, participants had to identify the previously shown pictograms within a set of 24 successively shown pictograms. During the encoding phase, fixations on the pictograms were recorded with an eye tracker. In the second half of the experiment, participants successively saw the 153

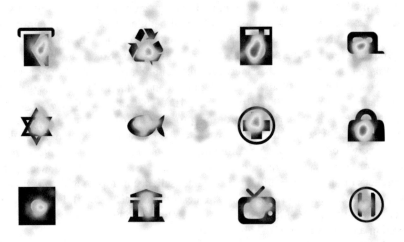

Fig. 8.2 Fixation heat map. In the study, participants saw and had to memorize sets of 12 landmark pictograms. Semantic salience of the pictograms was assessed based on the visual attention directed toward each pictogram, as reflected in the measured fixations. In the example above, the helipad pictogram (down right) was fixated less often than the postbox pictogram (top left) and was therefore scored as less semantically salient. Potential order effects in the stimuli (e.g., based on reading direction) were addressed by varying pictogram locations between trials

pictograms. Each pictogram was shown together with a continuous scale that was used to assess to what extent participants were certain to understand the meaning of the pictogram (meaningfulness).

The findings demonstrate that pictograms with a very low meaningfulness rating attracted more visual attention and were recognized more often. An explanation of this unexpected finding needs to consider the experimental design and the general source of salience, which is a contrast with surrounding stimuli (Claramunt and Winter 2007; Sorrows and Hirtle 1999). Most of the 153 pictograms were rated as having a relatively high meaningfulness. Therefore, the few pictograms with a low meaningfulness were the ones with the highest semantic contrast. Consequently, as the semantic contrast is assumed to direct visual attention, the few pictograms with a low meaningfulness should also be the ones with the highest semantic salience. As selective visual attention has been associated with improved object learning (Walther et al. 2005), this also explains the better memory performance of landmark pictograms with a low meaningfulness.

However, it is important to also consider that the perception of landmark pictograms in this experiment does not match the perception of landmark pictograms in their "natural" environment. In OSM and other maps, semantic pictograms are surrounded by unified representations with a low meaningfulness, for example, buildings, roads, or green spaces. Thus, pictograms with a high meaningfulness should have a higher semantic contrast within a map and therefore a higher semantic salience. Furthermore, due to the higher selective attention, landmark representations in maps with a higher meaningfulness are assumed to be more likely to be stored in a mental spatial model (Walther et al. 2005). Taken together, the study provides an approach for assessing the semantic salience of landmark pictograms and demonstrates that semantic salience affects the attraction of visual attention and the memory of map elements.

8.4.2 Visual Salience

Visual salience is probably the most investigated salience characteristic. It describes the visual contrast of an object to its surrounding objects and depends on parameters as illumination, size, color, texture, or shape (Clarke et al. 2013; Davoudian 2011; Duckham et al. 2010; Röser et al. 2011) and has been demonstrated to direct visual attention to stimuli (Wenczel et al. 2017). Commonly used examples of visually salient landmarks used for orientation and navigation are large or tall buildings, unique objects like statues, or buildings with an eccentric architecture or uncommon visual features (Klippel and Winter 2005). According to von Stülpnagel and Frankenstein (2015), visual salience affects the selection of spatial objects as landmarks for orientation. In other words, people look for visually salient objects that can be used as the spatial reference points required for making sense out of space and building mental spatial models (cf. Clarke et al. 2013).

Opposed to semantic and structural salience, the allocation of attention based on visual salience has been argued to be a stimulus-driven bottom-up process (Itti 2005; Ouerhani et al. 2004). This is supported by neurological findings demonstrating that the ability to perceive feature contrasts as the foundation for visual salience is located in the V1 area of the visual cortex (Li 2002). This means that the visual salience of visual stimuli is evaluated in an early and automatic stage likely before top-down processes affect the direction of visual attention. Its dependence on feature contrast means that visual salience is a context-dependent characteristic. An object is visually salient relative to its surrounding objects (Claramunt and Winter 2007; Klippel and Winter 2005). Thus, a tall building might not be visually salient if it is part of the skyline in a large city. On the contrary, a small building can be visually salient because it is surrounded by tall buildings.

Visual salience has been intensively investigated both for real-world objects like buildings or facades (e.g., Davoudian 2011; Franke 2021; Dong et al. 2020; Röser 2017; von Stülpnagel and Frankenstein 2015; Wenczel et al. 2017), but also for the perception of 2D stimuli, for example, computer interfaces or images (e.g., Clarke et al. 2013; Ouerhani et al. 2004; Sutherland et al. 2017). However, the effects of visual salience on the perception and processing of maps and landmark representations have received little attention in the literature so far.

The design of map elements is often not based on the visual features of the represented real-world objects. Streets, buildings, green spaces, and water bodies are represented by geometric shapes, and the colors of these shapes are determined by the map design guidelines (Dickmann 2018). Landmarks, on the other hand, as mentioned earlier, are usually represented as specific semantic pictograms based on the semantic categories they are assigned to (e.g., restaurants, shops, statues, etc.). In both cases, the individual visual characteristics are lost due to the type of representation. Thus, the visual salience of a real-world object does not match the visual salience of its map representation. Furthermore, opposed to real-world environments, it is easily possible to adjust the visual characteristics of object representations in a map and, consequently, the visual salience of specific map elements.

In a second study of our project, we explored how adjustments of map design can be used to systematically direct visual attention toward specific map areas. Furthermore, we investigated to what extent different map designs and the resulting visual salience differences of specific map regions affect spatial memory. The full study is described in detail in Keil et al. (2018). As study materials, we obtained maps from OSM and added routes to the maps. As a second stimulus condition, map areas offside the route (more than 10 pixels from the route) were displayed transparently (see Fig. 8.3). This was meant to reduce the visual salience of the areas offside the route and expected to direct visual attention toward the map areas close to the route. Participants saw the maps for 30 s and were asked to memorize the route (encoding phase). During this phase, fixations on the map areas were recorded with an eye tracker. Two different map areas of interest (AOIs) were defined. The first AOI contained the route and the area 10 pixels around the route, thus the area which was not transparent in the second stimulus condition (route AOI). The second

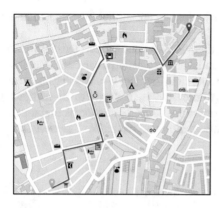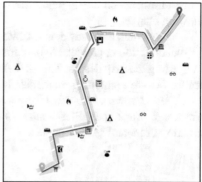

Fig. 8.3 Stimulus conditions. Maps were obtained from OSM (© OpenStreetMap contributors), and a route was added. For the second stimulus condition, areas offside the route (more than 10 pixels) were displayed transparently. Eye tracking was used to record fixations on the area close to the route and the area offside the route

AOI contained the rest of the map (offside route AOI). Each fixation was recorded according to the AOI it targeted at. After the encoding phase, participants were shown four versions of the previously presented map, either containing the correct route or a slightly manipulated route. For each map, they had to decide whether the displayed route matched the previously learned route.

The results show that significantly fewer fixations were directed at the map areas offside the route (offside route AOI) when these areas were transparent. Fixations on the area around the route (route AOI) did not differ significantly between the standard map and the transparent map. Thus, we were able to demonstrate that changing map design, and consequently, the visual salience of specific map elements affects the distribution of visual attention across the map. Interestingly, the fact that landmark representations offside the route were not displayed transparently did not significantly undermine the shift of visual attention toward the route. We argue that visual attention was in both conditions distinctively affected by the task requirement of memorizing the route. Thus, most fixations in the non-transparent map were already relatively close to the route and not on landmark representations far offside the route. Applying transparency only narrowed the area of fixations around the route. This interpretation is supported by the fact that route memory performance did not differ significantly between the original map and the transparent map. Participant route memory performance seems to have relied primarily on reference points close to the route. Therefore, making reference points offside the route less visible and less likely to receive attention did not affect memory performance. However, performance differences could potentially occur if the task is carried out with more time pressure. If a task requires map readers to make quick decisions or to capture map information quickly, directing visual attention toward relevant map areas could reduce distraction from less task-relevant map areas. The effects of task requirements on the direction of visual attention are further investigated in the following chapter.

Taken together, the study demonstrates that visual salience affects the distribution of visual attention in predictable patterns and that these patterns can be manipulated by adjusting the map design. This makes it possible to direct visual attention toward specific task-relevant map elements. In future studies, it needs to be addressed how different ways of guided visual attention based on landmark pictogram and map design affect spatial tasks such as orientation and navigation, as well as the formation of mental spatial models.

8.4.3 Structural Salience

Structural salience is a task- or context-dependent salience (Peebles et al. 2007). Previous research on structural salience focused primarily on the relative location of landmarks during navigation and the conceptualization of routes (Klippel and Winter 2005; Röser et al. 2011). Based on their relative location to the observer or a route, structurally salient landmarks can be divided into global and local landmarks (Elias and Paelke 2008).

Global landmarks in real-world settings are located far enough to only marginally change their relative location based on movements of the observer (Keil 2021). Thereby, they can act as beacons for assessing the general travel direction (Lynch 1960; Steck and Mallot 2000; Wenig et al. 2017). An important characteristic of a global landmark is its size, as it determines its remote visibility (von Stülpnagel and Frankenstein 2015). The distance from the observer can range between a few hundred meters (e.g., a tall building) and hundreds of kilometers (e.g., a mountain) or even light years (e.g., the north star). Global landmarks are not suitable to identify or memorize specific routes. Instead, they can be used to choose paths that lead to the general direction of the travel destination. This is reflected in a large flexibility of selected routes when people navigate based on global landmarks (Hurlebaus et al. 2008).

Local landmarks, on the other hand, are located close to the observer and/or to a specified route. Compared to local landmarks, global landmarks provide more precise information about the observer's location and support encoding of and navigation along a specific route (Hurlebaus et al. 2008; Ruddle et al. 2011; Steck and Mallot 2000). Furthermore, they are frequently used for communicating routes (Anacta et al. 2017). Local landmarks can be subdivided based on their relative location to specific fragments of a route. They can be located close to decision points, potential decision points, or along the route (see Fig. 8.4). Decision points are intersections where the travel direction needs to be adjusted. Landmarks located at decision points can be used as a reference marker for identifying, memorizing, or communicating a required turn (Millonig and Schechtner 2007). Potential decision points are locations where the travel direction could be adjusted but should not be adjusted, for example, an intersection where the route follows a straight direction.

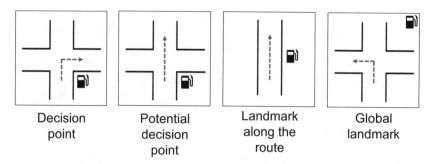

| Decision point | Potential decision point | Landmark along the route | Global landmark |

Fig. 8.4 Navigation-oriented landmark locations. Local landmarks close to a route can be located at locations where the travel direction needs to be adjusted (decision point), at locations where the travel direction could be adjusted (potential decision point), or where the travel direction cannot be adjusted (along the route). Global landmarks act as beacons and are located offside the route (figure adapted from Bauer 2018)

Landmarks along the route are located next to the route where the travel direction cannot be adjusted (Anacta et al. 2017; Elias and Paelke 2008). Landmarks at potential decision points or along the route are not necessarily required for conceptualizing a route. However, they can be used during navigation to ensure that the route is still followed correctly (Millonig and Schechtner 2007). At decision points, landmarks can further be subdivided based on their location relative to the turn direction of a route. Röser et al. (2012) found that landmarks are more likely to be perceived and used in way finding and thus are more structurally salient, if they are located in the direction of the turn. Furthermore, turning decisions have been found to be more likely to be correct if a local landmark is available in the direction of the turn (Albrecht and Stuelpnagel 2018).

Similar to semantic and visual salience, the question of how structural salience affects the distribution of visual attention in maps has received little attention yet. As it has been argued that structural salience is task-dependent (Peebles et al. 2007), we carried out three studies on the effects of different map-based spatial tasks on the distribution of visual attention across the map. The first two studies investigated the structural salience of landmark representations in map-based route memory tasks (for the complete study details, see Keil et al. 2020a). In both studies, participants had to memorize routes displayed in maps obtained from OSM. Eye tracking was used to assess the distribution of visual attention across the maps. Landmark pictograms were available in the maps according to the locations of landmark representations added by volunteers to these OSM maps. However, to control semantic salience of pictograms, the original pictograms were randomly replaced by a set of OSM pictograms with similar levels of meaningfulness as measured in the first study of our project (see Keil et al. 2019). As other map elements than landmark pictograms could also be used as spatial reference points for memorizing the route, we also addressed how the visual complexity of a map (number of spatial elements) affects the structural salience of landmark pictograms.

Therefore, the first of the two route memory studies compared rural maps with a moderate number of map elements and urban maps with a high number of map elements. In the second study, only urban maps were used, but in one condition, fractions of the original map were used, and these fractions of the map were stretched to the original map size to create a second map condition with less map elements per map display. The second study was meant to address the limitation of the first study that landmark pictograms and other map elements are usually less evenly distributed across rural maps compared to urban maps. This was argued to affect the distribution of visual attention across the maps.

Both studies found distinctive effects of the task on visual attention. Most fixations were directed at map areas around the to-be-learned routes and toward decision points of the route. Map areas and landmarks offside the route were only rarely fixated (see Fig. 8.5). Thus, we found clear evidence for the effects a specific task has on the distribution of visual attention (as assessed using an eye tracker) across a map. Furthermore, the controlled manipulation of visual map complexity in the second experiment provided additional insights in the relevance of spatial reference points for the formation of mental spatial models. In the stretched maps with reduced spatial reference points, landmarks farther offside the displayed route were fixated. This indicates that not enough spatial reference points close to the route were available for the map users to memorize the route. Thus, participants seem to have expanded their search area for suitable spatial reference points.

The third study on the structural salience of landmark representations in maps addressed location memory tasks (for the complete study details, see Keil et al.

Fig. 8.5 Fixations during a route-learning task. The fixation heat map demonstrates that visual attention was almost exclusively directed at the map areas around the to-be-learned route. Especially high fixation counts can be seen around landmark pictograms close to the route. These pictograms can be suggested to have a high structural salience. Maps were obtained from OSM (© OpenStreetMap contributors)

2020b). Participants were presented maps taken from the OSM project. The maps contained several landmark pictograms and a to-be-learned object location highlighted by a red pictogram. After a short distractor task, participants were presented the previously shown map again, but this time without the red pictogram. They were asked to click on the recalled location of the red pictogram using a computer mouse. During the encoding phase, fixations on landmark pictograms were recorded with an eye tracker (cf. Kuchinke et al. 2016; Dickmann et al. 2015).

The results of the third study on structural salience demonstrate how the task requirements of an object location memory task affect the structural salience of landmark pictograms in a map. According to these results, most visual attention was directed toward the landmark pictograms closest to the to-be-learned object location. Additionally, landmark pictograms were fixated more often if they were located closer to the (imaginary) horizontal and vertical cardinal axes of the to-be-learned object location (see Fig. 8.6). In agreement with assumptions of Rock (1997) and Tversky (1981), people appear to apply an imaginary coordinate system to perceived maps, either based on the viewing angle or the map borders. Landmark pictograms seem to receive privileged access to visual attention as spatial reference points if they are easy to conceptualize based on such an imaginary coordinate system, e.g., being directly above, below, left, or right to a to-be-learned object location.

Also, object location memory in this task was more accurate if the closest landmark pictogram was closer to the to-be-learned object location. Fixation rates steeply dropped toward the second- and third-closest landmark pictogram to the to-be-learned object location. Of interest is that there was no significant relation between the distance of these pictograms to the to-be-learned object location and the memory performance found in this study. This seems to indicate that a single landmark pictogram already can act as a spatial reference point for object location memory and that this task requirement directly affects (or better implies) the structural salience of this reference point and the low structural salience of other (potential) spatial reference points.

Taken together, the three studies provide new insights into the structural salience of landmark pictograms applying different map-based memory tasks. All three studies clearly indicate that task requirements affect how visual attention is distributed across maps. People appear to search for suitable reference points for memorizing locations or route nodes. Whether landmark representations are chosen as reference points depends on their distance and orientation relative to memorized locations and routes.

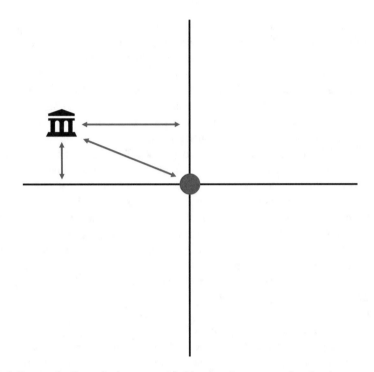

Fig. 8.6 Structural salience in the context of object location memory. Landmark representations in maps were fixated more often when they were located close to a to-be-learned object location and its (imaginary) horizontal and vertical cardinal axes

8.5 Conclusion

The reported studies of the sub-project "The Effects of Landmarks on Navigation Performance in VGI-based Maps" of the SPP 1894 provide new insights in how landmark representations in maps are perceived and how they affect the formation of mental spatial models. Previous research on landmark salience primarily focused on the parameters that affect the direction of visual attention toward real-world landmarks (e.g., Golledge 1991; Klippel and Winter 2005; Millonig and Schechtner 2007; Röser et al. 2012). The studies reported in this paper extend these findings by investigating the salience of landmark representations in maps and depict similarities and differences between real-world landmarks and landmark representations in maps.

Semantic salience has been argued to affect the likelihood of real-world landmarks to attract visual attention both based on general attributes, as well as on individual characteristics of the observer. General attributes include cultural and historical significance, as well as the purpose or function of a landmark (Claramunt and Winter 2007; Nothegger et al. 2004; Raubal and Winter 2002; Röser et al. 2011). Individual characteristics consider emotional or knowledge associations of an observer with a perceived landmark (Golledge 1991; Nuhn and Timpf 2017; Quesnot and Roche 2015).

In the context of landmark representations in maps, semantic salience seems only partially based on the general attributes and the individual associations with the represented landmarks. Rather, the type of representation also needs to be considered. As landmarks are often represented in maps as pictograms, specific associations with a unique landmark can be overridden. For example, a famous church with a unique architecture could be represented by the same pictogram as every other church. Thus, the unique associations related to this church could not be evoked by only seeing its map representation. Furthermore, intercultural and individual differences can affect the ability to understand pictograms used to represent landmarks and, consequently, their semantic salience (Spinillo 2012).

Our first reported study demonstrated that, similar to real-world landmarks, visual attention toward landmark representations in the form of pictograms is affected by their semantics respectively their meaningfulness. However, opposed to approaches of defining semantic salience in advance (based on obtained knowledge), we found that landmark representations can also attract visual attention if the conveyed semantic information is more difficult to understand than that of surrounding objects. In other words, visual attention is not directed based on the intrinsic semantic characteristics of a landmark representation (where more meaningfulness equals more semantic salience and visual attention) but is in some circumstances also based on the semantic contrast to other objects.

However, how semantic landmark pictograms in interaction with different map designs affect the distribution of visual attention remains unclear and needs to be addressed in future studies. Interestingly, although semantic salience has been argued to be affected by individual characteristics (Golledge 1991; Nuhn and Timpf 2017; Quesnot and Roche 2015), our study found no pronounced individual differences in the perceived meaningfulness of landmark pictograms or the direction of visual attention toward specific pictograms. This demonstrates that, as mentioned above, individual associations with specific landmarks get lost when the landmarks are represented as abstract (generalized) pictograms. Still, as the study was carried out with a homogeneous cultural group, future research needs to address to what extent cultural background affects the semantic associations with specific landmark pictograms and, consequently, the direction of visual attention toward these map elements.

The role of visual salience for the selection of spatial objects as landmarks has been emphasized repetitively (Klippel and Winter 2005; Röser 2017; von Stülpnagel and Frankenstein 2015). A commonly mentioned criterion of visual salience is the visual contrast of an object to its surroundings based on its visual characteristics, for example, its size, color, or illumination (Clarke et al. 2013; Davoudian 2011; Duckham et al. 2010; Röser et al. 2011). We argued that the form of landmark representation in maps (usually as pictograms) overwrites the visual characteristics of a real-world landmark and thereby also its visual salience. However, due to their accentuated design relative to other map elements, landmark pictograms in maps are also visually salient and attract visual attention (see example in Fig. 8.5). This is reflected in their common use as spatial reference points, despite the availability of other potential spatial reference points in the maps (see Keil et al. 2020a,b). Furthermore, opposed to real-world landmarks, the visual contrast of landmark representations and other reference points in maps to surrounding map elements and, consequently, their visual salience can easily be modified. As demonstrated in our second study, adding transparency to specific map areas directs visual attention away from these areas. Thus, by manipulating map design according to task requirements, visual attention can be systematically directed toward relevant map areas and spatial reference points as landmark representations (though the effects on memory are in need of future research).

However, it needs to be considered that not only the inherent visual characteristics of landmarks and landmark representations in contrast to their surroundings affect their visual salience and, consequently, their likelihood to attract visual attention. The general ability to perceive landmarks and their representations in maps also needs to be addressed. Concerning the selection of objects as landmarks, Röser et al. (2012) and Klippel and Winter (2005) stress the relevance of visibility, thus the ability to perceive the object from a specific viewpoint. Spatial objects might be hidden behind other spatial objects. If this is the case, they cannot be used as landmarks, even if they have a high visual contrast to their surroundings. The visibility of landmark representations in maps, on the other hand, depends not on viewpoints in real-world space but on the displayed map region and the selection of landmarks to be represented in a map. Consequently, during map-based navigation, visual attention could be attracted by landmark representations in the map that are not visible from the user's viewpoint. This could impair navigation performance, as only visible landmarks can be used for self localization. Therefore, future studies should address how visibility of landmarks can be assessed based on real-time tracking of map users and how this visibility information can be used to dynamically adjust which landmarks are represented in a map.

The third type of salience according to the approach of Sorrows and Hirtle (1999), structural salience, has been argued to be task- or context-dependent (Peebles et al. 2007). Landmarks have previously been argued to be structurally salient if they are located along a specific route and at locations that can be used to conceptualize a route (Klippel and Winter 2005; Millonig and Schechtner 2007; Röser et al. 2011). In three consecutive studies (see Keil et al. 2018, 2020a), we were able to demonstrate that, similar to real-world landmarks, landmark representations in maps are structurally salient based on their relative location to specific routes. The visualization of a task-relevant route directs visual attention to the map elements close to the route and its decision points. Furthermore, an additional study (Keil et al. 2020b) extended previous findings by demonstrating that landmark representations are also structurally salient and attract visual attention if they are located close to a to-be-learned object location and its (imaginary) cardinal axes. The effects of the cardinal axes of the map on the distribution of visual attention demonstrate a unique map-related characteristic of structural salience that cannot be applied to real-world space. Either based on the head orientation of the observer relative to a map or based on the (usually rectangular) shape of a map, up/down and left/right dimensions can induce such cardinal axes that are not induced by the perception of real-world space.

The reported findings emphasize the relevance of landmarks and landmark representations for location memory, as theorized by the mental spatial network of nodes and edges proposed by Werner et al. (2000). However, in contradiction to the allocentric mental spatial representation structure proposed by Werner et al. (1997, 2000), we found no evidence for a landmark configuration or framework used to conceptualize and memorize map space. In both the route memory and location memory tasks, visual attention was directed mainly toward the task-relevant landmarks close to the routes and object locations. Landmarks offside the routes and object locations received almost no visual attention. In the location memory task, already the second- and third-closest landmarks received significantly less visual attention, even if they were close to the to-be-learned object location. This demonstrates that participants did not explore the general spatial configuration of landmark representations. However, the lack of evidence in our experiments cannot be used to reject the assumption that landmark representations in maps are used to form a mental representation of maps based on nodes, as proposed by Werner et al. (2000). According to Millonig and Schechtner (2007), such allocentric mental spatial models are formed gradually during repeated interaction with spatial information. The short and clearly task-oriented perception of map information in our experiments might not have had sufficient power to detect or sufficient exploration time for the formation of an allocentric mental spatial model based on landmark configurations. In order to further explore the relevance of landmark representation configurations (landmark patterns) in maps on the formation of mental spatial models, future studies need to provide repeated interaction with maps and should be designed to support the formation of less task-specific mental spatial models, for example, by asking participants to draw sketch maps of the perceived maps. If, as assumed, landmark representation configurations are used as the first building block for the formation of mental spatial models (cf. Bestgen et al. 2017), identifying the ideal configuration characteristics and highlighting landmark representations in maps according to these characteristics could help to effectively and efficiently communicate spatial information to map users and support the formation of mental spatial models.

The reported studies provide first insight into salience characteristics of landmark representations in maps and their effects on the formation of mental spatial models. Based on these findings, better predictions of the distribution of visual attention across maps are possible. By systematically manipulating the salience of landmark representations, visual attention can be actively directed toward task-relevant map elements. The insights into the effects of directed visual attention on spatial tasks such as orientation or navigating along a selected route could be used to optimize automatic map generalization processes based on task requirements. In order to further exploit the potential benefits of the reported outcomes, future research needs to investigate how the systematic direction of visual attention toward specific map elements can improve performance on spatial tasks such as self-localization, orientation, and navigation in real-world environments. As a final step, the insights into the effects of directed visual attention on spatial tasks such as orientation or navigating along a selected route could be used to optimize automatic map generalization processes based on task requirements.

Acknowledgments This research was supported by the German Research Foundation DFG within Priority Research Program 1894 *Volunteered Geographic Information: Interpretation, Visualization and Social Computing* (VGIscience, Landmarks VGI, DI 771/11-1, KU 2872/6-1).

References

Albrecht R, Stuelpnagel Rv (2018) Memory for salient landmarks: empirical findings and a cognitive model. In: German Conference on Spatial Cognition. Springer, pp 311–325

Anacta VJA, Schwering A, Li R, Muenzer S (2017) Orientation information in wayfinding instructions: evidences from human verbal and visual instructions. GeoJournal 82(3):567–583

Bauer C (2018) Unterstützung der orientierung im innenbereich: analyse landmarkenbasierter karten-interfaces anhand des blickverhaltens der nutzer. PhD thesis, https://epub.uni-regensburg.de/37666/

Bestgen AK, Edler D, Kuchinke L, Dickmann F (2017) Analyzing the effects of vgi-based landmarks on spatial memory and navigation performance. KI-Künstliche Intelligenz 31(2):179–183

Buscher G, Cutrell E, Morris MR (2009) What do you see when you're surfing? Using eye tracking to predict salient regions of web pages. In: Proceedings of the SIGCHI Conference on Human Factors in Computing Systems, pp 21–30

Caduff D, Timpf S (2008) On the assessment of landmark salience for human navigation. Cogn Process 9(4):249–267

Claramunt C, Winter S (2007) Structural salience of elements of the city. Environ Plann B Plann Des 34(6):1030–1050

Clarke AD, Elsner M, Rohde H (2013) Where's wally: the influence of visual salience on referring expression generation. Front Psychol 4:329

Davoudian N (2011) Visual saliency of urban objects at night: impact of the density of background light patterns. Leukos 8(2):137–152

Dickmann F (2018) Kartographie. Westermann, Braunschweig

Dickmann F, Edler D, Bestgen AK, Kuchinke L (2015) Auswertung von heatmaps in der blickbewegungsmessung am beispiel einer untersuchung zum positionsgedächtnis. KN J Cartogr Geogr Inf 65(5):272–280

Dong W, Qin T, Liao H, Liu Y, Liu J (2020) Comparing the roles of landmark visual salience and semantic salience in visual guidance during indoor wayfinding. Cartogr Geogr Inf Sci 47(3):229–243

Duckham M, Winter S, Robinson M (2010) Including landmarks in routing instructions. J Locat Based Serv 4(1):28–52

Elias B (2003) Extracting landmarks with data mining methods. In: Kuhn W, Worboys MF, Timpf S (eds) Spatial information theory. Foundations of geographic information science. Springer, Berlin, Heidelberg, pp 375–389

Elias B, Paelke V (2008) User-centered design of landmark visualizations. In: Map-based mobile services. Springer, pp 33–56

Ferguson EL, Hegarty M (1994) Properties of cognitive maps constructed from texts. Mem Cognit 22(4):455–473

Fine MS, Minnery BS (2009) Visual salience affects performance in a working memory task. J Neurosci 29(25):8016–8021

Franke C (2021) Die salienz lokaler landmarken. doctoralthesis, Ruhr-Universität Bochum, Universitätsbibliothek. https://doi.org/10.13154/294-8103

Golledge RG (1991) Cognition of physical and built environments. In: Environment, cognition, and action. Oxford University Press, Oxford

Golledge RG (1993) Geographical perspectives on spatial cognition. In: Advances in psychology, vol 96. Elsevier, pp 16–46

Golledge RG (1999) Human wayfinding and cognitive maps. In: Golledge RG (ed) Wayfinding behavior. The Johns Hopkins University Press, Baltimore and London, pp 5–45

Herman JF, Kail RV, Siegel AW (1979) Cognitive maps of a college campus: a new look at freshman orientation. Bull Psychon Soc 13(3):183–186

Hurlebaus R, Basten K, Mallot HA, Wiener JM (2008) Route learning strategies in a virtual cluttered environment. In: International Conference on Spatial Cognition. Springer, pp 104–120

Itti L (2005) Models of bottom-up attention and saliency. In: Neurobiology of attention. Elsevier, pp 576–582

Itti L (2007) Visual salience. Scholarpedia 2(9):3327

Keil J (2021) The salience of landmark representations in maps and its effects on spatial memory. doctoralthesis, Ruhr-Universität Bochum, Universitätsbibliothek. https://doi.org/10.13154/294-8216

Keil J, Mocnik FB, Edler D, Dickmann F, Kuchinke L (2018) Reduction of map information regulates visual attention without affecting route recognition performance. ISPRS Int J Geo-Inf 7(12):469. https://doi.org/10.3390/ijgi7120469

Keil J, Edler D, Dickmann F, Kuchinke L (2019) Meaningfulness of landmark pictograms reduces visual salience and recognition performance. Appl Ergon 75:214–220. https://doi.org/10.1016/j.apergo.2018.10.008

Keil J, Edler D, Kuchinke L, Dickmann F (2020a) Effects of visual map complexity on the attentional processing of landmarks. Plos One 15(3):e0229575. https://doi.org/10.1371/journal.pone.0229575

Keil J, Edler D, Reichert K, Dickmann F, Kuchinke L (2020b) Structural salience of landmark pictograms in maps as a predictor for object location memory performance. J Environ Psychol 72:101497. https://doi.org/10.1016/j.jenvp.2020.101497

Klippel A, Winter S (2005) Structural salience of landmarks for route directions. In: International Conference on Spatial Information Theory. Springer, pp 347–362

Kuchinke L, Dickmann F, Edler D, Bordewieck M, Bestgen AK (2016) The processing and integration of map elements during a recognition memory task is mirrored in eye-movement patterns. J Environ Psychol 47:213–222. https://doi.org/10.1016/j.jenvp.2016.07.002

Li Z (2002) A saliency map in primary visual cortex. Trends Cogn Sci 6(1):9–16

Loomis JM, Klatzky RL, Golledge RG, Philbeck JW (1999) Human navigation by path integration. Wayfinding behavior: cognitive mapping and other spatial processes, pp 125–151

Lynch K (1960) The image of the city. MIT Press, Cambridge

Meilinger T, Hölscher C, Büchner SJ, Brösamle M (2006) How much information do you need? Schematic maps in wayfinding and self localisation. In: International Conference on Spatial Cognition. Springer, pp 381–400

Millonig A, Schechtner K (2007) Developing landmark-based pedestrian-navigation systems. IEEE Trans Intell Transp Syst 8(1):43–49

Nothegger C, Winter S, Raubal M (2004) Selection of salient features for route directions. Spat Cogn Comput 4(2):113–136

Nuhn E, Timpf S (2017) Personal dimensions of landmarks. In: Bregt A, Sarjakoski T, van Lammeren R, Rip F (eds) Societal geo-innovation. Lecture notes in geoinformation and cartography. Springer International Publishing, Cham, pp 129–143. https://doi.org/10.1007/978-3-319-56759-4_8

Nuhn E, Timpf S (2019) Prediction of landmarks using (personalised) decision trees. In: LBS 2019; Adjunct Proceedings of the 15th International Conference on Location-Based Services/Gartner, Georg; Huang, Haosheng, Wien

Ouerhani N, von Wartburg R, Hugli H, Muri R (2004) Empirical validation of the saliency-based model of visual attention. ELCVIA Electron Lett Comput Vis Image Anal 3(1):13–23. https://doi.org/10.5565/rev/elcvia.66

Peebles D, Davies C, Mora R (2007) Effects of geometry, landmarks and orientation strategies in the 'drop-off' orientation task. In: Winter S, Duckham M, Kulik L, Kuipers B (eds) Spatial information theory. Lecture notes in computer science, vol 4736. Springer, Berlin, Heidelberg, pp 390–405. https://doi.org/10.1007/978-3-540-74788-8_24

Peters D, Wu Y, Winter S (2010) Testing landmark identification theories in virtual environments. In: International Conference on Spatial Cognition. Springer, pp 54–69

Pilarczyk J, Kuniecki M (2014) Emotional content of an image attracts attention more than visually salient features in various signal-to-noise ratio conditions. J Vis 14(12):4–4

Presson CC, Montello DR (1988) Points of reference in spatial cognition: Stalking the elusive landmark. Br J Dev Psychol 6(4):378–381

Quesnot T, Roche S (2015) Quantifying the significance of semantic landmarks in familiar and unfamiliar environments. In: Fabrikant SI, Raubal M, Bertolotto M, Davies C, Freundschuh S, Bell S (eds) Spatial information theory. Lecture notes in computer science, vol 9368. Springer International Publishing, Cham, pp 468–489. https://doi.org/10.1007/978-3-319-23374-1_22

Raubal M, Winter S (2002) Enriching wayfinding instructions with local landmarks. In: International Conference on Geographic Information Science. Springer, pp 243–259

Richter KF, Winter S (2014) Landmarks: GIScience for Intelligent Services. Springer International Publishing, New York

Rock I (1997) Orientation and form. In: Rock I (ed) Indirect perception, MIT Press / Bradford Books series in cognitive psychology. MIT Press, Cambridge, MA, pp 133–150

Röser F (2017) A cognitive observer-based landmark-preference model. KI-Künstliche Intelligenz 31(2):169–171

Röser F, Hamburger K, Knauff M (2011) The giessen virtual environment laboratory: human wayfinding and landmark salience. Cogn Process 12(2):209–214

Röser F, Krumnack A, Hamburger K, Knauff M (2012) A four factor model of landmark salience– a new approach. In: Proceedings of the 11th International Conference on Cognitive Modeling (ICCM), Technische Universität Berlin, pp 82–87

Ross T, May A, Thompson S (2004) The use of landmarks in pedestrian navigation instructions and the effects of context. In: International Conference on Mobile Human-Computer Interaction. Springer, pp 300–304

Ruddle RA, Payne SJ, Jones DM (1997) Navigating buildings in "desk-top" virtual environments: experimental investigations using extended navigational experience. J Exp Psychol Appl 3(2):143

Ruddle RA, Volkova E, Mohler B, Bülthoff HH (2011) The effect of landmark and body-based sensory information on route knowledge. Mem Cognit 39(4):686–699

Sadeghian P, Kantardzic M (2008) The new generation of automatic landmark detection systems: Challenges and guidelines. Spat Cogn Comput 8(3):252–287

Siegel AW, White SH (1975) The development of spatial representations of large-scale environments. Adv Child Dev Behav 10:9–55

Sorrows ME, Hirtle SC (1999) The nature of landmarks for real and electronic spaces. In: Freksa C, Mark DM (eds) Spatial information theory. Cognitive and Computational Foundations of Geographic Information Science. Springer, Berlin, Heidelberg, pp 37–50. https://doi.org/10.1007/3-540-48384-5_3

Spinillo CG (2012) Graphic and cultural aspects of pictograms: an information ergonomics viewpoint. Work 41(Supplement 1):3398–3403

Steck SD, Mallot HA (2000) The role of global and local landmarks in virtual environment navigation. Presence 9(1):69–83

Sutherland MR, McQuiggan DA, Ryan JD, Mather M (2017) Perceptual salience does not influence emotional arousal's impairing effects on top-down attention. Emotion 17(4):700–706

Tversky B (1981) Distortions in memory for maps. Cogn Psychol 13(3):407–433 von Stülpnagel R, Frankenstein J (2015) Configurational salience of landmarks: an analysis of sketch maps using space syntax. Cogn Process 16(1):437–441

Walther D, Rutishauser U, Koch C, Perona P (2005) Selective visual attention enables learning and recognition of multiple objects in cluttered scenes. Comput Vis Image Underst 100(1-2):41–63

Wenczel F, Hepperle L, von Stülpnagel R (2017) Gaze behavior during incidental and intentional navigation in an outdoor environment. Spat Cogn Comput 17(1-2):121–142

Wenig N, Wenig D, Ernst S, Malaka R, Hecht B, Schöning J (2017) Pharos: Improving navigation instructions on smartwatches by including global landmarks. In: Rogers Y, Jones M, Tscheligi M, Murray-Smith R (eds) Proceedings of the 19th International Conference on Human-Computer Interaction with Mobile Devices and Services. ACM, New York, NY, pp 1–13. https://doi.org/10.1145/3098279.3098529

Werner S, Krieg-Brückner B, Mallot HA, Schweizer K, Freksa C (1997) Spatial cognition: The role of landmark, route, and survey knowledge in human and robot navigation. In: Jarke M, Pasedach K, Pohl K (eds) Informatik '97 Informatik als Innovationsmotor, Informatik aktuell. Springer, Berlin, Heidelberg, pp 41–50. https://doi.org/10.1007/978-3-642-60831-5_8

Werner S, Krieg-Brückner B, Herrmann T (2000) Modelling navigational knowledge by route graphs. In: Freksa C, Brauer W, Habel C, Wender KF (eds) Spatial cognition II. Lecture notes in computer science, vol 1849. Springer, Berlin, Heidelberg, pp 295–316. https://doi.org/10.1007/3-540-45460-8_22

Zhu Y (2012) Enrichment of routing map and its visualization for multimodal navigation. Dissertation, Technische Universität München, München

Chapter 9
Addressing Landmark Uncertainty in VGI-Based Maps: Approaches to Improve Orientation and Navigation Performance

Julian Keil, Frank Dickmann, and Lars Kuchinke

Abstract Landmarks, salient spatial objects, play an important role in orientation and navigation. They provide a spatial reference frame that helps to make sense of complex environments. Landmark representations in maps support map matching and orientation, because matching landmarks to their map representations provides information about spatial directions and distances. However, effective landmark-based map matching demands sufficiently accurate georeferencing of the landmarks represented in a map, because spatial inaccuracies of landmark representations cause distortions of the spatial reference frame and derived directions and distances. The requirement of accurate landmark georeferencing imposes difficulties on the use of maps based on Volunteered Geographic Information (VGI) for map matching. Differences of the motivation, competence, and available apparatus of volunteers can cause great variations of the data quality in VGI-based maps, including spatial accuracy of landmark representations. In a series of experiments, we investigated and quantified to what extent spatial inaccuracies of landmark representations in VGI-based maps affect map matching. Based on the findings, we were able to identify critical thresholds for spatial landmark inaccuracies. Furthermore, we explored potential ways to sustain successful map matching at higher degrees of spatial landmark inaccuracies. Through visual communication of spatial uncertainties, we were able to make map users more resilient to potential inaccuracies and sustain successful map matching.

Keywords Landmarks · Spatial inaccuracy · Uncertainty · Visualization · Map matching · Orientation

J. Keil (✉) · F. Dickmann
Ruhr-Universität Bochum, Bochum, Germany
e-mail: julian.keil@rub.de; frank.dickmann@rub.de

Lars Kuchinke
International Psychoanalytic University Berlin, Berlin, Germany
e-mail: lars.kuchinke@ipu-berlin.de

© The Author(s) 2024
D. Burghardt et al. (eds.), *Volunteered Geographic Information*,
https://doi.org/10.1007/978-3-031-35374-1_9

9.1 Introduction

In unfamiliar environments, people tend to use maps for orientation and navigation (Roskos-Ewoldsen et al. 1998). By matching spatial representations in maps to real-world objects, people identify their own location and obtain spatial information about orientation and route directions that are necessary for effective navigation (Kiefer et al. 2014). In the context of such a map matching process, landmarks, salient spatial objects with a fixed geographic location (Anacta et al. 2017; Bestgen et al. 2017; Claramunt and Winter 2007), have been discussed to play an important role (see Chap. 8). In maps, landmarks are often represented as pictograms that communicate semantic information about the nature or purpose of the represented landmark (Keil 2021). Furthermore, Peebles et al. (2007) found that people tend to use single landmarks and their map representations to match 2D maps to 3D spaces.

In recent years, due to the widespread use of mobile Internet and smartphones, paper maps have been increasingly replaced by online map services. The advantages of these online maps are their convenient availability and their ability to record and display one's own position in real time and to calculate and display routes in real time. Furthermore, they are—in comparison to printed maps—(often) more up to date, since these digital maps do not have to be reprinted after each modification. A particular phenomenon that has arisen in connection with online maps are maps based on Volunteered Geographic Information (VGI), with maps based on the OpenStreetMap (OSM) project as the most prominent representatives of this map category. Opposed to "traditional" commercial or official maps, VGI-based maps are created in a process that allows and even encourages the participation of map users (Goodchild 2007). Volunteers can add or modify map content and thereby contribute to the map creation process.

The use of VGI data can provide substantial benefits, in particular in regard to the quality of a map. In areas with numerous volunteers, these volunteers are able to make corrections to the map at very short notice if local conditions change, for example, when a road is closed or a new building is built (Barrington-Leigh and Millard-Ball 2017; Olteanu-Raimond et al. 2017). Thus, the availability of VGI data has improved geographic information (Flanagin and Metzger 2008) and the way such information is spread and processed. Volunteers also share the role of map users of this geographic information. Hence, the data provided clearly relies on individual experiences and thus shares a natural, implicit advantage over commercial products. As a result, the involvement of volunteers in map creation can lead to the mapping of spatial elements that are less relevant from the point of view of a public or commercial authority but are very relevant for certain groups of map users. For example, OSM contains many hiking trails that are not mapped in official maps (See et al. 2017). However, the source of these advantages of VGI-based maps is directly linked to disadvantages in terms of data quality (cf. Bégin et al. 2013).

Overall, there is an ongoing and thorough discussion on quality issues of such spatial data, mainly in comparison to commercial products (e.g., Degrossi et al. 2018; Flanagin and Metzger 2008; Senaratne et al. 2017; Zhang and Malczewski

2017). In regions with only a few active volunteers who provide VGI data, the maps are usually much less detailed (Rousell and Zipf 2017), and map errors tend to get fixed late, if at all. It is also known that interindividual differences between volunteers affect the quality of VGI-based maps. These differences include personal motivations, skills and mapping expertise, as well as the technical equipment used, for example, devices to record GPS data (Van Exel et al. 2010). These between-region and interindividual differences result in a pronounced heterogeneity of data quality (c.f. Chap. 3) and completeness available in VGI-based maps (Girres and Touya 2010). Thus, data quality and data characteristics are of a strongly heterogeneous nature in the case of VGI. Most map readers, however, do not question data quality when, for example, using OSM or are even not aware of the fact that maps are based on OSM. They are not aware of the very different characteristics of VGI as opposed to traditional or commercial datasets (Skopeliti et al. 2017). In contrast, Schiewe and Schweer (2013) report "a rather high degree of awareness of uncertainty problems" in OSM users. But this awareness circulates around completeness and up-to-dateness of the data, while localization errors and thematic inaccuracies remain unaware (Schiewe and Schweer 2013). As a result, it can be assumed that map readers treat every available landmark in the same way, independent of its representational quality.

In the context of successful map matching, two potential problems arise from the described disadvantages of VGI-based maps. First, in some cases, there may not be a sufficient number of spatial reference points represented in certain map areas that would be necessary for successful map matching. And second, localization errors of important spatial reference points to be used as landmarks in navigation and orientation (see Fig. 9.1) could potentially lead to unsuccessful map matching, i.e., elements in real space not being recognized in the map or landmarks represented in the map are not identified in real space.

As part of the SPP 1894 (Volunteered Geographic Information (VGI): Interpretation, Visualization and Social Computing) of the DFG, the sub-project on "The Effects of Landmark Uncertainty in VGI-based Maps: Approaches to Improve Wayfinding and Navigation Performance" carried out by the Ruhr-Universität Bochum (RUB) and the International Psychoanalytical University Berlin (IPU) addresses these presumed effects of spatial landmark inaccuracies on map matching, orientation, and navigation performance. The first aim was to assess and quantify to what extent spatial inaccuracies of landmark representations affect map matching and, consequentially, orientation and navigation (see Sect. 9.2). In a second step, approaches for reducing the assumed negative effects of landmark inaccuracies in maps on map matching are being developed and evaluated (see Sect. 9.3).

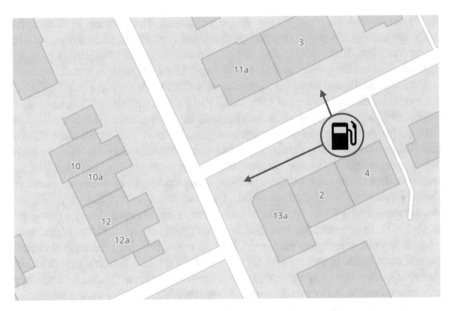

Fig. 9.1 Inaccuracies of landmark representations in maps. Due to data quality issues intrinsic to VGI data, individual landmark representations in VGI-based maps can be spatially more or less inaccurate. For example, the gas station represented in the map above may also be located on the other side of the road or on a different location along the road. If spatial inaccuracies are too high, map users may experience difficulties when trying to match map representations to the represented real-world environment (© OpenStreetMap contributors)

9.2 Effects of Landmark Inaccuracies on Map Matching

In a first experiment of the SPP 1894 sub-project, we aimed to investigate and quantify how spatial inaccuracies of landmark representations in maps affect the ability of map users to match the map to the represented 3D environment. For this purpose, we created a virtual 3D environment and a digital map that allowed us to fully control the locations of a landmark building and its pictogram representation in the map (see Fig. 9.2).

The locations of both the 3D landmark and the landmark representation were fully randomized, independent of each other along the road. Consequentially, the spatial inaccuracy of the landmark representations was different in each trial. After each trial, participants used a continuous scale to respond to what extent they perceived (as pedestrians) the map as matching the 3D environment. For full details on the study design, see Keil et al. (2022a).

The results demonstrated a pronounced and significant nonlinear relation between the spatial inaccuracy of the landmark representation in the 2D map and the perceived match between the 3D environment and the map (see Fig. 9.3). A tipping point was observed at approximately 10 meters of spatial inaccuracy, i.e., 10-meter walking distance in a virtual 3D environment. Maps with less spatial

Fig. 9.2 Stimulus design. Participants saw a 3D environment containing a landmark building and a corresponding map representation. Random spatial inaccuracies were applied to the landmark representation in the map (here, landmark building matching the map representation)

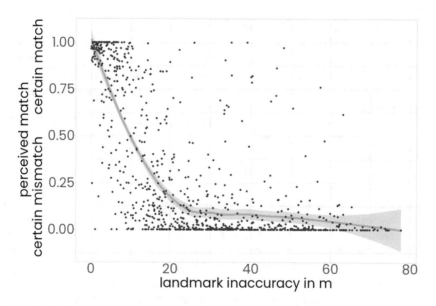

Fig. 9.3 Relation between inaccuracies of the landmark representation in the map and the perceived match between the 3D environment and the map. Values of one represent a certain match, values of zero represent a certain mismatch, and values between zero and one represent uncertainty concerning a match or mismatch. If spatial inaccuracies of the landmark representation were too high, maps were perceived as not matching the represented 3D environment

inaccuracy were mainly rated as matching the 3D environment. Maps with more than 10 meters of spatial inaccuracy were mainly rated as not matching the 3D environment.

These findings suggest that inaccuracies of a landmark's map representation of more than 10 meters for a pedestrian observer and map user are recognized. Map users then seem to start to mismatch map and 3D space representations, meaning that they are getting unable to match a corresponding map to the represented 3D space, although in the present experiment, only one map element (the landmark pictogram) is spatially inaccurate. However, some issues concerning the generalizability of these findings still need to be considered. First, in this experiment, only one fixed map scale was applied. However, most modern digital maps support dynamic adjustments of the visualized map scale. Selecting smaller map scales results in smaller misplacements of spatially inaccurate landmark representations in terms of pixels or millimeters. Consequentially, spatial landmark inaccuracies could be more difficult to recognize if a smaller map scale is used. Hence, how specific spatial inaccuracies of landmark representations affect map matching also needs to be quantified with different map scales in future experiments (see Keil et al. 2022a).

A second limitation concerning the generalizability of the findings is that the experiment was carried out with controlled virtual 3D environments. Although full experimental control is important to isolate the effect of the spatial inaccuracy of the landmark pictogram on map matching, it needs to be considered that people who carry out a map matching task are usually confronted with complex real-world environments containing numerous spatial elements that can act as helpful spatial reference points or as distractors. For example, an unusual route shape could support the map matching process or visually highly salient spatial objects that attract visual attention but are not represented in the map could disturb the map matching process. How the perception of complex real-world environments affects the map matching process still needs to be investigated. Still, one could assume that the effects observed in this experiment could be less pronounced in real-world environments because the salience of a single landmark and its map representation is less pronounced.

Finally, maps often represent more than one landmark in a map section. Due to the heterogeneity of VGI-based maps (Girres and Touya 2010), only some of these landmark representations may be significantly spatially inaccurate. If this is the case, other landmark representations could still be used to maintain successful map matching. Thus, how exactly one or some spatially inaccurate landmark representations affect map matching if other spatially more accurate landmark representations are available needs to be addressed in future studies.

Despite these mentioned limitations, the findings demonstrate that spatial inaccuracy of landmark representations in maps can jeopardize successful map matching and most likely also orientation and navigation performance. Therefore, finding ways for reducing the impact of spatial landmark pictogram inaccuracies on the map matching process seems to be a relevant topic for further research (cf. next paragraph).

9.3 Visualizing Spatial Uncertainty

In a second experiment, we aimed to assess ways to reduce the negative effect of spatial inaccuracies of landmark representations on map matching. According to Padilla et al. (2021), uncertainty concerning data quality is an issue of most data and can affect all kinds of decision-making. However, people may not be aware of these data quality issues. Thus, if map users do not consider during navigation that some landmark representations may be more or less spatially accurate, they could interpret a map as representing another spatial environment and not the currently perceived one. Pang et al. (1997) argue that by visualizing data uncertainty, people are provided with important information of data quality and can therefore make more informed decisions. However, according to Mason et al. (2016), there is a "lack of comprehensive and generalizable empirical studies across the entire domain of uncertainty visualization." Therefore, we investigated the effects of visualizing uncertainty concerning the correct location of map-based landmark representations on map matching. For this purpose, the uncertainty visualization variables transparency and size were selected and manipulated based on suggestions of MacEachren (1992) and MacEachren et al. (2005). Furthermore, an uncertainty area visualization already used by the commercial map provider Google Maps to visualize uncertainty of GPS locations was investigated and compared to the other visualizations as well (see Fig. 9.4).

The stimulus design was the same as in the first experiment, with one exception. In addition to the control condition with the unmodified landmark pictogram, three different visualizations for spatial landmark uncertainty were compared in a within-subject design. In four sets of twelve trials, participants either saw a map with an unmodified landmark pictogram (control condition), a transparent pictogram, a pictogram with modified size, or a pictogram with a circular transparent uncertainty area (see Fig. 9.4). The level of transparency, the size of the landmark pictogram, or the uncertainty area was linked to the spatial inaccuracy of the landmark pictogram in the map. Again, participants used a continuous scale ranging from 0 to 1 after

Fig. 9.4 Visualizations for spatial uncertainty. Pictogram transparency (left), size (middle), and circular (transparent) uncertainty areas (right) were used to visualize uncertainty concerning the correctness of the landmark pictogram location

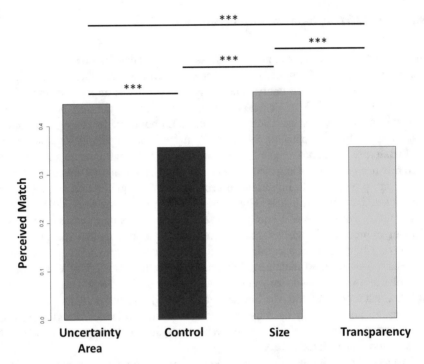

Fig. 9.5 Perceived match per landmark pictogram condition. Participants significantly less often perceived a mismatch between the 3D environment and the map representation (mismatches represented by low perceived match scores) when potential spatial inaccuracies of landmark representations were visualized by increasing the landmark pictogram size or by adding a circular uncertainty area

each trial to indicate to what extent they perceived the map as matching the 3D environment. For more study details, see Keil et al. (2022b).

The results demonstrate that participants are less likely to perceive a mismatch between the 3D environment and the map (mismatches represented by low perceived match scores), if the size of the landmark pictogram was modified or if an uncertainty area was added around the landmark pictogram to illustrate uncertainty (see Fig. 9.5). However, adding transparency to the landmark pictogram did not lead to fewer perceived mismatches between the 3D environments and the maps compared to the control condition. Figure 9.6 shows that if pictogram size or uncertainty areas are used to communicate uncertainty concerning the correct location of the landmark representation, a larger landmark inaccuracy was accepted by the participants to still be perceived as a match between the 3D environment and the map. Based on this result, it seems that in map matching (i.e., comparison of map and 3D space), the size of the pictogram or the size of the uncertainty area placed around the pictogram are particularly suitable for depicting spatial uncertainty. These two pictogram variants seem to support the processing of uncertainty by

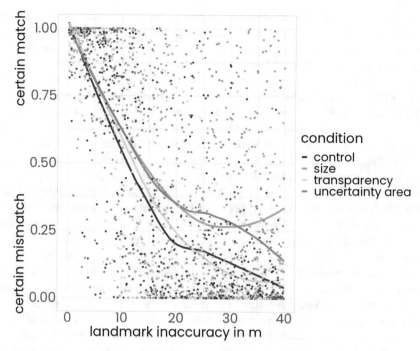

Fig. 9.6 Relation between inaccuracies of the landmark representation in the map and the perceived match between the 3D environment and the map per landmark pictogram condition. When uncertainty concerning the correct location of the landmark representation in the map was visualized by increasing the pictogram size or adding a circular uncertainty area, higher values of spatial landmark pictogram inaccuracy were required for participants to perceive a certain mismatch between the map representation and the represented 3D environment

map users significantly better than, for example, the also tested transparency representations of pictograms.

Based on these initial findings, we conclude that uncertainty visualization can be used to reduce the negative effects of spatial inaccuracies of landmark representations in maps on map matching. Especially, modifying the size of pictograms as suggested by MacEachren et al. (2005) and adding transparent uncertainty areas around the landmark pictogram proved to communicate spatial uncertainty effectively and to improve successful map matching. It seems likely that an explanation for the fact that adding transparency to landmark pictograms did not prove to be effective in communicating spatial uncertainty needs to consider the difficulties to registering different transparency levels of our map users in the experiment, as only one landmark pictogram was visible simultaneously.

Concerning the generalizability of these findings, two issues need to be addressed. First, Fig. 9.6 reveals a data artifact for extremely high landmark inaccuracy values in the size condition (green line). Opposed to the other visualizations, uncertainty increased when spatial inaccuracy values were extremely high. This particular artifact might be explained by the fact that pictogram size in this condition was directly linked to the spatial inaccuracy of the landmark pictogram. Thus, if the spatial inaccuracy was extremely high, the pictogram was extremely large and covered large areas of the map. However, if the pictogram gets too large, it will make the map unreadable, and map matching will become impossible. This artifact demonstrates that, as also stressed by Kinkeldey et al. (2014), new ways to visualize uncertainty call for studies to investigate their usability. Potential usability issues of the addressed uncertainty visualizations need to be identified in further extensive series of usability tests to ensure that the general map reading ability is not impaired. For example, as a consequence of the identified data artifact, maximum values for the pictogram size could be defined.

The second issue is related to the values used to control the intensity of the uncertainty visualization. In this experiment, the spatial inaccuracy of the landmark representations was used to control the transparency or size of landmark pictograms or the uncertainty area around the pictogram. In a real application scenario, however, these values would not be available. If precise information concerning the correct landmark location would be available, then in most cases, the errors could easily be corrected, and an uncertainty visualization would not be required. Instead, measures for uncertainty based on the available metadata related to VGI contributions need to be developed, or average inaccuracy values need to be estimated and applied to the uncertainty visualization. For example, an uncertainty score could be calculated based on the number and spread of VGI contributions linked to a specific landmark pictogram, the number of contributions of a contributor, or the number of corrections linked to the contributions of a contributor. The development of such uncertainty value calculations and their evaluation in applied user studies are subject to future research.

9.4 Conclusion and Outlook

The studies reported above provide first insights into how spatial inaccuracies of landmark representations in maps affect map matching. Our first study demonstrates that spatially inaccurate landmark representations reduce the likelihood of successful map matching. This could create problems for real-world orientation and navigation, especially in VGI-based maps, due to their great heterogeneity in terms of data quality (Girres and Touya 2010; Rousell and Zipf 2017). In a second study, we were able to show that by visualizing uncertainty concerning the spatial accuracy of landmark representations, the negative effects on map matching can be reduced. Especially modifying the size of landmark pictograms as suggested by MacEachren et al. (2005) or adding uncertainty areas as used in Google Maps proved to maintain successful map matching at higher levels of spatial landmark inaccuracies. We argue that it is this visualization of spatial uncertainty that provides the information required for more informed decision-making (Padilla et al. 2021; Pang et al. 1997).

Fig. 9.7 Map matching with 360-degree photos. Due to the higher number of real-world distractors and spatial reference points in the maps obtained from OSM (© OpenStreetMap contributors), effects of spatial landmark inaccuracies in the maps are assumed to be less pronounced compared to experimentally fully controlled virtual environments

In future studies, the generalizability of the identified relation between spatial inaccuracies of landmark representations and map matching to real-world environments needs to be investigated. As previously argued, it is assumed that the negative effect on map matching will be less pronounced in real-world environments compared to the virtual 3D environment used in the first experiment of our SPP 1894 sub-project. The higher complexity of real-world environments should provide both more spatial reference points that are represented in maps and further distractors (like persons, building facades, cars, etc.) that are not represented in maps. Both the additional spatial reference points and the distractors are expected to compete with the landmark representations for visual attention. In consequence, this distribution of attention across more objects is assumed to reduce the relevance of the landmark representations for map matching. To test this assumption, it is necessary to conceptually extend the experiment design of the two studies reported above (see Keil et al. 2022a,b). Instead of using virtual 3D scenes in the map matching task, 360-degree images of real-world scenes with landmarks that are also included in the corresponding OSM map section may be tested (see Fig. 9.7).

In addition, to investigate the effect of map scale on the relationship between spatial inaccuracy of landmark representations and map matching, in future studies, different map scales should be used. It is expected that the results of such studies will not only provide a deeper understanding of the role of individual landmark representations for map matching in the context of spatial inaccuracies in real-world environments. They will also provide new findings on the effects of different map scales and potential usability issues of the used uncertainty visualizations in widespread maps as OSM.

Furthermore, potential ways to quantify spatial uncertainty based on available OSM metadata need to be explored, developed, and tested. This will allow visualizing different levels of uncertainty based on the suggested data quality levels of different OSM map regions. Finally, the effects of the uncertainty visualizations on orientation ability and navigation performance (i.e., real-world map matching) also need to be investigated with regard to different target groups. The results can not only provide detailed information on how spatial inaccuracies of landmark representations in maps affect map matching. They will also show how successful map matching can be maintained by providing map users with information about the uncertainty of map content to support more informed decision-making.

Acknowledgments This research was supported by the German Research Foundation DFG within Priority Research Program 1894 *Volunteered Geographic Information: Interpretation, Visualization and Social Computing* (VGIscience, Landmark Uncertainty, DI 771/11-2, KU 2872/6-2).

References

Anacta VJA, Schwering A, Li R, Muenzer S (2017) Orientation information in wayfinding instructions: evidences from human verbal and visual instructions. GeoJournal 82(3):567–583

Barrington-Leigh C, Millard-Ball A (2017) The world's user-generated road map is more than 80% complete. PloS One 12(8):e0180698

Bégin D, Devillers R, Roche S (2013) Assessing volunteered geographic information (vgi) quality based on contributors' mapping behaviours. Int Arch Photogramm Remote Sens Spat Inf Sci 2013:149–154

Bestgen AK, Edler D, Kuchinke L, Dickmann F (2017) Analyzing the effects of vgi-based landmarks on spatial memory and navigation performance. KI-Künstliche Intelligenz 31(2):179–183

Claramunt C, Winter S (2007) Structural salience of elements of the city. Environ Plann B Plann Des 34(6):1030–1050

Degrossi LC, Porto de Albuquerque J, Santos Rocha Rd, Zipf A (2018) A taxonomy of quality assessment methods for volunteered and crowdsourced geographic information. Trans GIS 22(2):542–560

Flanagin AJ, Metzger MJ (2008) The credibility of volunteered geographic information. GeoJournal 72(3):137–148

Girres JF, Touya G (2010) Quality assessment of the french openstreetmap dataset. Trans GIS 14(4):435–459

Goodchild MF (2007) Citizens as sensors: the world of volunteered geography. GeoJournal 69(4):211–221

Keil J (2021) The salience of landmark representations in maps and its effects on spatial memory. doctoralthesis, Ruhr-Universität Bochum, Universitätsbibliothek. https://doi.org/10.13154/294-8216

Keil J, Edler D, Dickmann F, Kuchinke L (2022a) Uncertainties in spatial orientation: Critical limits for landmark inaccuracies in maps in the context of map matching. KN J Cartogr Geogr Inf, 1–12. https://doi.org/10.1007/s42489-022-00105-7

Keil J, Edler D, Kuchinke L, Dickmann F (2022b) Visualization of spatial uncertainty improves map matching. Abstr ICA 5:55. https://doi.org/10.5194/ica-abs-5-55-2022. https://www.abstr-int-cartogr-assoc.net/5/55/2022/

Kiefer P, Giannopoulos I, Raubal M (2014) Where am i? Investigating map matching during self-localization with mobile eye tracking in an urban environment. Trans GIS 18(5):660–686

Kinkeldey C, MacEachren AM, Schiewe J (2014) How to assess visual communication of uncertainty? A systematic review of geospatial uncertainty visualisation user studies. Cartogr J 51(4):372–386

MacEachren AM (1992) Visualizing uncertain information. Cartogr Perspect (13):10–19

MacEachren AM, Robinson A, Hopper S, Gardner S, Murray R, Gahegan M, Hetzler E (2005) Visualizing geospatial information uncertainty: what we know and what we need to know. Cartogr Geogr Inf Sci 32(3):139–160

Mason JS, Klippel A, Bleisch S, Slingsby A, Deitrick S (2016) Special issue introduction: approaching spatial uncertainty visualization to support reasoning and decision making. Spat Cogn Comput 16(2):97–105

Olteanu-Raimond AM, Hart G, Foody GM, Touya G, Kellenberger T, Demetriou D (2017) The scale of vgi in map production: a perspective on european national mapping agencies. Trans GIS 21(1):74–90

Padilla L, Kay M, Hullman J (2021) Uncertainty visualization. Wiley, New York, pp 1–18. https://doi.org/10.1002/9781118445112.stat08296. https://onlinelibrary.wiley.com/doi/abs/10.1002/9781118445112.stat08296

Pang AT, Wittenbrink CM, Lodha SK, et al (1997) Approaches to uncertainty visualization. Vis Comput 13(8):370–390

Peebles D, Davies C, Mora R (2007) Effects of geometry, landmarks and orientation strategies in the 'drop-off' orientation task. In: International Conference on Spatial Information Theory. Springer, pp 390–405

Roskos-Ewoldsen B, McNamara TP, Shelton AL, Carr W (1998) Mental representations of large and small spatial layouts are orientation dependent. J Exp Psychol Learn Mem Cogn 24(1):215

Rousell A, Zipf A (2017) Towards a landmark-based pedestrian navigation service using osm data. ISPRS Int J Geo-Inf 6(3):64

Schiewe J, Schweer MK (2013) Vertrauen im rahmen der nutzung von karten. KN J Cartogr Geogr Inf 63(2):59–66

See L, Estima J, Pődör A, Arsanjani JJ, Bayas JCL, Vatseva R (2017) Sources of VGI for mapping. Ubiquity Press, London, pp 13–35. https://doi.org/10.5334/bbf.b

Senaratne H, Mobasheri A, Ali AL, Capineri C, Haklay M (2017) A review of volunteered geographic information quality assessment methods. Int J Geogr Inf Sci 31(1):139–167

Skopeliti A, Antoniou V, Bandrova T (2017) Visualisation and communication of vgi quality. Mapping and the citizen sensor. Ubiquity Press, London, pp 197–222

Van Exel M, Dias E, Fruijtier S (2010) The impact of crowdsourcing on spatial data quality indicators. In: Proceedings of the GIScience 2010 Doctoral Colloquium, Zurich, Switzerland, pp 14–17

Zhang H, Malczewski J (2017) Accuracy evaluation of the canadian openstreetmap road networks. Int J Geospat Environ Res 5(2). https://dc.uwm.edu/ijger/vol5/iss2/1/

Chapter 10
Improvement of Task-Oriented Visual Interpretation of VGI Point Data

Martin Knura and Jochen Schiewe

Abstract VGI is often generated as point data representing points of interest (POIs) and semantic qualities (such as accident locations) or quantities (such as noise levels), which can lead to geometric and thematic clutter in visual presentations of regions with numerous VGI contributions. As a solution, cartography provides several point generalization operations that reduce the total number of points and therefore increase the readability of a map. However, these operations are applied rather general and could remove specific spatial pattern, possibly leading to false interpretations in tasks where these spatial patterns are of interest. In this chapter, we want to tackle this problem by defining task-oriented sets of map generalization constraints that help to maintain spatial pattern characteristics during the generalization process. Therefore, we conduct a study to analyze the user behavior while solving interpretation tasks and use the findings as constraints in the following point generalization process, which is implemented through agent-based modeling.

Keywords Point generalization · Constraints · Agent-based modeling

10.1 Introduction

As shown by the variety of different aspects and applications which are observed in this book, the volume and relevance of Volunteered Geographic Information (VGI) have immensely increased in recent years. In many cases, this VGI data is generated and visualized as point data, e.g., representing the location of a point of interest (POI), an event, or a data source. However, utilizing VGI data needs to take some specific characteristics into account in comparison to geospatial data acquired and processed in the "traditional" way. In particular, VGI "is produced by heteroge-

M. Knura (✉) · J. Schiewe
HafenCity University Hamburg, Hamburg, Germany
e-mail: martin.knura@hcu-hamburg.de; jochen.schiewe@hcu-hamburg.de

© The Author(s) 2024
D. Burghardt et al. (eds.), *Volunteered Geographic Information*,
https://doi.org/10.1007/978-3-031-35374-1_10

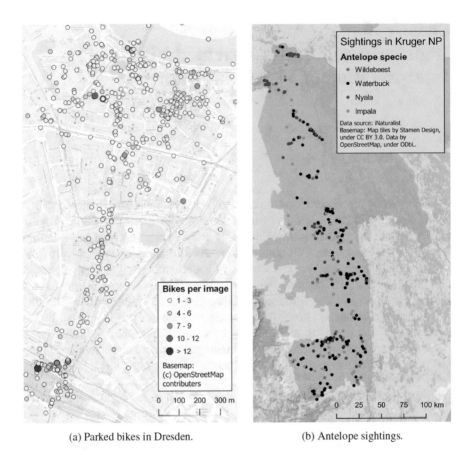

(a) Parked bikes in Dresden.   (b) Antelope sightings.

Fig. 10.1 Examples for VGI point data with point clutter. (**a**) Parked bikes in the city of Dresden, detected on Flickr images between 2004 and 2014 according to Knura et al. (2021). (**b**) Sightings of selected antelope species in Kruger National Park uploaded on the platform iNaturalist

neous contributors, using various technologies and tools, having different levels of details and precision, serving heterogeneous purposes, and a lack of gatekeepers" (Senaratne et al. 2017), leading to an enormous volume and heterogeneity within the data. All of these characteristics could harm the usability of the data, especially when it comes to the visual presentation and exploration of very dense and even overlapping point markers or symbols (see Fig. 10.1a and b), commonly known as geometric point clutter (Moacdieh and Sarter 2015).

As a solution to this clutter problem, cartography provides several point generalization operations such as selection, aggregation, or displacement, which rearrange or reduce the total number of points and therefore increase the readability of a map. However, these operations are applied rather general and could remove a specific spatial pattern, possibly leading to false interpretations in tasks where these spatial patterns are of interest.

The aim of the TOVIP project is to tackle this problem by defining a set of cartographic constraints—i.e., conditions a generalized map should satisfy—that preserve these spatial patterns throughout the whole generalization process. The first research question we want to answer in this chapter is therefore:

- What is the minimum set of constraints and constraint measures that should be used to evaluate interpretation tasks based on VGI point visualizations, such as pattern identification, pattern comparison, or relation seeking?

Different cartographic constraints often describe contradicting aspects with no optimal solution, as it is not possible during map generalization to maintain all information—i.e., fulfill all information preservation constraints—while keeping the map readable, i.e., fulfill all legibility constraints. Constraint-based generalization is therefore an optimization task, which tries to find a solution that satisfies as many constraints as good as possible, and has been implemented in recent years through multi-agent systems (Duchêne et al. 2018). We want to contribute to this research and define our second research question as:

- Is it possible to optimize the task-oriented generalization using an agent-based modeling approach?

The following chapter describes the workflow to answer the research questions as follows: in Sect. 10.2, we introduce the cartographic concept of constraint-based generalization, on which the TOVIP project is based upon. Section 10.3 summarizes the results of a user study, which analyzed the user behavior while working with spatial patterns in point data sets. In Sect. 10.4, we translate the findings of the previous section into measurable constraints that could be utilized in map generalization practice. In Sect. 10.5, we apply these constraints in an agent-based generalization model. That followed, we discuss our findings in Sect. 10.6 before concluding in Sect. 10.7.

10.2 Constraint-Based Map Generalization

Cartography provides a variety of different point generalization operations—and various combinations between them—to solve the aforementioned clutter problem. As an example, a *simplification* describes a straight reduction of source points based on geometric criteria (e.g., only points which have a minimum distance to their neighbors are preserved; (Slocum et al. 2009)). When semantic criteria are used, a *selection* operation could take place. For example, points can be selected based on respective information filtering methods (Huang and Gartner 2012) or scale-dependent (Gröbe and Burghardt 2021). *Aggregation* takes place when multiple points are replaced by a single aggregator marker. Most frequently, points are grouped through clustering with a respective initialization method (e.g., random, k-means, Voronoi-based; (Yan and Weibel 2008)), while alternatives, for example, use heat maps (Meier 2016), or geometry objects (Zahtila and Knura 2022) to aggregate point data. A different solution to overcome clutter, but with the possibility to preserve the cardinality (i.e., the overall number of points) of the dataset, is to *displace* the points. During the displacement operation, an iterative workflow of overlap detection, relocation, and re-evaluation is executed (Mackaness and Purves 2001). Furthermore, if the preservation of cardinality and the original topology is of interest, a *spatial distortion* based on pixels (Keim et al. 2004) or point density (Bak et al. 2009) could help.

An operational system for point generalization must implement a workflow to trigger and orchestrate the individual point generalization operations described above. First implementations used a rule-based approach where a predefined set of well-defined and unambiguous rules guided the generalization process (Beard 1991). Each rule thereby states what has to be done in a process at a certain condition, so each condition was connected to a specific action (Harrie and Weibel 2007). The problem that occurred with this approach was that the enormous variety of spatial and non-spatial characteristics that exist in the world and therefore in maps led to a number of rules which were not possible to handle anymore. This leads to the constraint-based approach, which focused on the requirements that the final map should fulfil instead of providing a set of isolated generalization operations, leaving more flexibility within the generalization process on how to reach these results. According to Beard (1991), these constraints can be classified into aspects related to position, topology, shape, structural, functional, and legibility. Furthermore, it is necessary to introduce respective measures for these constraints, which are grouped by Mackaness and Ruas (2007) into either internal or external and either micro, meso, or macro.

If the constraints are defined in a complete and measurable way, there are different techniques available for implementation. Looking at optimizing single generalization methods, there is considerable work done, for example, regarding the displacement operation by applying least squares adjustment (Sester 2000), simulated annealing (Ware and Jones 1998), or snakes (Burghardt 2005). For more complex processes, agent-based modeling has shown great success in terms of applicability (Duchêne et al. 2018). In this approach, agents represent autonomous map objects trying to minimize a given cost function, which is based on the fulfillment of the constraint measures. As a result, the whole complexity of the generalization workflow is distributed to a set of relatively simple interacting agents.

The agent-based modeling approach is also used in the TOVIP project. Regarding the aim of TOVIP—defining a set of constraints that optimizes the generalization workflow designed for visual interpretation tasks where specific spatial patterns are of interest—it is necessary to consider two potentially contradictory aspects: On the final map, the aforementioned spatial patterns have to be visible (*preservation constraints*), while the map must still be readable by the users (*legibility constraints*). Describing constraints that preserve the relevant information during the generalization process is often done with object-specific measures, e.g., preserving the area of a polygon before and after generalization (Harrie and Weibel 2007). On the other hand, legibility constraints ensure the readability of the map, for example, by avoiding any spatial conflict—i.e., display clutter—and showing objects in a suitable degree of detail according to the scale of the map. A respective list of analytical legibility measures, such as the number of vertices or the object line length, was developed by Stigmar and Harrie (2011). For the aim of the TOVIP project, it is now of interest to find the minimum set of preservation and legibility constraints that allow the interpretation of specific spatial pattern even after generalization.

10.3 User Behavior When Interpreting VGI Point Data

Developing constraints that support users while interpreting specific tasks implies profound knowledge of their behavior while doing so. Before defining constraints that support interpretation tasks, it is therefore necessary to analyze the behavior of users working with VGI point data sets. We conduct a user study where participants have to perform different interpretation tasks—like finding clusters within a dataset, comparing point densities, or finding areas with a specific point distribution—using a novel method that combines postal questionnaires, think-aloud interviews, and techniques from visual analytics. A more detailed overview on the technical aspects and the execution of the user study, including a detailed description of the analysis of the think-aloud interviews, is given by Knura and Schiewe (2021). In this chapter, we want to summarize the results of the study (see also Knura and Schiewe 2022), focusing on the impact of the user behavior on the definition of a minimum set of constraints as described above.

10.3.1 Task-Solving Strategies

We analyzed the strategies of the participants by dividing the overall task-solving process into three sequential actions: (1) finding a start position, (2) obtaining information, and (3) decision-making. Apart from a task where the participants have to find a similar pattern compared to a given reference, the point density of a cluster—as a combination of proximity and cardinality of points—was the most important factor when selecting a starting position on the map, followed by the point color. For the process of obtaining information, point density was again the most important factor, as more dense clusters were described and analyzed earlier and more often. Moreover, density was the main evaluation measure in comparison tasks and during decision-making. Although we had different categories of synoptic interpretation tasks, which—in contrast to elementary tasks—include pattern identification, pattern comparison, and relation seeking (see Andrienko and Andrienko 2006), the task-solving strategies did not differ significantly between different kinds of tasks. As a first result of the study, we state that point density has the biggest impact on the task-solving behavior of the participants and has to be addressed in the first place when defining constraints.

10.3.2 Influence of Point Data Cardinality and Background Map

A key factor that could have an impact on the user behavior during visual interpretation is the map complexity. There are numerous definitions and concepts of map complexity in cartography (see Touya et al. 2016). As most of them distinct between the intellectual complexity, which relates to the cognitive process of map reading, and the graphical complexity, which relates to the visual perception of individual map objects, we vary the maps for some of the tasks with respect to these two categories. To learn more about the influence of the intellectual complexity on user behavior, we varied the data cardinality—i.e., the number of points—for two of the tasks. Although we recognized some minor differences in the behavior between the user groups, the overall task execution strategy remains unchanged with a higher data cardinality. For analyzing the impact of the graphical complexity, we varied the background map source between Google Maps, Bing Maps, and Stamen Terrain. This time, we identified both an implicit and explicit influence from the background map. Implicitly, because participants frequently identified clusters which were visually supported by the background map and explicitly because they refer to the characteristics of the background map when explaining their strategies. But again, and despite the influence of the background map on the reasoning, the overall task-solving strategies described in the section above remain unchanged between different levels of graphical complexity.

10.3.3 Implications for Constraints Supporting Interpretation Tasks

Following the results of our study, there are two main aspects that have to be considered while defining constraints for map generalization. First, it is of major importance to preserve the original pattern proportions during the generalization process. Our study revealed that the point density had the biggest impact on the task-solving process, and participants discussed both interrelations between clusters with different density, as well as between different classes of points within the same cluster. Information preservation constraints regarding the point density should therefore:

- Retain the proportion of points between areas with different densities
- Preserve the ranking of densities between different areas
- Preserve proportions between classes while maintaining at least one point per class
- Preserve Gestalt law rules regarding similarity and proximity of clusters

The second aspect to consider is the use of cartographic techniques to guide the interpretation of points. The use of specific colors to draw attention is common in cartography, and this can be applied to other map objects with the aim of lowering the graphical complexity. Respective constraints could ensure to:

- Use cartographic style elements where pattern preservation is difficult to manage
- Optimize the guiding effect of the background map (e.g., preservation of other map objects in close proximity to point clusters)

These constraints could be categorized as both preserving information and legibility, and they address not only the location and visibility of the point symbols but also their style and the surrounding map areas.

10.4 Defining Constraints and Measures for Spatial Pattern Interpretation

The previous section revealed the importance of preserving point densities during generalization. In this section, we collect a list of different approaches—both from the literature and own experiments—to define constraints and respective measures, which can help to preserve point densities and spatial pattern and test them on exemplary point distributions. The aim is thereby to find a minimum set of constraints that fit best to the list of requirements described above. We thereby focus on information preservation constraints regarding the point distributions. Constraints related to cartographic techniques are a key aspect of our future work.

10.4.1 Measures Describing Spatial Pattern and Densities

When defining measures for spatial pattern and densities, we follow the categorization of Mackaness and Ruas (2007), who distinguish between macro-measures that deal with all point objects of interest, micro-measures that deal with individual characteristics of objects (i.e., points), and meso-measures that deal with the specific properties of different groups of objects (i.e., point clusters). The authors also distinguish between internal and external measures, which states if a measure can be calculated based on a single dataset (internal) or is a relation between two datasets (e.g., before and after a generalization operation; external).

10.4.1.1 Macro-Measures

Macro-measures are able to describe the entirety of information and respective characteristics in a single value. One of the most basic macro-measures is the radical law (Töpfer and Pillewizer 1966), which estimates how many features should be maintained at a smaller scale in the generalization process. It is defined as:

$$n_D = n_S \sqrt{\frac{m_S}{m_D}}, \tag{10.1}$$

where n is the number of objects of the derived ($_D$) resp. source ($_S$) map, and m is the scale denominator. In the context of this project, it is worth noting that the calculation should be based on a readable map, i.e. without any point clutter. If this measure is calculated from a map with point clutters, the calculated value should be interpreted as the **maximum number of objects** on the derived map. Even more basic is the measure that describes the **amount of information** N_i as the number of all map objects (Harrie and Stigmar 2010), calculated as:

$$N_i = \sum_{i=1}^{n} \sum_{j=1}^{m_i} O_{ij}. \tag{10.2}$$

For objects other than points, this measure can be expanded with the number of object points, calculating the overall measure as the sum of all object points of all map objects.

Beside measures that deal with the amount of information in general, global measures can also describe a specific characteristic of the dataset or the map. In the same work, Harrie and Stigmar (2010) defined an index to characterize the **spatial distribution of points** I_{SDP} based on Voronoi regions. The index is calculated as:

$$I_{SDP} = \frac{\sum_{i=1}^{NP} P_{SDP,i} \log P_{SDP,i}}{\log \frac{1}{NP}}, \tag{10.3}$$

where $P_{SDP,i}$ is the relative size of the Voronoi region for a point i and NP is the number of points. I_{SDP} converges to 1 the more even the sizes of the Voronoi regions are. Zhang et al. (2009) use the **Voronoi region size as the variable of interest in Moran's I** to discern if point distributions are clustered, dispersed, or random:

$$I = \frac{N}{W} \frac{\sum_{i=1}^{N} \sum_{j=1}^{N} w_{ij}(x_i - \bar{x})(x_j - \bar{x})}{\sum_{i=1}^{N} (x_i - \bar{x})^2}, \tag{10.4}$$

where N is the number of spatial units indexed by i and j, x is the size of the Voronoi region A_V, $\bar{x}$ is the mean of x, w_{ij} is a matrix of spatial weights with zeroes on the diagonal (i.e., $w_{ii} = 0$), and W is the sum of all w_{ij}.

10.4.1.2   Micro-Measures

Micro-measures describe characteristics of individual objects and therefore can take the local neighborhood into account. Analog to the macro-measures before, calculating the size of the Voronoi region of a point A_V is used as a fundamental metric to describe local density. Based on this, Zhang et al. (2008) calculate the *object-oriented density* OD as:

$$OD = \frac{1}{A_V}. \tag{10.5}$$

A higher object-oriented density implies a smaller Voronoi region and therefore a higher point density in the local neighborhood. Vice versa, a small object-oriented density indicates a bigger Voronoi area and a more dispersed distribution around that point.

Besides density measures, qualitative and quantitative information about the points in close proximity are also of interest when point generalization operations like selection are used. Therefore, Delauney triangulations are often used to identify "natural" neighbors in point distributions (Sadahiro 1997). Applying this tessellation to a point data set provides a *list of neighbors* for each point, and micro-measures like the *number of natural neighbors*, the *mean neighbor distance*, and the existence of *local extreme values* can be calculated (see Fig. 10.2). Delauney triangulation also helps to define clusters, so the *cluster affiliation* can also be defined in this way.

All the measures introduced in this section are internal because they can be calculated solely based on one dataset. However, it is possible to compare the measures of an individual point to measures of the same point during the

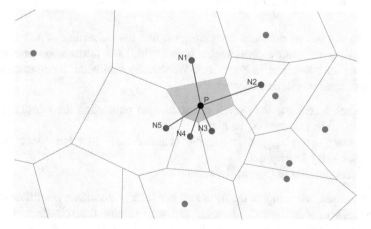

Fig. 10.2 Example of micro-measures for point P. Natural neighbors N1 to N5 in red, other points in gray. Voronoi region A_V for point P in light blue

generalization process, and note the amount of change as an additional external measure. In the same way, the ***distance to the origin location*** of a point is of interest when displacement operations take place during the generalization.

10.4.1.3 Meso-Measures

Compared to macro- and micro-measures, meso-measures are not bound to a predefined number or list of points they describe. The first step to calculating meso-measures is therefore to define which points are members of the group of interest. In the TOVIP project, we focus on spatial pattern, and so the definition of clusters is relevant for further processing. Clustering can be made—among many other techniques—by cluster algorithms such as k-means and HDBscan, based on Delauney triangulation (Sadahiro 1997), or by using grids (Yan et al. 2021). These clusters can then be described by meso-measures such as the ***number of group members***, the ***existence of different point categories***, and the ***mean distance*** between members or between members and the group centroid. Comparing the measures of the respective clusters, it is also possible to define ***cluster rankings***. Furthermore, according to the findings of the user study presented in Sect. 10.3, measures regarding the shape and the orientation of the clusters can be of interest. Common methods to represent ***the shapes of point clusters*** are convex hulls or alpha shapes (Edelsbrunner et al. 1983). The ***orientation of a cluster*** can be described by the minimum rotated rectangle, a technique which is usually used for building orientation (Duchêne et al. 2003). Furthermore, all macro-measures defined in Sect. 10.4.1.1 can also be applied on clusters with a defined border.

10.4.2 Deriving a Minimum Set of Constraints

Based on the list of different measures, we test the suitability of the measures to control the different aspects which help to fulfil the information preservation constraints we developed in Sect. 10.3. We thereby subdivide the constraints and respective measures into three groups:

1. Measures describing the overall distribution of points and the density ranking between different areas of the map
2. Measures preserving pattern-specific characteristics like hot spots, extreme values, cluster density, etc.
3. Measures describing Gestalt law rules

Furthermore, we compare the measures and their performance on different point distributions to identify redundancies, and we examine the robustness on point cardinality, which is essential when applied in map generalization operations. We create a series of experimental point distributions with 100, 200, 500, and 1000 points and different characteristics: a regular and a random distribution, distributions

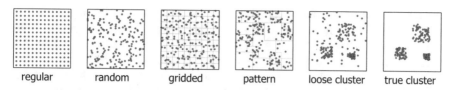

regular   random   gridded   pattern   loose cluster   true cluster

Fig. 10.3 Point distributions with different characteristics, using the example of 200 points. Gray borders in the background of the gridded and pattern distributions indicate the areas in which the same number of points were randomly distributed

where we predefined regular (gridded distribution) and irregular areas (pattern distribution) to control the density, and distributions with loose and clear clusters (see Fig. 10.3). Furthermore, we tested the behavior of micro-measures on points in different VGI datasets to evaluate their utility.

10.4.2.1 Preserve the Overall Distribution of Points and the Density Ranking Between Areas

The first subgroup of measures combines the first two constraints on information preservation in Sect. 10.3.3 and can be controlled through a combination of macro- and meso-measures. We tested both the Voronoi-based Moran's I and the spatial distribution of points with our series of different artificial point distributions. Table 10.1 shows the calculated measures and the standard deviation over the different point cardinalities. We can see that the spatial point distribution measure is to a certain degree stable toward the point cardinality and is smaller when the points

Table 10.1 Results for point distribution measures. Values with (*) signs indicate that there was a small deviation to the given number of points because of distribution characteristics (e.g., 196 instead of 200 points for the regular distributions)

Point distribution measure

Number of points	regular	random	gridded	pattern	l. cluster	t. cluster
100	1.0000	0.9673	0.9924	0.9493*	0.8624	0.7623*
200	0.9994*	0.9718	0.9904	0.9405*	0.8603	0.7481*
500	0.9987*	0.9789	0.9847	0.9394*	0.8664	0.7079*
1000	0.9991*	0.9805	0.9832	0.9391*	0.8704	0.6658*
std	*0.0005*	*0.0061*	*0.0044*	*0.0048*	*0.0044*	*0.0434*

Voronoi-based Moran's I

Number of points	regular	random	gridded	pattern	l. cluster	t. cluster
100	0.4683	0.1707	0.0032	0.1819*	0.5665	0.2733*
200	0.4372*	0.1919	−0.0196	0.2126*	0.623	0.4653*
500	0.4690*	0.1575	0.1995	0.2988*	0.6378	0.4990*
1000	0.4817*	0.2907	0.1809	0.3841*	0.2891	0.5187*
std	*0.0189*	*0.0603*	*0.1151*	*0.0911*	*0.1629*	*0.1126*

are more clustered. Moran's I on Voronoi regions is more sensitive to variations of point cardinality, as it uses the area size as the variable of interest—whose value decreases with more points. As a result, we decided to use the point distribution measure to control the overall distribution and the distribution within clusters. That also includes the fact that the measures for amount of information and cluster size are variables within the calculation of the spatial point distribution on the macro- and meso-level. In contrast, the cluster ranking measure has no overlap with other measures and is therefore inevitable.

10.4.2.2   Preserve Pattern-Specific Characteristics

Pattern-specific measures are a crucial part of the goals of the TOVIP project. If local extreme values and the existence of different point categories are of interest in an interpretation task, it is mandatory to preserve these points and therefore control them with related measures. As this measure requires a Delauney triangulation to define the neighborship, respective measures that are based on this can be performed with low additional effort, even if not compulsory. As an example, the mean distance to neighbors can be calculated this way. As an alternative, the distance to all points within the predefined cluster can be used to decide which points are overlapping and thus should be a controlling measure. If a displacement operation is implemented, the distance to the origin location of a point can be of interest. For the other pattern-specific measures, we did not find a unique behavior in which we see an additional utility for our model.

10.4.2.3   Preserve Gestalt Law Rules

The maximum number of points can be utilized as a target value for the generalization process, although it is not mandatory if all legibility measures are satisfied. Measures related to the shape and orientation of clusters are utilizing common techniques from the field of geospatial analysis, such as calculating the minimum bounding rectangle, the convex hull, or the alpha shape of a point set. We compared the different approaches on different data sets and decided to use the convex hull to describe the shape, as it needs no additional parameter compared to the alpha shape and is more detailed than the rectangular bounding box. If the orientation is of interest, the longer side of the minimum bounding rectangle can be utilized.

Table 10.2 shows the selected measures which we initially implemented in the agent-based model, together with additional measures that could be relevant for certain tasks and were also recognized. Nevertheless, because most of the measures are defined in code blocks outside the actual agent-based model, it is possible to adopt measures from other scale levels during model optimization.

Table 10.2 Subgroups of measures and selected measures in column "Set". (X) indicates the measure is not mandatory in certain applications

	Macro	Meso	Micro	Set
Overall distribution of points/cluster rankings				
amount of information	X			
spatial distribution of points	X	X		X
Moran's I on Voronoi regions	X	X		
cluster density ranking	X			X
cluster size		X		
Pattern-specific characteristics				
local extreme values			X	X
point category preservation		X		X
number of natural neighbors		X	X	
mean distance to cluster members		X	X	X
mean neighbor distance			X	
object-oriented density			X	
distance to the origin location			X	(X)
Gestalt Law				
maximum number of points	X			(X)
shape of a cluster		X		X
orientation of a cluster		X		(X)

10.5   Application Using Agent-Based Modeling

The set of constraints and respective measures developed in the previous section is a key component for the implementation of a map generalization process, which preserves spatial patterns. We apply the constraint-based approach using agent-based modeling, which is a powerful method for controlling complex processes (Harrie and Weibel 2007). As the model is currently in the final phase of development, this section will focus on the architecture and parametrization we implemented: First, we introduce the software framework we use and explain the different components within the model. In the second part, we describe the integration of global map specifications and the translation of measures to a satisfaction scale, which helps the agents to better evaluate their fulfillment of constraints. Evaluation of the model results will be part of our future work.

10.5.1 Software and Components

We implement our agent-based model[1] using the Mesa framework (Kazil et al. 2020). Mesa is an open-source framework for creating agent-based models written in Python. It includes four core components (Model, Agent, Schedule and Space), along with additional components for analysis and visualization. Thereby, the *Model* class is the core class for creating the environment of the model using the *Space* class, initializing the agents which are objects of the *Agent* class, and orchestrating the running model through the *Scheduler* class. Applied to the process of map generalization, our model has a map area which is implemented through a continuous space—providing a high flexibility for different map scales— and contains map agents which represent objects that generalize themselves by performing generalization operations, according to the perception of their current state and their fulfillment of given constraint measures. Besides micro-agents, which represent the individual points, our model also contains meso-agents, which are generated within the model initialization and control the pattern preservation.

(Map) agents and the implementation of their decision-making process are the most complex part of an agent-based model. Duchêne et al. (2018) decomposed the "brain" of map agent into three main components: capacities, mental representation, and procedural knowledge. We followed this approach and used these components in our model (see Fig. 10.4). The capacities of our agents include the ability to perceive their surrounding space, to evaluate themselves, and to perform generalization operations. The updating process of the first two capacities is thereby provided by the *Model* class, which performs several spatial analysis operations on the totality of map objects after each simulation step and transmits the calculated measures back to the individual micro- and meso-agents. The mental representation of the agents compares their current state with the goals they are aiming at—i.e., the fulfillment of map constraints—and calculates their satisfaction. It also memorizes all previous actions the agents took and the respective outcome of it. Finally, the procedural knowledge component is the decision-making unit of the agents. Based on the agent's constraint satisfaction and the knowledge of the past steps, it decides which operation the agent should execute in the next step.

Besides the core functions for agent-based modeling, Mesa offers functionalities for data analysis and model visualization. The *DataCollector* class of Mesa is able to record, store, and export all relevant data of the agents for further analysis. It allows us to control the mechanisms of the model, as well as tuning the decision-making process of the agents. Via the visualization components, Mesa also provides a browser-based visualization of the running model, but until now, we haven't implemented a respective function in our model yet. Instead, we set up and run the model via Jupyter Notebook and present the generalized map in an interactive browser map.

---

[1] Our source code is online: https://gitlab.com/g2lab/tovip.

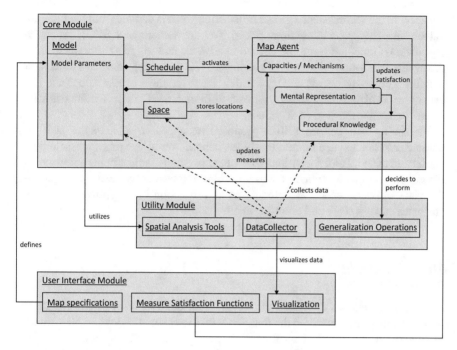

Fig. 10.4 Components of our model for agent-based map generalization

10.5.2   Global Map Specifications and Measure Satisfaction

Besides the specific constraints we defined in order to support interpretation tasks, there are also global map specifications and characteristics which are important for the process of generalization in general and for the point generalization in particular and which have to be defined in advance. For example, it is necessary to know the scale of the source map and if this map satisfies all legibility constraints regarding the point symbols (i.e., the source map has no point clutter and is readable). It is also required to define the target scale of the map and the (pixel) size of the point symbols. Moreover, it is of interest if the point data set contains different classes and, if it does, the respective scale of measurement. While these global map specifications are determined in most of the use cases for point generalization (e.g., the target map scale via predefined zoom levels), they can also be changed in the model setup.

Furthermore, the "brain" of the map agent requires determining a predefined behavior regarding the task of translating a list of measures into a value representing the satisfaction of an agent at its current status. The common workflow for this task consists of two steps (Touya 2012): First, the measures get translated into a Likert-like satisfaction scale, which ranges from 1 ("unacceptable") to 8 ("perfect"). Each measure thereby has its own method for translation, which has to be defined in advance. In the second step, a global satisfaction value is derived from the individual values, utilizing principles from Social Welfare Orderings (SWO). Again, the specific SWO method has to be chosen in advance and triggers different strategies the agents will follow. For example, using a SWO that emphasizes low values with a higher weight, the agents will try to minimize the number of low values and focus less on maximizing other values, while a utilitarian SWO fosters strategies where agents maximize the sum of all values.

Taillandier and Gaffuri (2012) proposed an approach to help the user with parameterization using a human-machine dialogue. We utilize this approach for our model and offer a guided user interface where parameter adjustments get visualized via samples on a map. It allows the user to adjust the satisfaction scales by modifying the class dividers, which are predefined as a function of respective global map attributes such as scale and point cardinality.

10.6 Discussion

The previous sections described the workflow of defining a set of constraints and respective measures based on the findings from a user study and the implementation in an agent-based model. This already answers our research questions to a certain degree. In this section, we want to further discuss implications that occur with the results of the previous sections.

In Sect. 10.4.2, we define a set of constraints containing measures that control the spatial distribution of points, the ranking between clusters, the shapes of clusters, and the distances between the points of a cluster. Furthermore, the preservation of local extreme values and all point categories should be added if their existence is of interest in the interpretation task. Taking the different measure scales of two of the constraints into account, there are only six to eight measures, which can be used to control the generalization process. Still, this requires at least six different predefined parameter adjustments to translate measures into satisfaction values. The complexity of the parametrization process has been identified as one of the main drawbacks of the agent-based approach (Duchêne et al. 2018), and this is also the case in our model. As an example, defining a function to evaluate the macro-measure of point distribution is little intuitive, as it needs to define class dividers for narrow-value ranges, which differences are hard to visualize. However, six parameters and a user-friendly way to adjust them are still feasible in our opinion.

Manual parameter adjustment is one reason which makes it difficult to transfer our approach of point generalization to other applications. The second reason is the time-consuming calculation of measures that rely on rather complex geospatial operations. On-the-fly point generalization (Jabeur et al. 2006) is therefore not possible with our approach, and the computing time depends heavily on the number of points to generalize. A solution to this problem could be the integration of novel learning techniques (Touya et al. 2019). If a model can learn how to generalize points while preserving the right information, it could predict a generalized point set on-the-fly.

10.7   Summary and Outlook

The generation of VGI data in general, and of points in particular, has shown an immense increase in recent years. As one of the main properties of VGI is its enormous volume and heterogeneity within the data, it leads to dense clutters when it is presented on maps. The cartographic solution to this problem is point generalization: rearranging or reducing the number of points. If this is applied rather general, specific spatial pattern could be eliminated—which is a major problem when these patterns were of high importance and subject of interest to the user. This chapter presents a workflow to resolve this problem by defining a set of constraint that can be used to control the generalization workflow. We developed the list of constraints based on a user study and applied them by implementing an agent-based model for point generalization.

VGI point data is often produced in multiple scale levels and over longer periods of time. In our future work, we want to factor this and develop our model further by adding functionalities for *multi-scale views*, which requires consideration of scale transitions, and *multi-temporal representations*, where the cognitive workload related to animations must also be considered. Furthermore, we plan to improve the usability of our agent-based model by developing a more intuitive user interface, which would allow more users to apply the findings of our model to their objectives. A third task we plan to work on in the near future is the integration of other cartographic techniques such as point color, point symbolization, and others in our generalization model. The overall plan is thereby to stepwise add functionalities for broader applications of map generalization into our system.

Acknowledgments This research was supported by the German Research Foundation DFG within Priority Research Program 1894 *Volunteered Geographic Information: Interpretation, Visualization and Social Computing* (VGIscience, TOVIP, SCHI 1008/11-1).

References

Andrienko N, Andrienko G (2006) Exploratory analysis of spatial and temporal data. Springer, New York. https://doi.org/10.1007/3-540-31190-4

Bak P, Schaefer M, Stoffel A, Keim DA, Omer I (2009) Density equalizing distortion of large geographic point sets. Cartogr Geogr Inf Sci 36(3):237–250. https://doi.org/10.1559/152304009788988288

Beard K (1991) Constraints on rule formation. Map generalization: making rules for knowledge representation, pp 121–135

Burghardt D (2005) Controlled line smoothing by snakes. GeoInformatica 9(3):237–252. https://doi.org/10.1007/s10707-005-1283-3

Duchêne C, Touya G, Taillandier P, Gaffuri J, Ruas A, Renard J (2018) Multi-agents systems for cartographic generalization: feedback from past and on-going research. Research report, IGN (Institut National de l'Information Géographique et Forestière); LaSTIG, équipe COGIT. https://hal.archives-ouvertes.fr/hal-01682131

Duchêne C, Bard S, Barillot X, Ruas A, Trévisan J, Holzapfel F (2003) Quantitative and qualitative description of building orientation. In: ICA Generalisation Workshop, April 2003

Edelsbrunner H, Kirkpatrick D, Seidel R (1983) On the shape of a set of points in the plane. IEEE Trans Inf Theory 29(4):551–559. https://doi.org/10.1109/TIT.1983.1056714

Gröbe M, Burghardt D (2021) Scale-dependent point selection methods for web maps. KN J Cartogr Geogr Inf 71(3):143–154. https://doi.org/10.1007/s42489-021-00079-y

Harrie L, Stigmar H (2010) An evaluation of measures for quantifying map information. ISPRS J Photogramm Remote Sens 65(3):266–274. https://doi.org/10.1016/j.isprsjprs.2009.05.004. Theme issue "Visualization and exploration of geospatial data"

Harrie L, Weibel R (2007) Modelling the overall process of generalisation. In: Mackaness WA, Ruas A, Sarjakoski LT (eds) Generalisation of geographic information. International Cartographic Association, Elsevier Science B.V., Amsterdam, pp 67–87. https://doi.org/10.1016/B978-008045374-3/50006-5

Huang H, Gartner G (2012) A technical survey on decluttering of icons in online map-based mashups. Online Maps with APIs and WebServices

Jabeur N, Boulekrouche B, Moulin B (2006) Using multiagent systems to improve real-time map generation. In: Lamontagne L, Marchand M (eds) Advances in artificial intelligence. Springer, Berlin, Heidelberg, pp 37–48

Kazil J, Masad D, Crooks A (2020) Utilizing python for agent-based modeling: The mesa framework. In: Thomson R, Bisgin H, Dancy C, Hyder A, Hussain M (eds) Social, cultural, and behavioral modeling. Springer International Publishing, Cham, pp 308–317

Keim DA, Panse C, Sips M, North SC (2004) Pixel based visual data mining of geo-spatial data. Comput Graph 28(3):327–344. https://doi.org/10.1016/j.cag.2004.03.022

Knura M, Schiewe J (2021) Map evaluation under covid-19 restrictions: a new visual approach based on think aloud interviews. Proc ICA 4:60. https://doi.org/10.5194/ica-proc-4-60-2021

Knura M, Schiewe J (2022) Analysis of user behaviour while interpreting spatial patterns in point data sets. KN J Cartogr Geogr Inf 72(3):229–242. https://doi.org/10.1007/s42489-022-00111-9

Knura M, Kluger F, Zahtila M, Schiewe J, Rosenhahn B, Burghardt D (2021) Using object detection on social media images for urban bicycle infrastructure planning: A case study of dresden. ISPRS Int J Geo-Inf 10(11). https://doi.org/10.3390/ijgi10110733

Mackaness WA, Purves RS (2001) Automated displacement for large numbers of discrete map objects. Algorithmica 30(2):302–311. https://doi.org/10.1007/s00453-001-0007-9

Mackaness WA, Ruas A (2007) Evaluation in the map generalisation process. In: Mackaness WA, Ruas A, Sarjakoski LT (eds) Generalisation of geographic information. International Cartographic Association, Elsevier Science B.V., Amsterdam, pp 89–111. https://doi.org/10.1016/B978-008045374-3/50007-7

Meier S (2016) The marker cluster:: A critical analysis and a new approach to a common web-based cartographic interface pattern. Int J Agric Environ Inf Syst 7:28–43. https://doi.org/10.4018/IJAEIS.2016010102

Moacdieh N, Sarter N (2015) Display clutter: A review of definitions and measurement techniques. Hum Factors 57(1):61–100. https://doi.org/10.1177/0018720814541145. pMID: 25790571

Sadahiro Y (1997) Cluster perception in the distribution of point objects. Cartogr Int J Geogr Inf Geovis 34(1):49–62. https://doi.org/10.3138/Y308-2422-8615-1233

Senaratne H, Mobasheri A, Ali AL, Capineri C, Haklay MM (2017) A review of volunteered geographic information quality assessment methods. Int J Geogr Inf Sci 31(1):139–167. https://doi.org/10.1080/13658816.2016.1189556

Sester M (2000) Generalization based on least squares adjustment. GeoInformatica 6:233–261

Slocum T, McMaster R, Kessler F, Howard H (2009) Thematic cartography and geovisualization. Prentice Hall series in geographic information science. Pearson Prentice Hall, Hoboken

Stigmar H, Harrie L (2011) Evaluation of analytical measures of map legibility. Cartogr J 48(1):41–53. https://doi.org/10.1179/1743277410Y.0000000002

Taillandier P, Gaffuri J (2012) Designing generalisation evaluation function through human-machine dialogue. CoRR abs/1204.4332. http://arxiv.org/abs/1204.4332

Touya G (2012) Social welfare to assess the global legibility of a generalized map. In: Geographic information science. Springer Berlin, Heidelberg, pp 198–211. https://doi.org/10.1007/978-3-642-33024-7_15

Touya G, Hoarau C, Christophe S (2016) Clutter and map legibility in automated cartography: A research agenda. Cartogr Int J Geogr Inf Geovis 51(4):198–207. https://doi.org/10.3138/cart.51.4.3132

Touya G, Zhang X, Lokhat I (2019) Is deep learning the new agent for map generalization? Int J Cartogr 5(2-3):142–157. https://doi.org/10.1080/23729333.2019.1613071

Töpfer F, Pillewizer W (1966) The principles of selection. Cartogr J 3(1):10–16. https://doi.org/10.1179/caj.1966.3.1.10

Ware JM, Jones CB (1998) Conflict reduction in map generalization using iterative improvement. GeoInformatica 2(4):383–407. https://doi.org/10.1023/A:1009713606524

Yan H, Weibel R (2008) An algorithm for point cluster generalization based on the voronoi diagram. Comput Geosci 34(8):939–954. https://doi.org/10.1016/j.cageo.2007.07.008

Yan X, Chen H, Huang H, Liu Q, Yang M (2021) Building typification in map generalization using affinity propagation clustering. ISPRS Int J Geo-Inf 10(11):732. https://doi.org/10.3390/ijgi10110732

Zahtila M, Knura M (2022) Visualizing point density on geometry objects: Application in an urban area using social media vgi. KN J Cartogr Geogr Inf 72(3):187–200. https://doi.org/10.1007/s42489-022-00113-7

Zhang X, Ai T, Stoter J (2008) The evaluation of spatial distribution density in map generalization. In: ISPRS 2008: Proceedings of the XXI congress: Silk road for information from imagery: the International Society for Photogrammetry and Remote Sensing, 3–11 July, Beijing, China. Comm. II, WG II/2. International Society for Photogrammetry and Remote Sensing (ISPRS), Beijing, pp 181–187. https://www.isprs.org/proceedings/XXXVII/congress/2_pdf/2_WG-II-2/03.pdf

Zhang X, Ai T, Stoter JE (2009) A voronoi-like model of spatial autocorrelation for characterizing spatial patterns in vector data. In: 2009 Sixth International Symposium on Voronoi Diagrams, pp 118–126. https://doi.org/10.1109/ISVD.2009.19

Part III
Active Participation, Social Context, and Privacy Awareness

Part III of the book addresses the human impact associated with VGI. The influence of humans is manifold during acquisition, processing, and interpretation of VGI. People can take both active and passive roles in the generation of VGI, for example, simple unintentional actions such as georeferencing pictures or messages in social media or intentional involvement of volunteers in citizen science and crowdsourced projects. Potentials for active participation can be released event-driven in the short term, e.g., in the case of catastrophic events, but also result in continuous participation over a long period of time, e.g., while monitoring animals, plants, or recording weather and climate phenomena. Specialist and local knowledge is thereby used to build up free knowledge archives and geodata collections.

Humans not only play a central role in the generation of VGI but can also become the subject of research themselves when human behavior is studied based on the published content. This includes studies of who is involved in the creation of VGI, e.g., differentiated by age, gender, or previous education. Furthermore, a distinction has been made between the large majority, which makes contributions available independently often at short notice and a rather small user group, which contributes content very persistently and thus also determines future directions of VGI projects. Motives of all groups are the subject of research and are just as interesting as approaches on how participation can be promoted.

The first chapter in this section describes the use of wearable sensors worn by volunteers to record their exposure to environmental stressors on their daily journeys. With that the behavior of individuals and the effect of customized recommendations are examined. The following chapter investigates the collective behavior of people using location-based social media and movement data from football matches. In addition to a conceptual behavioral model, various generic methods and application-related workflows for visual analysis are presented. The third chapter in this section deals with the motivation of volunteers who provide important support in information gathering, filtering, and analysis of social media in disaster situations. Since humans play a central role in the generation of VGI, special precautions are also required with regard to the protection of privacy. In the case of active participation, the consent of the user can be obtained. This is not possible in the case of passive, unintentional participation. The last chapter in this section describes a set of concepts that take privacy by design into account to process social media and make it impossible to identify individual persons.

Risks and negative side effects must be examined more closely in future research. This includes the invasion of masses when promoting sensitive places via social media, the emergence of filter bubbles when information is sorted based on user preferences, or the manipulative use of information about users who contribute content on social media. Selected, simple approaches for considering bias and for normalizing the social media data are already taken into account in the various chapters.

Chapter 11
Environmental Tracking for Healthy Mobility

Anna Maria Becker, Carolin Helbig, Abdelrhman Mohamdeen, Torsten Masson, and Uwe Schlink

Abstract Environmental stressors in city traffic are a relevant health threat to urban cyclists and pedestrians. These stressors are multifaceted and include noise pollution, heat, and air pollution such as particulate matter. In the present chapter, we describe the use of wearable sensors carried by volunteers to capture their exposure to environmental stressors on their everyday routes. These wearable sensors are becoming increasingly important to capture the spatial and temporal distribution of environmental factors in the city. They also offer the unique opportunity to provide individualized feedback to the person wearing the sensor as well as possibilities to visualize different stressors in their temporal and spatial distribution in a virtual reality environment. We used the option of providing individualized feedback on personal exposure levels in two randomized controlled field studies. In these experiments, we studied the psychological health-related outcomes of carrying a wearable sensor and receiving feedback on one's individual exposure levels.

Keywords Wearable sensors · Air pollution · Noise · Heat · Mobility · Health

11.1 Introduction

Volunteered geographic information (VGI) that is collected by laypeople can be utilized by other citizens and enables scientific investigations that can be seen as a special form of citizen science (Connors et al. 2012; Goodchild 2007). User-generated content such as VGI has proliferated in recent years and is a form of crowdsourcing, where information is drawn together from several non-

A. M. Becker (✉) · T. Masson
Department of Social Psychology, Wilhelm-Wundt-Institute for Psychology, Leipzig, Germany
e-mail: anna.becker@uni-leipzig.de; torsten.masson@uni-leipzig.de

C. Helbig · A. Mohamdeen · U. Schlink
Department of Urban and Environmental Sociology, Helmholtz Centre for Environmental Research, UFZ, Leipzig, Germany
e-mail: carolin.helbig@ufz.de; mahmoud.mohamdeen@ufz.de; uwe.schlink@ufz.de

© The Author(s) 2024
D. Burghardt et al. (eds.), *Volunteered Geographic Information*,
https://doi.org/10.1007/978-3-031-35374-1_11

expert individuals (Elwood et al. 2012). Besides many other applications, VGI has been used in environmental monitoring (Connors et al. 2012) and public transport planning (Giuffrida et al. 2019). Multi-level data mashups—the integration of different forms of information—are becoming increasingly important to layer information in respect to specific locations (Elwood et al. 2012). While some VGI is socially embedded and subjective (Elwood et al. 2012), sensor data can be objectively geotagged.

The aim of this chapter is to demonstrate how VGI can take advantage of smart sensors for the assessment of environmental data that is relevant for human health. In the approach described here, we develop a specific design of VGI techniques that can support city planners and urban authorities as well as individuals who are vulnerable to elevated levels of environmental pollutants and heat. We hypothesize that VGI can improve health protection, which is particularly important for susceptible and vulnerable individuals. The following sections outline the methods and the design of our studies and summarize the results.

11.2   Measuring Environmental Stressors

Epidemiological studies have consistently demonstrated links between exposure to particulate matter and adverse health effects (World Health Organization 2016; Stafoggia et al. 2022). To protect human health, air quality monitoring is regulated by European Directives (European Parliament 2008). Personal exposure to environmental stressors is multifactorial and includes exogenous factors contributing to human health risks (Schlink and Ueberham 2020), such as air temperature, air humidity, air pollutants (gasses, particulate matter), and noise. In Leipzig, traffic is a major source of intra-city PM10 and NOx emissions, while small combustion plants are a source of other emissions (Stadt Leipzig 2018).

Due to the individual differences in daily activities, each person has very individual exposure patterns, which obviously cannot be adequately captured by measurements at a few monitoring stations in the city (improved approach; see Steininger et al. 2020) but require person-specific measurements (Dias and Tchepel 2018; Hinwood et al. 2007). However, air pollution data from a terrestrial monitoring station can only be considered representative of the surrounding district. Such measurements are strongly influenced by the location of the stations, and the dispersion and dilution of air pollutants are affected by meteorological and local conditions (e.g., urban structure; built-up areas, street canyons, traffic networks, and green areas). Therefore, it does not adequately capture the spatial variability of locally sourced pollutants and cannot be an indicator of human exposure (Adams and Kanaroglou 2016; Dionisio et al. 2013). The activities of individuals and air pollution levels exhibit a large degree of spatiotemporal dynamics. This emphasizes the necessity to investigate near-real-time measurements and to develop methods for data assessment (Yang et al. 2022).

With advances in technology in the fields of electrical engineering and wireless networks, "low-cost" air quality sensors have been developed to make air quality monitoring more accessible and portable (Castell et al. 2017; Snyder et al. 2013). In the past several years, evidence has begun to emerge about the usefulness of low-cost sensors. Studies have shown various applications, for example, the classification of emissions sources. Low-cost sensors have a limited ability to detect nanoparticle real-time emissions from residential sources (Wang et al. 2020). However, low-cost sensors were found to perform well in both ambient (field) and controlled (laboratory) conditions (Connolly et al. 2022).

11.3 Citizen Science and VGI in the Context of Environmental Tracking

Including the general public in the production of scientific knowledge is becoming increasingly important (Eitzel et al. 2017). From collecting data (e.g., about local wildlife sightings), processing large amounts of data on private computers, or contributing tedious work by cataloging pictures of animal colonies or galaxies to a deeper involvement in identifying research questions and procedures, laypeople can contribute to scientific progress (Strasser et al. 2019). While citizen science can arise from grassroots movements with an activist agenda (Ottinger 2010), it is most often initiated by scientific and educational institutions (Eitzel et al. 2017). Citizen science can not only shift workload to interested citizens and create unique monitoring opportunities (Aceves-Bueno et al. 2015; Verplanke et al. 2016). It can also serve in educating laypeople about the scientific process and increase public interest and trust in science (Strasser et al. 2019). The use of wearable sensors to collect data on environmental stressors around the city can be understood in the framework of citizen science, as laypersons collect data and investigate their own exposure to pollution as they explore their surroundings. This immersion in the scientific process can not only aid in data collection but also help citizens to identify the least polluted areas in their city. Platforms that integrate data from citizen scientists have been developed to share and visualize environmental information from different sources (Lautenschlager et al. 2018). A plethora of studies indicates that volunteers have a high interest in their personal exposure to environmental stressors, though behavior change to avoid these stressors is often hard to implement (see Becker et al. 2021 for a review).

11.4 A Health-Psychological Perspective

Tracking environmental stressors with a wearable sensor makes it possible to give individualized feedback to the users. Thereby, they receive information on potentially harmful exposure levels to air pollution, heat, and noise pollution. As these stressors (e.g., particulate matter) cannot be perceived directly, this is new information that may elicit fear in the user. Fear appeals have been studied by psychologists in the past, showing that effects of fear appeals on behavior change (e.g., smoking cessation) are strongest when the negative consequences of a harmful behavior are seen as personally relevant and likely to occur (Rosenstock 1974; Rogers 1975). Choosing highly polluted routes in the city can be framed as a harmful behavior to one's health, and messaging should motivate people to choose less polluted routes, e.g., smaller streets with less traffic. Two influential theories in health psychology, namely, the health belief model (Rosenstock 1974) and protection motivation theory (Rogers 1975, 1983; Maddux and Rogers 1983), have studied health-related behavior change such as smoking cessation or physical exercise. Both theories have in common that they predict a healthy behavior change when negative health outcomes of a current behavior are seen as severe and a person feels susceptible to these negative outcomes. In our case, this may be that people perceive the health effects of particulate matter to be serious and that it is likely that they will be affected by these, if they do not lower their exposure. Additionally, persons must see a behavior change as useful to mitigate the risk of illness and feasible to implement (Rogers 1983; Maddux and Rogers 1983). Perceived psychological costs or barriers will reduce the likelihood of behavior change (Rogers 1983). For example, if a person is not willing to take a longer route to work or to use a side street with a less comfortable surface to cycle on, this will reduce the likelihood of them changing their route. In more technical terms, protection motivation theory posits that both threat appraisal (harmful outcomes of the current situation are severe and likely to occur) and coping appraisal (alternative behaviors will be effective in mitigating the threat and are feasible to implement) contribute to health-related behavior change (Rogers 1983; Maddux and Rogers 1983). Research on protection motivation theory has found that both threat appraisal and coping appraisal must interact to elicit a problem-focused coping response (Babcicky and Seebauer 2019; Rippetoe and Rogers 1987). If a person only perceives high threat but has a low coping appraisal, they might choose emotion-focused coping strategies rather than problem-focused coping. These emotion-focused strategies, such as denial, fatalism, or wishful thinking, will reduce the emotional burden of the threat but not result in healthy behavior change (Babcicky and Seebauer 2019; Rippetoe and Rogers 1987). Protection motivation is most commonly applied to health behaviors (Plotnikoff and Trinh 2010; Prentice-Dunn et al. 2009). However, it can also be applicable to other behavioral safety measures, for example, in relation to flood events (Babcicky and Seebauer 2019), climate change (Bagagnan et al. 2019), the adoption of electric vehicles (Bockarjova and Steg 2014), household water management (Bryan et al. 2019), earthquake preparedness (Mulilis and Lippa 1990), and wildfire protection (Dupéy and Smith 2019).

11.5 Visualizing Environmental Stressors

The amount and variety of data are increasing in all research fields, which means that analyzing these large, complex datasets has become a challenging task (Avazpour et al. 2019). That includes integrating data from multiple heterogeneous sources, normalizing these data, and providing a unified view of these data sets to users (Tian and Li 2019). Collecting, integrating, aligning, and efficiently extracting information from heterogeneous and autonomous data sources are considered a major challenge (Fusco and Aversano 2020). In this context, visualization is an important tool that makes it possible to analyze complex data. Scientific visualization assists scientists in analyzing data by transforming data into geometric representations, thus supporting the analysis of complex data (Gershon and Eick 1995; Defanti and Brown 1991). In addition, virtual reality (VR) is recognized as a powerful human-computer interface (Burdea and Coiffet Philippe 2003). Users can immerse themselves in a virtual world and manipulate it by changing their viewpoint and interacting (Brooks 1999). VR environments are a promising tool for scientists to visualize their large and complex data sets and to control the behavior of virtual objects using interaction functions (Simpson et al. 2000). The combination of the power of a VR system and the human ability to detect interesting patterns and inconsistencies in the data makes VR a suitable tool to solve future scientific visualization tasks (van Dam et al. 2002). VR is also used in the context of urban planning in some prototypical projects, for example, in knowledge transfer on the value of urban greenery (Mokas et al. 2021), the investigation of the perceived safety level of cyclists (Nazemi et al. 2021), different planning phases of high-rise building construction (Lu et al. 2021), green landscape planning and design (Pei 2021), and the evaluation of urban spaces (Luigi et al. 2015; Zhang and Zhang 2021). One way to combine methods of visualization and VR, as well as to implement analysis methods and make them available via a graphical user interface (GUI), is to use the Unity Game Engine (Helbig et al. 2015). Its use in scientific projects has increased in recent years, e.g., for integration of 3D building model data (Keil et al. 2021), visualizing 3D point clouds captured by drones (Weißmann et al. 2022), implementing a walkable virtual city model (Schmohl et al. 2020), and simulating secure hazardous transportation (Yang et al. 2020). Another argument for the use of game engines is the possibility to integrate audio data in a straightforward way, which increases the degree of immersion considerably (Hruby 2019; Berger and Bill 2019; Rafiee et al. 2017) and is not available in conventional visualization tools for geodata. In the presented project, we implemented a visualization and analysis application for mobile sensing data and questionnaire data in combination with other urban data such as 3D city model, traffic data and noise maps from the city of Leipzig, and weather station data to explore the data with scientists as well as decision makers and citizens.

11.6 Implementation: Two Field Experiments Using Wearable Sensors

We conducted two field experiments with wearable sensors in order to explore the spatiotemporal distribution of environmental stressors in the city of Leipzig. Another aim of these studies was to test the psychological effects of carrying wearable sensors and receiving feedback about one's personal exposure levels. Lastly, one of our aims was to implement innovative ways to integrate and visualize sensor data and subjective experiences of participants with the sensors.

11.6.1 Field Study 1

11.6.1.1 Design and Procedure of Study 1

To test the psychological effects of carrying the sensors, we conducted experiments as randomized controlled trials, including a measurement group that carried the wearable sensors and a control group that did not receive the sensor kit. The first field experiment was run from July to September 2020. After registering on our website, participants were allocated to a week of participation. Within each study week, participants were randomly allocated to either the measurement group or the control group. Persons in the measurement group carried the measurement kit for three days and received individualized feedback about their exposure to heat, particulate matter, and noise after the measurement phase. Participants in the control group did not carry the measurement kits but received questionnaires including the same questions as in the measurement group. This allowed us to compare the answers of those in the control group to those in the measurement group and identify the effects of the intervention. On Fridays before their allocated study week, all participants received a first questionnaire. Persons in the measurement group then picked up the measurement device on Monday and used it for three consecutive days to measure their exposure on everyday routes by bike or as pedestrians. They returned the measurement kit on Thursdays. The next day, all participants received a link to the second questionnaire (both those who carried the measurement kit and those in the control group). We then retrieved the data from the measurement kits and compiled individual feedback reports for those who wore the sensors. One week after the measurement ended, each person from the measurement group received their feedback report and a third questionnaire. The feedback report included general information about particulate matter, heat, and noise as environmental stressors. The report also showed graphs indicating the accumulated stressor levels throughout the measurement phase. A color-coded legend indicated the magnitude of exposure with reference to noise levels (silent room–pain threshold) or temperature ranges (no temperature stress–extreme temperature stress). Approximately 2 to 4 months after their participation, both the measurement and the control group were sent the link

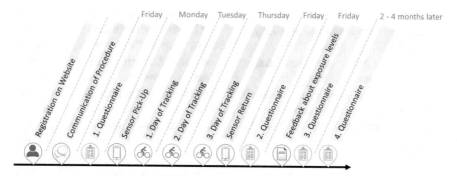

Fig. 11.1 Experimental procedure for the measurement group in study 1

to a follow-up questionnaire (with some outliers submitting after 1.6 to 4.7 months) (Fig. 11.1).

11.6.1.2 Sensors Used for Measuring Environmental Stressors in Study 1

The measurement system consisted of three different devices: (1) smartphone Motorola, Moto G3, launched in July 2015 and operated with android system; (2) Little Environmental Observatories (LEO) Electrochemical sensors widely used for the detection of toxic gasses at the parts per million (ppm) level and for oxygen in levels of percent of volume (% vol)—toxic gas sensors are available for a wide range of gasses, including NO2, NO, and O3 (Robinson et al. 2018), providing the current temperature and relative humidity (Ateknea Solutions Catalonia S.A.; https://ateknea.com); and (3) Dylos-DC1100, an air quality monitor that measures particulate matter (PM) number concentration to provide a continuous assessment of ambient suspended particles. The unit counts particles in two size ranges: large and small. According to the manufacturer, large particles have diameters between 2.5 and 10 micrometers, μm (i.e., PM10); small particles have diameters from 0.5 μm up to 2.5 μm (i.e., PM1 and PM2.5).

11.6.2 Field Study 2

11.6.2.1 Design and Procedure of Study 2

The second field experiment was conducted from September 2021 to August 2022. Similar to study 1, there was a control group and a measurement group. However, we extended the design of the field experiment by providing participants with daily feedback about their personal exposure as well as suggestions for alternative routes as a strategy to strengthen participants' coping capacity. After signing up through our website and being allocated to a study week, participants were randomly assigned to their group. Both groups received a first questionnaire, before the measurement group was given the wearable sensor and smartphone kit. Participants in the measurement group used the sensor and smartphone for 3 days on their everyday routes. They received feedback about their exposure in the smartphone app every evening. The app allowed them to (a) see the measured levels of heat, noise, and particulate matter along the routes they had used during the day and to (b) receive an alternative route suggestion for each route they had recorded that day. These alternative route suggestions were based on a set of routing preferences optimized for low pollution and developed in collaboration with the VGI-Routing team (see Chap. 3). Participants in the measurement group were asked to fill in a short daily questionnaire after looking at their feedback and the alternative route suggestions. After their study week, both the measurement group and the control group received the link to another questionnaire. Approximately 3 to 6 months after their study week, all participants received a follow-up questionnaire (with some outliers submitting after 2 to 8.7 months).

11.6.2.2 Sensors Used for Measuring Environmental Stressors in Study 2

The system included a smartphone and the PAM AS520 wearable air quality monitor (www.atmosphericsensors.com), which can assess air quality in the wearer's immediate environment. The PAM uses electrochemical cells ("Alphasense" sensors with 2 electrodes), a laser-optical particle monitor, temperature and humidity sensors, and a high-sensitivity GPS receiver and 3-axis accelerometer. A built-in microphone records ambient noise. At night, the PAM is stored in a docking station where the battery is charged. The docking station is equipped with a modem and SIM card that transmits the data stored on the device to the project's central database for further processing (Fig. 11.2).

11.6.3   The Questionnaires

Multiple questionnaires were used to assess psychological variables in our field experiments. Questionnaires of each participant were linked through an identifier code that was indicated at each measurement point. This code was also used to identify each participant's sensor measurements and connect this information to the questionnaire data. In the first questionnaire, participants were asked for their preferred mode of transport (public transport, car, bike, or by foot) and which aspects were particularly important to them when choosing a route as a cyclist or pedestrian (e.g., speed, green space, etc.) (adapted from Ueberham et al. 2019). Participants were also asked about the stage of action they were in regarding healthy routing choices with low pollution levels (adapted from Olsson et al. 2018). To

Fig. 11.2 The sensor and smartphone kit used in Study 2 (Foto by André Künzelmann, UFZ)

capture the main predictors of protection motivation theory (Rogers 1975, 1983), we measured health risk perception for particulate matter, noise, and heat (e.g., *particulate matter, noise, and heat on my daily routes have very negative effects for my health*), as well as coping appraisals regarding these three environmental stressors (e.g., *I can reduce my exposure to environmental stressors in street traffic*) (partly adapted from Ueberham et al. 2019). As our main dependent variable, we measured individual protection intentions to change one's everyday routes in a way that would avoid environmental stressors (e.g., avoiding streets with a lot of car traffic). We further measured collective action intentions to fight for lower levels of environmental stressors in city traffic (e.g., by signing petitions), as well as willingness to pay for a service providing daily information on current low-pollution routes. We measured a range of other constructs in the questionnaire, including emotion-focused coping strategies that may come into play when facing a threat while coping appraisals are low (Rippetoe and Rogers 1987) and routing behavior habits when traveling to school/university, when going shopping, and during leisure time (Verplanken and Orbell 2003). Further measures included, among others, social norms (e.g., whether participants thought that others in their city were avoiding environmental stressors on everyday routes); identification with the city of Leipzig, cyclists, pedestrians, and car drivers; derogation of car drivers; general health concerns (Fahrenberg et al. 2003); and preference for technology use.

11.7 Results of the Field Experiments

11.7.1 Results Regarding the Sensor Measurements

The two field experiments (FE) were designed to monitor individual's experiences on a trajectory based real-time (FE$_1$; $\Delta t = 5$ sec. and FE$_2$; $\Delta t = 10$ sec.) for a given population (i.e., volunteer cyclists and pedestrians) during their daily life activities. These specific conditions of personal exposure are used to develop a new perception of the spatial categories of the road network in Leipzig, where our population activities took place. The framework of the developed system includes (1) mobile sampling using low-cost air quality sensors along road segments of individual's routes traveled (first field experiment), (2) empirically identified classes of microenvironments for road segments in Leipzig, and (3) recognition of classes developed by applying supervised machine learning (ML) models (e.g., Random Forest, RF; logistic regression, LR; support vector machines, SVM; and K-nearest neighbors, KNN). The outdoor microenvironment is categorized into three categories, each of which is virtually differentiated by buildings, road structure, traffic rate, vehicle fleet number, and driving mode with OSM-based field categories (such as highway, bicycle, and pedestrian lane), helping to reduce the multiple factors affecting personal exposure in urban data areas such as weather conditions, land use, and traffic. The three categories are (1) main, characterized by detached blocks and broad road structure, e.g., avenues, semi-continuous pollutant line source from fleet flow of vehicles with large number and high rate of dissipation, which also depends on weather conditions, i.e., wind factors; (2) secondary, spaced by linear semi-tall buildings, secondary streets (e.g., residential streets and parking areas, i.e., idle driving mode), discrete emissions from vehicle critical dual periods (i.e., daily back-and-forth movements of the workforce), and large eddy currents which counteract or reduce aerodynamic dispersal for pollutants in street canyons; and (3) green, vegetation cover, e.g., forests and parks. Several non-parametric ML algorithms were also tested with the same selected urban stressors (10 features, PM2.5, PM10, NO, NO2, O3, temperature, specific humidity, light intensity, speed, and noise) to determine which ones are suitable for our data set. The results showed that the performance difference between the individual models (RF, LG, SVM, and KNN) was significant, while RF models performed best over others, achieving >90% accuracy. Model confusion from multiple experimental samples is less than 10%, and this can be explained in terms of the following: (1) the high degree of overlap between classes is one of the main challenges affecting the accuracy of the classifier (Deberneh and Kim 2021), and (2) the developed categories are strongly coherent in a heterogeneous urban environment, so sensor response time and recording interval must also be considered when interpreting the results. In addition, the time to transfer a cyclist or pedestrian from one category to another is always much less than the response time of the sensors (Ueberham and Schlink 2018). This poses a great challenge in the crossings and adjacent parts between the three classes.

11.7.2   Results Regarding the Questionnaires

11.7.2.1   Statistical Analysis

To identify effects of our intervention, we computed linear mixed-effect models with random intercepts to estimate within-participant changes from pretest to follow-up for our outcome measures while also assessing differences between the intervention group and the control group. The mixed models included time (pretest, posttest, test after receiving exposure feedback, follow-up) as well as the group (intervention vs. control) and their interaction term to predict each outcome variable. When using an additional moderator, we entered time, group, the moderator variable, as well as the two-way and three-way interaction terms into the mixed model.

11.7.2.2   Results of Study 1: Descriptive Analysis

One hundred and eighty-two participants completed the pretest questionnaire, 167 participants completed the posttest questionnaire, and 121 participants completed the follow-up questionnaire. The datasets were merged based on an identifier code, generated by each participant at the start of each questionnaire, resulting in a final sample of 109 participants ($N_{intervention}$ = 56, $N_{control}$ = 53; 59.89% of the pretest sample). Sixty-one participants identified as female and 48 identified as male. Ages ranged from 19 to 67 years (M = 36.33, SD = 9.68). Most participants (72.5%) had a university degree. 6.4% reported having a respiratory health condition such as asthma, and 29.4% reported having allergies. Overall, participants rated their health as good (Mdn = 6 on a seven-point scale ranging from 1, very bad, to 7, very good) and reported medium levels of health concerns (M = 4.04, SD = 1.00). All scales were 7-point Likert scales. Participants rated their previous knowledge about particulate matter, heat, and noise pollution as medium (for PM, Mdn = 3; heat, Mdn = 3; noise, Mdn = 3). Satisfaction with the measurement kit was medium (Mdn = 4.5). 57.1% of the participants rated the handling of the measurement kit as at least somewhat easy, indicating that the ease of use was only medium (Mdn = 3). More than 70% reported a high or very high tracking frequency throughout the measurement phase (Mdn = 6). Participants reported that they used the tracker on typical everyday routes (Mdn = 7). More than 74% of the respondents at least somewhat agreed that wearable sensors can help people reduce their personal exposure to environmental health risks (Mdn = 5), and 63% at least somewhat agreed that wearable sensors may support behavior change aimed at reducing personal exposure (Mdn = 5). Following the individualized feedback report, participants rated their exposure to particulate matter (Mdn = 5) and noise (Mdn = 5) as medium to high while rating the heat exposure as lower (Mdn = 3.00). The feedback did not greatly differ from participants' expectations, as participants indicated it being neither much higher nor lower than expected (PM, Mdn = 4.00; noise, Mdn = 4.00;

heat, $Mdn = 4.00$). Participants perceived the feedback to be mostly representative for their everyday exposure (PM, $Mdn = 6.00$; noise, $Mdn = 6.00$; heat, $Mdn = 5.00$).

11.7.2.3 Results of Study 1: Mixed Model Results

For individual action intentions, results showed no significant main effects of time and group and, more importantly, no significant interaction effect of time and group. In other words, our results did not indicate that participation in the measurement group (vs. control group) would increase respondents' action intentions to protect themselves against environmental health risks. However, results of exploratory analysis, including routing behavior habits as an additional moderator in the analysis, revealed an interesting pattern of results. First, participation in the measurement group (vs. control group) increased action intentions from pretest to posttest for respondents with weak (but not strong) routing behavior habits. This initial increase, however, was not stable throughout the intervention period. At the follow-up measurement point, we found no differences in action intentions between respondents with weak and strong routing behavior habits. For perceptions of environmental health risks, our findings indicate that participation in the measurement group (vs. control group) led to a significant increase in the perception of particulate matter health risks from pretest to posttest. Importantly, increased health risk perceptions for particulate matter were retained throughout the follow-up period, indicating a robust intervention effect. There were no intervention effects on health risk perceptions for noise and heat.

11.7.2.4 Results of Study 2: Descriptive Analysis

The pretest questionnaire was completed by 267 eligible participants, 225 completed the posttest questionnaire, and lastly 151 eligible participants completed the follow-up questionnaire. After matching participants' codes, the final sample with complete data sets consisted of 136 participants ($N_{intervention} = 67$, $N_{control} = 69$; 50.9% of pretest sample). Eighty-eight participants identified as female, 45 identified as male, and three participants identified as diverse. Ages ranged from 18 to 70 years ($M = 29.76$, $SD = 10.43$). Approximately half of the respondents had a university degree. Regarding health condition, 7.4% reported a respiratory health condition such as asthma, and 27.9% reported having allergies. Overall, participants rated their health as good to very good ($Mdn = 6$) and reported medium levels of health concerns ($M = 4.15$, $SD = 1.08$). Satisfaction with the measurement kit was high ($Mdn = 6$). Approximately 85% of the participants rated the handling of the measurement kit as somewhat easy, easy, or very easy ($Mdn = 2$), and more than 90% reported a high or very high tracking frequency throughout the measurement phase ($Mdn = 6$). More than 77% of the respondents at least somewhat agreed that wearable sensors can help people reduce their personal exposure to environmental health risks ($Mdn = 5$), and more than 62% at least somewhat agreed that wearable sensors may support behavior change aimed at reducing personal exposure ($Mdn = 5$).

11.7.2.5   Results of Study 2: Mixed Model Results

For individual action intentions, results showed no significant interaction effect of time and group. While inspection of simple effects showed a trend indicating that participation in the measurement group (vs. control group) increased action intentions from pretest to posttest, the intervention effect was not significant. In contrast to Study 1, we found no effects of routing behavior habits on changes in action intentions. Regarding perceptions of environmental health risks, our findings indicate a robust intervention effect on perceived particulate matter risk. Specifically, results revealed that participation in the measurement group (vs. control group) led to a significant increase in perceived PM health risks from pretest to posttest. Importantly, increased levels of perceived PM health risk were retained throughout the follow-up period. For perceived heat and noise health risks, results indicate no significant treatment effects.

11.7.3   The Visualization and Analysis Application and Achieved Results

As part of the project, a visualization and analysis application was implemented that allows the data from the measurement campaign to be evaluated in combination with data on building structure, weather, GI, and traffic (Helbig et al. 2022). The application was implemented with the Unity Game Engine, which provides a development environment for computer games and other interactive 3D graphic applications and enables us to meet the following requirements: (1) Provide a user interface and interaction functionality, (2) performance (avoid long loading times), (3) Integrate various data sources/data types, (4) 3D visualization, (5) integrate analysis methods, and (6) presentation on PC and VR environment. By using mobile sensors within the measurement campaign of our project, we get data from two different perspectives: (1) the individual exposure to environmental stressors for a participant and (2) the distribution of environmental stressors within the urban area. In the application, methods must be provided that enable the analysis to be carried out from both perspectives. This is possible by displaying individual routes or the routes of individual participants in the 3D city model, as well as in a 2D representation. The distribution of all measurement points over the urban space can also be displayed. In addition, a hot and cold spot analysis of the distribution of the values can be carried out. The application also offers a range of functions that allow a comprehensive analysis, such as different perspectives (perspective and orthographic), showing/hiding of data layers, display of individual routes (incl.

Fig. 11.3 Analysis and visualization of the trajectories of mobile sensor data

extracting GPS outliers), different coding of values (by color and/or size), chained filtering by parameters, and space-time cube visualization (Fig. 11.3).

The analysis of the data with the help of the application showed that paths, especially on which many cyclists move and which also allow faster speeds due to their condition, are polluted with high nitrogen oxide levels due to their location directly on main roads. The evaluation so far also shows the temperature differences that occur between parks and heavily sealed surfaces, especially on hot days. There, the surroundings heat up strongly, especially in the second half of the day, and can be poorly ventilated with dense development at the same time.

11.8 Conclusions

Our study was conducted in the frame of the ExpoAware project[1] and supported by the German Research Foundation DFG within the Priority Research Program 1894 Volunteered Geographic Information: Interpretation, Visualization and Social Computing. The research developed an integrated design that demonstrated (1) the feasibility of individuals applying advanced VGI techniques to explore the urban environmental conditions and provide useful information for city planners and officials; (2) the potential of VGI for the assessment of personal exposure that is highly relevant for adverse health effects, especially in vulnerable persons; and (3) the ability to study protective behavior of individuals and to develop actions for behavior changes and individual adaptation strategies. From our results of the urban measurements, we conclude that the proposed novel microenvironment classifications successfully reduced the complexity of urban data, and the results of the random forest models emphasized the validity of the hypothesized classification method. Nonparametric methods such as random forest are promising approaches

---

[1] https://www.vgiscience.org/projects/expoaware.html.

when the model is complex and can benefit from a large number of training examples. The study also highlights the benefits and potential of low-cost sensors in categorizing street-level personal exposure in urban areas. The field experiments are an important contribution to improving the framework conditions for cycling and walking and make it more attractive. For a sustainable and resilient city, the mobility transition and thus the reduction of motorized individual transport and at the same time the promotion of active mobility are a decisive component. By integrating the results of mobile measurements into long-term planning, urban planners can reduce the level of environmental stressors in cities and better react to extreme events (e.g., heat waves) in the short term. Very short-term reactions to elevated exposure can be made by the individuals themselves. In our study, the application of personal sensors increased participants' health risk perceptions (with regard to particulate matter) and was also able to temporarily elevate their action intentions to protect themselves against environmental health risks, albeit only for participants with weak routing habits in the first study. With the help of Unity, we were able to combine methods from 2D and 3D visualization, as well as implement analysis methods and make them available via GUI. Unity supports various platforms and enables us to use it in different contexts, ranging from PC for individual analysis to VR environments for collective analysis and presentation. We claim that our approach should be used for implementing analysis and visualization tools in future projects because (1) it has the potential to become a modular system for applications by reusing and further developing methods, (2) it enables the combination of modern 3D visualization with various analysis methods, and (3) it supports presentation in VR environments and thereby facilitates multidisciplinary, research in collaborative projects, and projects with high interdisciplinary (Helbig et al. 2022).

Acknowledgments This research was funded by the Deutsche Forschungsgemeinschaft (DFG, German Research Foundation)—ExpoAware, 424979005.

References

Aceves-Bueno E, Adeleye AS, Bradley D, Tyler Brandt W, Callery P, Feraud M, Garner KL, Gentry R, Huang Y, McCullough I, Pearlman I, Sutherland SA, Wilkinson W, Yang Y, Zink T, Anderson SE, Tague C (2015) Citizen science as an approach for overcoming insufficient monitoring and inadequate stakeholder buy-in in adaptive management: Criteria and evidence. Ecosystems 18(3):493–506. https://doi.org/10.1007/s10021-015-9842-4

Adams MD, Kanaroglou PS (2016) Mapping real-time air pollution health risk for environmental management: Combining mobile and stationary air pollution monitoring with neural network models. J Environ Manag 168:133–141. https://doi.org/10.1016/j.jenvman.2015.12.012

Avazpour I, Grundy J, Zhu L (2019) Engineering complex data integration, harmonization and visualization systems. J Ind Inf Integr 16:100103. https://doi.org/10.1016/j.jii.2019.08.001

Babcicky P, Seebauer S (2019) Unpacking protection motivation theory: evidence for a separate protective and non-protective route in private flood mitigation behavior. J Risk Res 22(12):1503–1521. https://doi.org/10.1080/13669877.2018.1485175

Bagagnan A, Ouedraogo I, M Fonta W, Sowe M, Wallis A (2019) Can protection motivation theory explain farmers' adaptation to climate change decision making in the gambia? Climate 7(1):13. https://doi.org/10.3390/cli7010013

Becker AM, Marquart H, Masson T, Helbig C, Schlink U (2021) Impacts of personalized sensor feedback regarding exposure to environmental stressors. Curr Pollut Rep 7:579–593. https://doi.org/10.1007/s40726-021-00209-0

Berger M, Bill R (2019) Combining vr visualization and sonification for immersive exploration of urban noise standards. Multimodal Technol Interact 3(2):34. https://doi.org/10.3390/mti3020034

Bockarjova M, Steg L (2014) Can protection motivation theory predict pro-environmental behavior? explaining the adoption of electric vehicles in the netherlands. Glob Environ Change 28:276–288. https://doi.org/10.1016/j.gloenvcha.2014.06.010

Brooks FP (1999) What's real about virtual reality? IEEE Comput Graph Appl 19(6):16–27. https://doi.org/10.1109/38.799723

Bryan K, Ward S, Barr S, Butler D (2019) Coping with drought: Perceptions, intentions and decision-stages of south west england households. Water Resour Manag 33(3):1185–1202. https://doi.org/10.1007/s11269-018-2175-2

Burdea GC, Coiffet Philippe (2003) Virtual reality technology. Wiley-IEEE Press, New York

Castell N, Dauge FR, Schneider P, Vogt M, Lerner U, Fishbain B, Broday D, Bartonova A (2017) Can commercial low-cost sensor platforms contribute to air quality monitoring and exposure estimates? Environ Int 99:293–302. https://doi.org/10.1016/j.envint.2016.12.007

Connolly RE, Yu Q, Wang Z, Chen YH, Liu JZ, Collier-Oxandale A, Papapostolou V, Polidori A, Zhu Y (2022) Long-term evaluation of a low-cost air sensor network for monitoring indoor and outdoor air quality at the community scale. Sci Total Environ 807(Pt 2):150797. https://doi.org/10.1016/j.scitotenv.2021.150797

Connors JP, Lei S, Kelly M (2012) Citizen science in the age of neogeography: Utilizing volunteered geographic information for environmental monitoring. Ann Assoc Am Geogr 102(6):1267–1289. https://doi.org/10.1080/00045608.2011.627058

Deberneh HM, Kim I (2021) Prediction of type 2 diabetes based on machine learning algorithm. Int J Environ Res Public Health 18(6). https://doi.org/10.3390/ijerph18063317

Defanti TA, Brown MD (1991) Visualization in scientific computing. Advances in Computers, vol 33. Elsevier, Amsterdam, pp 247–307. https://doi.org/10.1016/S0065-2458(08)60168-0. https://www.sciencedirect.com/science/article/pii/S0065245808601680

Dias D, Tchepel O (2018) Spatial and temporal dynamics in air pollution exposure assessment. Int J Environ Res Public Health 15(3). https://doi.org/10.3390/ijerph15030558

Dionisio KL, Isakov V, Baxter LK, Sarnat JA, Sarnat SE, Burke J, Rosenbaum A, Graham SE, Cook R, Mulholland J, Özkaynak H (2013) Development and evaluation of alternative approaches for exposure assessment of multiple air pollutants in atlanta, georgia. J Exposure Sci Environ Epidemiol 23(6):581–592. https://doi.org/10.1038/jes.2013.59

Dupéy LN, Smith JW (2019) Close but no cigar: how a near-miss wildfire event influences the risk perceptions and mitigation behaviors of residents who experienced a recent, nearby wildfire. https://digitalcommons.usu.edu/extension_curall/1959

Eitzel MV, Cappadonna JL, Santos-Lang C, Duerr RE, Virapongse A, West SE, Kyba CCM, Bowser A, Cooper CB, Sforzi A, Metcalfe AN, Harris ES, Thiel M, Haklay M, Ponciano L, Roche J, Ceccaroni L, Shilling FM, Dörler D, Heigl F, Kiessling T, Davis BY, Jiang Q (2017) Citizen science terminology matters: Exploring key terms. Citiz Sci Theory Pract 2(1):1. https://doi.org/10.5334/cstp.96

Elwood S, Goodchild MF, Sui DZ (2012) Researching volunteered geographic information: spatial data, geographic research, and new social practice. Ann Assoc Am Geogr 102(3):571–590. https://doi.org/10.1080/00045608.2011.595657

European Parliament (2008) Directive 2008/50/EC of the European Parliament and of the Council of 21 May 2008 on ambient air quality and cleaner air for Europe. Consolidated text: https://eur-lex.europa.eu/eli/dir/2008/50/2015-09-18

Fahrenberg J, Hampel R, Selg H (2003) Das Freiburger Persönlichkeitsinventar FPI. Hogrefe, Oxford

Fusco G, Aversano L (2020) An approach for semantic integration of heterogeneous data sources. PeerJ Comput Sci 6:e254. https://doi.org/10.7717/peerj-cs.254

Gershon N, Eick S (1995) Foreword in proc. ieee symp. information visualization, infovis 95. IEEE CS Press, Washington, DC, pp vii–viii

Giuffrida, Le Pira, Inturri, Ignaccolo (2019) Mapping with stakeholders: An overview of public participatory gis and vgi in transport decision-making. ISPRS Int J Geo-Inf 8(4):198. https://doi.org/10.3390/ijgi8040198

Goodchild MF (2007) Citizens as sensors: the world of volunteered geography. GeoJournal 69(4):211–221

Helbig C, Bilke L, Bauer HS, Böttinger M, Kolditz O (2015) Meva–an interactive visualization application for validation of multifaceted meteorological data with multiple 3d devices. PloS one 10(4):e0123811. https://doi.org/10.1371/journal.pone.0123811

Helbig C, Becker AM, Masson T, Mohamdeen A, Sen ÖO, Schlink U (2022) A game engine based application for visualising and analysing environmental spatiotemporal mobile sensor data in an urban context. Front Environ Sci 10. https://doi.org/10.3389/fenvs.2022.952725

Hinwood AL, Rodriguez C, Runnion T, Farrar D, Murray F, Horton A, Glass D, Sheppeard V, Edwards JW, Denison L, Whitworth T, Eiser C, Bulsara M, Gillett RW, Powell J, Lawson S, Weeks I, Galbally I (2007) Risk factors for increased btex exposure in four australian cities. Chemosphere 66(3):533–541. https://doi.org/10.1016/j.chemosphere.2006.05.040

Hruby F (2019) The sound of being there: Audiovisual cartography with immersive virtual environments. KN J Cartogr Geogr Inf 69(1):19–28. https://doi.org/10.1007/s42489-019-00003-5

Keil J, Edler D, Schmitt T, Dickmann F (2021) Creating immersive virtual environments based on open geospatial data and game engines. KN J Cartogr Geogra Inf 71(1):53–65. https://doi.org/10.1007/s42489-020-00069-6

Lautenschlager F, Becker M, Steininger M, Hotho A (2018) Everyaware gears: A tool to visualize and analyze all types of citizen science data. In: Burghardt D, Chen S, Andrienko G, Andrienko N, Purves R, Diehl A (eds) Proceedings of VGI Geovisual Analytics Workshop, colocated with BDVA 2018, KOPS

Lu X, Tomkins A, Hehl-Lange S, Lange E (2021) Finding the difference: Measuring spatial perception of planning phases of high-rise urban developments in virtual reality. Comput Environ Urban Syst 90:101685. https://doi.org/10.1016/j.compenvurbsys.2021.101685

Luigi M, Massimiliano M, Aniello P, Gennaro R, Virginia PR (2015) On the validity of immersive virtual reality as tool for multisensory evaluation of urban spaces. Energy Procedia 78:471–476. https://doi.org/10.1016/j.egypro.2015.11.703

Maddux JE, Rogers RW (1983) Protection motivation and self-efficacy: A revised theory of fear appeals and attitude change. J Exp Soc Psychol 19(5):469–479. https://doi.org/10.1016/0022-1031(83)90023-9

Mokas I, Lizin S, Brijs T, Witters N, Malina R (2021) Can immersive virtual reality increase respondents' certainty in discrete choice experiments? A comparison with traditional presentation formats. J Environ Econ Manag 109:102509. https://doi.org/10.1016/j.jeem.2021.102509

Mulilis JP, Lippa R (1990) Behavioral change in earthquake preparedness due to negative threat appeals: A test of protection motivation theory. J Appl Soc Psychol 20(8):619–638. https://doi.org/10.1111/j.1559-1816.1990.tb00429.x

Nazemi M, van Eggermond MAB, Erath A, Schaffner D, Joos M, Axhausen KW (2021) Studying bicyclists' perceived level of safety using a bicycle simulator combined with immersive virtual reality. Accid Anal Prev 151:105943. https://doi.org/10.1016/j.aap.2020.105943

Olsson LE, Huck J, Friman M (2018) Intention for car use reduction: Applying a stage-based model. Int J Environ Res Public Health 15(2). https://doi.org/10.3390/ijerph15020216

Ottinger G (2010) Buckets of resistance: Standards and the effectiveness of citizen science. Sci Technol Hum Values 35(2):244–270. https://doi.org/10.1177/0162243909337121

Pei L (2021) Green urban garden landscape design and user experience based on virtual reality technology and embedded network. Environ Technol Innov 24:101738. https://doi.org/10.1016/j.eti.2021.101738

Plotnikoff RC, Trinh L (2010) Protection motivation theory: is this a worthwhile theory for physical activity promotion? Exerc Sport Sci Rev 38(2):91–98. https://doi.org/10.1097/JES.0b013e3181d49612

Prentice-Dunn S, McMath BF, Cramer RJ (2009) Protection motivation theory and stages of change in sun protective behavior. J Health Psychol 14(2):297–305. https://doi.org/10.1177/1359105308100214

Rafiee A, van der Male P, Dias E, Scholten H (2017) Developing a wind turbine planning platform: Integration of "sound propagation model–gis-game engine" triplet. Environl Modell Softw 95:326–343. https://doi.org/10.1016/j.envsoft.2017.06.019

Rippetoe PA, Rogers RW (1987) Effects of components of protection-motivation theory on adaptive and maladaptive coping with a health threat. J Pers Soc Psychol 52(3):596–604. https://doi.org/10.1037/0022-3514.52.3.596

Robinson JA, Kocman D, Horvat M, Bartonova A (2018) End-user feedback on a low-cost portable air quality sensor system-are we there yet? Sensors (Basel, Switzerland) 18(11). https://doi.org/10.3390/s18113768

Rogers RW (1975) A protection motivation theory of fear appeals and attitude change1. J Psychol 91(1):93–114. https://doi.org/10.1080/00223980.1975.9915803

Rogers RW (1983) Cognitive and physiological processes in fear appeals and attitude change: A revised theory of protection motivation. In: Cacioppo J, Petty R (eds) Social psychophysiology. Guilford Press, New York

Rosenstock IM (1974) Historical origins of the health belief model. Health Educ Monogr 2(4):328–335. https://doi.org/10.1177/109019817400200403

Schlink U, Ueberham M (2020) Perspectives of individual-worn sensors assessing personal environmental exposure. Engineering (Beijing, China) https://doi.org/10.1016/j.eng.2020.07.023

Schmohl S, Tutzauer P, Haala N (2020) Stuttgart city walk: A case study on visualizing textured dsm meshes for the general public using virtual reality. PFG J Photogramm Remote Sens Geoinf Sci 88(2):147–154. https://doi.org/10.1007/s41064-020-00106-z

Simpson RM, LaViola JJ, Laidlaw DH, Forsberg AS, van Dam A (2000) Immersive vr for scientific visualization: a progress report. IEEE Comput Graph Appl 20(6):26–52. https://doi.org/10.1109/38.888006

Snyder EG, Watkins TH, Solomon PA, Thoma ED, Williams RW, Hagler GSW, Shelow D, Hindin DA, Kilaru VJ, Preuss PW (2013) The changing paradigm of air pollution monitoring. Environ Sci Technol 47(20):11369–11377. https://doi.org/10.1021/es4022602

Stadt Leipzig (2018) Luftreinhalteplan für die stadt leipzig - fortschreibung 2018. https://www.luft.sachsen.de/download/luft/LRP_Leipzig-2018_Fassung_14-5-2019.pdf

Stafoggia M, Oftedal B, Chen J, Rodopoulou S, Renzi M, Atkinson RW, Bauwelinck M, Klompmaker JO, Mehta A, Vienneau D, Andersen ZJ, Bellander T, Brandt J, Cesaroni G, de Hoogh K, Fecht D, Gulliver J, Hertel O, Hoffmann B, Hvidtfeldt UA, Jöckel KH, Jørgensen JT, Katsouyanni K, Ketzel M, Kristoffersen DT, Lager A, Leander K, Liu S, Ljungman PLS, Nagel G, Pershagen G, Peters A, Raaschou-Nielsen O, Rizzuto D, Schramm S, Schwarze PE, Severi G, Sigsgaard T, Strak M, van der Schouw YT, Verschuren M, Weinmayr G, Wolf K, Zitt E, Samoli E, Forastiere F, Brunekreef B, Hoek G, Janssen NAH (2022) Long-term exposure to low ambient air pollution concentrations and mortality among 28 million people: results from seven large european cohorts within the elapse project. Lancet Planet Health 6(1):e9–e18. https://doi.org/10.1016/S2542-5196(21)00277-1

Steininger M, Kobs K, Zehe A, Lautenschlager F, Becker M, Hotho A (2020) Maplur: Exploring a new paradigm for estimating air pollution using deep learning on map images. ACM Trans Spatial Algorithms Syst 6(3). https://doi.org/10.1145/3380973

Strasser BJ, Baudry J, Mahr D, Sanchez G, Tancoigne E (2019) "citizen science"? Rethinking science and public participation. Sci Technol Stud, 52–76. https://doi.org/10.23987/sts.60425

Tian C, Li G (2019) A framework for the data integration of earthquake events. IEEE Access 7:172628–172637. https://doi.org/10.1109/ACCESS.2019.2957024

Ueberham M, Schlink U (2018) Wearable sensors for multifactorial personal exposure measurements - a ranking study. Environ Int 121(Pt 1):130–138. https://doi.org/10.1016/j.envint.2018.08.057

Ueberham M, Schlink U, Dijst M, Weiland U (2019) Cyclists' multiple environmental urban exposures - comparing subjective and objective measurements. Sustainability 11(5):1412. https://doi.org/10.3390/su11051412

van Dam A, Laidlaw DH, Simpson RM (2002) Experiments in immersive virtual reality for scientific visualization. Comput Graph 26(4):535–555. https://doi.org/10.1016/S0097-8493(02)00113-9

Verplanke J, McCall MK, Uberhuaga C, Rambaldi G, Haklay M (2016) A shared perspective for pgis and vgi. Cartogr J 53(4):308–317. https://doi.org/10.1080/00087041.2016.1227552

Verplanken B, Orbell S (2003) Reflections on past behavior: A self-report index of habit strength 1. J Appl Soc Psychol 33(6):1313–1330. https://doi.org/10.1111/j.1559-1816.2003.tb01951.x

Wang Z, Delp WW, Singer BC (2020) Performance of low-cost indoor air quality monitors for pm2.5 and pm10 from residential sources. Build Environ 171:106654. https://doi.org/10.1016/j.buildenv.2020.106654

Weißmann M, Edler D, Rienow A (2022) Potentials of low-budget microdrones: Processing 3d point clouds and images for representing post-industrial landmarks in immersive virtual environments. Front Rob AI 9:886240. https://doi.org/10.3389/frobt.2022.886240

World Health Organization (2016) Ambient air pollution: A global assessment of exposure and burden of disease 118:1–131. ISBN: 9789241511353. https://www.who.int/publications/i/item/9789241511353

Yang L, Zhang F, Kwan MP, Wang K, Zuo Z, Xia S, Zhang Z, Zhao X (2020) Space-time demand cube for spatial-temporal coverage optimization model of shared bicycle system: A study using big bike gps data. J Transp Geogr 88:102861. https://doi.org/10.1016/j.jtrangeo.2020.102861

Yang Q, Liu G, Gonella F, Chen Y, Liu C, Zhao H, Yang Z (2022) Assessing the temporal-spatial dynamic reduction in ecosystem services caused by air pollution: A near-real-time data perspective. Resour Conserv Recycl 180:106205. https://doi.org/10.1016/j.resconrec.2022.106205

Zhang RX, Zhang LM (2021) Panoramic visual perception and identification of architectural cityscape elements in a virtual-reality environment. Future Gener Comput Syst 118:107–117. https://doi.org/10.1016/j.future.2020.12.022

Chapter 12
Extraction and Visually Driven Analysis of VGI for Understanding People's Behavior in Relation to Multifaceted Context

Dirk Burghardt, Alexander Dunkel, Eva Hauthal, Gota Shirato, Natalia Andrienko, Gennady Andrienko, Maximilian Hartmann, and Ross Purves

Abstract Volunteered Geographic Information in the form of actively and passively generated spatial content offers great potential to study people's activities, emotional perceptions, and mobility behavior. Realizing this potential requires methods which take into account the specific properties of such data, for example, its heterogeneity, subjectivity, and spatial resolution but also temporal relevance and bias.

The aim of the chapter is to show how insights into human behavior can be gained from location-based social media and movement data using visual analysis methods. A conceptual behavioral model is introduced that summarizes people's reactions under the influence of one or more events. In addition, influencing factors are described using a context model, which makes it possible to analyze visitation and mobility patterns with regard to spatial, temporal, and thematic-attribute changes. Selected generic methods are presented, such as extended time curves and the co-bridge metaphor to perform comparative analysis along time axes. Furthermore, it is shown that emojis can be used as contextual indicants to analyze sentiment and emotions in relation to events and locations.

D. Burghardt (✉) · A. Dunkel · E. Hauthal
Institute of Cartography, Faculty of Environmental Sciences, TU Dresden, Dresden, Germany
e-mail: dirk.burghardt@tu-dresden.de; alexander.dunkel@tu-dresden.de; eva.hauthal@tu-dresden.de

G. Shirato · N. Andrienko · G. Andrienko
Fraunhofer Institute IAIS, Sankt Augustin, Germany
e-mail: gota.shirato@iais.fraunhofer.de; natalia.andrienko@iais.fraunhofer.de; gennady.andrienko@iais.fraunhofer.de

M. Hartmann · R. Purves
University Zurich, Zurich, Switzerland
e-mail: maximilianchristoph.hartmann@geo.uzh.ch; ross.purves@geo.uzh.ch

© The Author(s) 2024
D. Burghardt et al. (eds.), *Volunteered Geographic Information*,
https://doi.org/10.1007/978-3-031-35374-1_12

Application-oriented workflows are presented for activity analysis in the field of urban and landscape planning. It is shown how location-based social media can be used to obtain information about landscape objects that are collectively perceived as valuable and worth preserving. The mobility behavior of people is analyzed using the example of multivariate time series from football data. Therefore, topic modeling and pattern analyzes were utilized to identify average positions and area of movements of the football teams.

Keywords Location-based social media · Football analytics · Visual analytics · Reactions · Behavior · Context · Emoji · Bias

12.1 Introduction

The rise of so-called Volunteered Geographic Information (VGI) has brought with it fundamental changes not only in the nature of geographic data but also its production and accessibility. These changes have implications across the board for those carrying out research requiring geographic data and most profoundly for research exploring how humans interact with and are affected by changes in their environment. The creation of user-generated spatial content is diverse, whether through the usage of a wide variety of sensors during activities in the real world or the use of social media platforms for information exchange in the digital world. Realizing the potential of VGI requires methods which take account of the specific properties of such data, for example, its heterogeneity, quality, subjectivity, spatial resolution, and temporal relevance. Of particular interest in this research are two types of information—first, information about people's reactions and behavior and, second, modeling different types of contexts, both derived from location-based social media (LBSM) and football match data.

Location-based social media respective geosocial media are extensively used for expressing and exchanging thoughts, opinions, ideas, and feelings (publicly or within a particular group of people)—thus reactions to an event or related to a theme. A definition of reactions is given by Dunkel et al. (2019) consisting of an identifier to a referent event and four facets (spatial, temporal, thematic, and social) describing the reaction. A series of reactions from an actor (where an actor might be an individual, a group, or an organization) can be seen as a manifestation of this actor's behavior (Luckmann 2013). Behaviors convey actions and provide information about doing and refraining. Activities in geosocial media are intentional, purposive, and subjectively meaningful, and they are usually targeted at others, i.e., they are social actions. The creation and production of such information is an expression of human behavior and as such influenced strongly by events and context. Key to any framework seeking to analyze behavior is a definition of dimensions through which space, time, thematic attributes, and events can be described. Moreover, modelling behavior and context requires methods to aggregate and generalize data across spatial and temporal scales.

The chapter is structured as follows. Section 12.2 describes the state-of-the art research related to modeling of reactions and behavior, context modelling, use of emojis, and consideration of privacy. Section 12.3 introduces conceptual models for describing behavior and context. Section 12.4 presents generic methods for accounting for representativeness and bias in LBSM, for comparative visual analysis, and using emojis as contextual indicators. Section 12.4 presents application-related workflows for activity analyzes in the area of landscape and urban planning, as well as the exploration of people's mobility behavior in football games.

12.2 Related Work

Geosocial media is a special case of Volunteered Geographic Information (VGI) (Goodchild 2007), which enables users to communicate and connect with each other by sharing location-based information. Geovisual analysis of this information enables the study of events, geographic phenomena, and related human reactions for a variety of application cases, e.g., in the areas of urban and landscape planning, tourism, health, transport, or disaster management. Human responses manifest themselves within and outside of social network communication in a variety of ways as emotional expressions, opinions, and thoughts expressed or actions taken (Hauthal et al. 2019). Thus, Volunteered Geographic Information has added a new dimension to traditional geospatial data acquisition for human activity research (Li et al. 2020). Activities in Geosocial Media are intentional, purposive, and subjectively meaningful, and they are usually targeted at others, i.e., they are social actions. From a technical point of view, they can be subdivided accordingly Davis (2016) into origination (creating own, original content, e.g., tweeting, posting), acknowledgment (reactions to content, e.g., liking, favorite), associating (interaction with content, e.g., replying, commenting, mentioning, following), amplification (spreading content, e.g., retweeting, sharing), and action (moving beyond content, e.g., signing up for a newsletter, buying something, or going to a demonstration).

The research presented within this paper aims at analyzing reactions as a component of behavior, incorporating external context to better allow events and activities to be related and compared. Key to any framework seeking to analyze reactions to events is a definition of dimensions through which both reactions and events can be described. As pointed out by Teitler et al. (2008), these dimensions form the core of a description of an event and include not only ways of describing (what, who, where, when) but also explaining (how and why). Answering these questions can be seen as a way of characterizing the context of an event and associated behaviors (Dunkel et al. 2019).

As the project has been focusing on the concept of individual and collective behaviors (particularly, behaviors of social media users), it was appropriate to review the existing literature on the topic of spatial behavior analysis. The most intensive studies have been conducted in the area of sports analytics, where the researchers focus on the behaviors of players and teams in sport games. Duarte et al. (2012) propose to consider sport teams as superorganisms. "Superorganism" is defined in sociobiology as a group of individuals self-organized by division of labor and united by a system of communication. The division of labor can be characterized by the "areas of responsibility" of the players on the pitch. The communications can be represented by graphs of interactions, e.g., passing networks. Fonseca et al. (2013) use Voronoi diagrams to study the spatial interaction behavior in terms of continuously changing players' arrangement on the pitch. Gudmundsson and Horton (2017) published a survey of approaches that use spatiotemporal data from team sports. This includes approaches from social network analysis, which are applied to passing networks and transition networks. Specifically for football (soccer), Memmert et al. (2017) present approaches to analysis of collective organization (distribution on the pitch, maintenance of distances between players); inter-player coordination; inter-team and inter-line coordination; correspondences between team formations as an aspect of inter-team interaction; formations of tactical groups, e.g., offense and defense; etc. Many of the existing approaches are specific for sports games and can hardly be applied to other types of data. In our project, we strive to develop more general concepts and techniques, and we also strive to consider behavior dynamics rather than derive summary characteristics of particular individual or group behaviors.

Emojis play a special role in the analysis of reactions, especially feelings and opinions. Their use in geosocial media is just as popular as the use of hashtags (Bai et al. 2019; Highfield and Leaver 2016), and as a language-independent characters, they have the advantage of avoiding error-prone language processing (Hu et al. 2013; Kralj Novak et al. 2015). The use of emojis is influenced by regional and cultural context (Kejriwal et al. 2021), so emojis have great potential to characterize spatial context but also analyze activities related to places and events (Hauthal et al. 2021).

Volunteered Geographic Information is based on people's willingness to collect and share content with the public community. It should be noted that this data is always related to individuals and could therefore be sensitive in terms of privacy and ethical issues. Olteanu et al. (2019) demands that an ethical approach must respect individual autonomy. In the case of purposeful active participation in VGI projects, this can be ensured by confirming active consent (by a declaration of informed consent). This is more critical in terms of analyzing millions of social media posts, even when users post the content publicly and agree to the terms of use for third parties to use it. Williams et al. (2017) therefore calls for user expectations to be taken into account, to perceive changed contexts, for example, in the form of "context collapse" (Crawford and Finn 2015), and to be respectful when combining potentially sensitive personal data. Through an "aggregation effect" (Solove 2013), privacy-relevant insights can be gained from different data sources without the contributing user being aware of it. An ethical research approach requires a balance between social and individual interests—the boundary of privacy is not rigid but depends on the topic, place, time, and user characteristics. Cartographers and geoscientists have a responsibility here to develop flexible methods that protect user privacy while taking these contextual factors into account (Burghardt and Dunkel 2022).

12.3 Theoretical and Conceptual Foundations

In the first phase of the project, we developed a conceptual model underpinning the extraction, analysis, and visualization of *events* and *reactions* to events in location-based social media (Dunkel et al. 2019). For instance, a message or post published on a social media platform could be considered as a reaction, related to either simple events (e.g., a tweet that is observable or a single rumble of thunder) or preceding and ongoing chains of events (e.g., the Brexit). The key feature of our conceptual model and its implementation is the integration of spatial, temporal, thematic, and social dimensions combined with an explicit link between events and reactions.

The second phase of the project focused on the concept of individual and collective *behavior*. Among multiple existing definitions of the concept of behavior (Henriques and Michalski 2020), a simple definition suitable for our purposes can be "an organism's activities in response to external or internal stimuli" (American Psychological Association n.d.). From our perspective, it is important that a behavior unfolds over time, i.e., it covers a certain time interval, unlike a singular reaction, which can be viewed as an instant event. Like a reaction, a behavior belongs to a certain *actor*, who may be an individual, a group, an organization, or even the population of a country. However, a behavior may include multiple reactions to the same or different events. A behavior may also include *movements* of the actor in space and changes of various characteristics, which can be expressed by thematic attributes. Hence, like the concept of reaction, the concept of behavior involves social (who?), temporal (when?), spatial (where?), and thematic (how?) facets. However, the temporal facet extends to a time interval, and the spatial and thematic aspects can no longer be expressed by a singular location in space and a combination of values of thematic attributes but need to be represented by a time-referenced sequence of locations (i.e., a trajectory in space) and time series of values of the thematic attributes.

Besides, as already mentioned, a behavior may include multiple reactions to one or more events, and the reactions themselves are conceptualized as events. Even more generally, a behavior may also include not only reactions but also other kinds of events in which the actor participates, for example, actions or decisions.

Hence, we conceptualize a behavior as a tuple $B = (p, T = [t_1, t_2], S(t), A(t), E^T = \{e_1, \ldots, e_N\})$, where p is an actor, T is a time interval, $S(t)$ is a function representing changes of the spatial position of p over time, $A(t)$ is a function representing changes of the thematic attributes of p over time, and E is a set of events involving p, including but not limited to p's reactions to other events. In $S(t)$ and $A(t)$, the variable t takes values from T. The times of all events in E^T are contained in T or at least overlap with T.

Since a behavior is a complex dynamic (i.e., changing over time) entity, study and understanding of behaviors require abstraction by which elementary locations, characteristics, and events that took place at different time moments are united and treated together as units, which are called *patterns*. Collins et al. (2018) proposed the following definition of a pattern: "a representation of a collection of items of any kind as an integrated whole with specific properties that are not mere compositions of properties of the constituent items." A more formal definition of a data pattern has been later proposed by Andrienko et al. (2021b). In brief, a pattern is a combination of relationships between data items. For time-related data, the pattern-forming relationships include temporal ordering, temporal distances, and relationships between temporally arranged spatial locations (direction and distance), attribute values (equality or difference, sign and amount of difference), and sets of events (equality, inclusion, overlap, disjointedness). Such relationships can also exist between patterns, which enable joining two or more patterns into more complex patterns. For example, a pattern of increase of values of a numeric variable followed by a pattern of decrease can be joined in a single pattern of a peak in the numeric value variation.

A behavior can thus be represented as a combination of patterns reflecting changes of the spatial location (i.e., movement patterns, such as quick movement straight to the north), changes of thematic attributes (e.g., increase in frequency and duration of sport activities), and/or changes in the set of relevant events (e.g., end of exams and beginning of a holiday).

A behavior unfolding during a time interval of substantial length may need to be described in terms of complex patterns composed of temporally arranged simpler patterns, i.e., sequences of simpler patterns. Such complex structures are hard to analyze and compare. Therefore, it is practically useful to divide behaviors into parts, which may be called *episodes*. An episode is an excerpt from a behavior B having the same structure as B, i.e., $EP = (p, T' = [t'_1, t'_2], S(t), A(t), E^{T'} = \{e_1, \ldots, e_N\})$, but $T' \subseteq T$, the variable t in $S(t)$ and $A(t)$ takes values from T', and $ET' \subseteq ET$ consists of those events that existed during the interval T'. The idea behind dividing a behavior into episodes or extracting episodes from a behavior is that an episode is relatively short and can be described in terms of simpler patterns. Then, analysis of a single behavior consists of comparison of the constituent episodes, i.e., revealing similarities and differences between the patterns in the episodes, which can be followed by detecting re-occurrences of similar patterns and investigating temporal relationships between the re-occurrences. Comparative analysis of two or more different behaviors involves comparisons of their episodes, which may include, in particular, comparison of co-occurring episodes from different behaviors as well as detection of similar asynchronous episodes in different behaviors.

Any behavior and, consequently, any episode of a behavior occurs in a certain *context*, i.e., a combination of circumstances that change over time. A context can be conceptually modelled as a tuple $C = (T, S, A^C(t, s), A^C(t), E^{CT}, B^{CT})$, where T is a time interval, S is a part of space, $A^C(t, s)$ represents various dynamic attributes whose values are associated with different spatial locations (e.g., weather, fuel prices), $A^C(t)$ are dynamic attributes considered as global (i.e., their values refer to S as a whole), E^{CT} is a set of events that occurred during the interval T (e.g., political events or epidemic outbreaks), and B^{CT} is a set of behaviors of various actors other than the actor(s) whose behaviors are analyzed. Context-aware analysis of behaviors and their constituent episodes includes determining relationships between patterns in the behaviors/episodes in focus and patterns occurring in the context attributes, context events, and context behaviors. The analysis can be directed from the context to the behaviors (e.g., what behavior patterns occurred when the air temperature was high or after pandemic lockdowns) or from the behaviors to the context (e.g., what events or attribute development patterns preceded the occurrence of a given pattern in one or more behaviors).

Let us illustrate the concepts using an example of a football game. Each player of a team is an actor having his/her behavior happening in the course of the game, i.e., T is the time of the game. The player's behavior includes movements of the player $S(t)$, his/her physical condition $A(t)$, and game events E involving the player, such as passes, attacks, shots on goal, goals, etc. Besides the individual behaviors of the players, there is a collective behavior of a whole team, including team movements, changes of team width, depth and relative arrangements of the players, and game events involving one or more players. The context of the team's behavior includes the behavior of the opponent team, the game events that have already happened, the weather and lighting conditions, the team's ranks, the goals set by the coach, etc. The context of the individual players' behaviors includes, additionally to what was listed above, the individual behaviors and skills of their teammates and of the opponents.

Since behaviors and their contexts are inherently complex (dynamic and multifaceted), analysis of behaviors and their relationships to contexts is a very challenging problem that cannot be tackled without simplification. Possible operations for achieving include *decomposition* (e.g., behaviors are divided into episodes), *abstraction* (e.g., unification of multiple items into patterns), *selection* (focusing on particular facets and their components, particular combinations of contextual circumstances, or particular types of patterns), and *aggregation* (representation of sets of items by summaries).

The concepts and ideas presented above are illustrated in Figs. 12.1, 12.2, 12.3, and 12.4 by example of football data. These data are more suitable for illustration purposes than data from location-based social media due to their high quality and absence of legal and ethical issues.

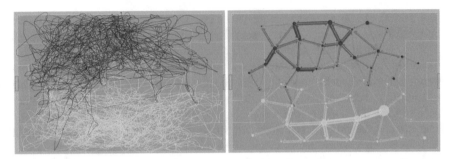

Fig. 12.1 An illustration of behavior simplification by means of abstraction and aggregation. The image on the left shows the trajectories of two players during a football game. To increase the level of abstraction, the points of the trajectories were grouped and abstracted to areas on the pitch. the segments of the trajectories were transformed into transitions between the areas which, in turn, were aggregated into flows between the areas. The resulting spatial network is a simplified representation of the spatial facet of the behavior

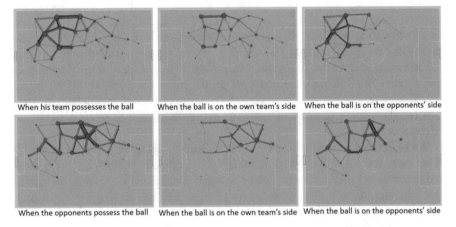

When his team possesses the ball   When the ball is on the own team's side   When the ball is on the opponents' side

When the opponents possess the ball   When the ball is on the own team's side   When the ball is on the opponents' side

Fig. 12.2 An illustration of the decomposition of a player's behavior during a football game into several behaviors occurring in different contexts. The behaviors are represented in the form of spatial networks resulting from abstraction and aggregation of original data, as shown in Fig. 12.1

12.4   Generic Methods That Support Studies of Reactions and Behaviors

12.4.1   *Representativity and Bias in Location-Based Social Media*

In the second part of this chapter, we discuss a range of analyzes based on location-based social media. Before we describe these analyzes in more detail, it is important to consider issues related to representativity and bias with respect to such data sources. It is important to firstly remember that all analyzes are subject to these

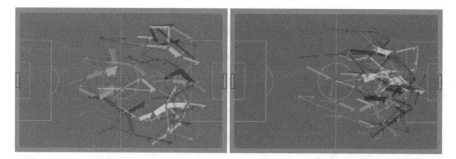

Fig. 12.3 Abstracted and aggregated representation of collective behavioral patterns of two teams in a selected subset of episodes. The left image corresponds to a team in possession of the ball and the right image to a defending team

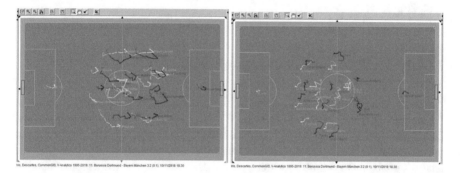

Fig. 12.4 Abstracted movement behaviors of players of two teams in two sets of episodes: first 12 seconds after gaining the ball by the yellow team (left) and after gaining the ball by the red team (right). The movements are abstracted by taking the average players' positions from all episodes at the time moments $t + 0, t + 1, \ldots, t + 12$ seconds, where t is the time moment of the ball possession change

issues—for example, if we generalize patterns found in the football data described above, it is important to consider which teams are more likely to be monitored and generate such data and to what extent these represent footballers more generally.

However, in location-based social media, these issues are more obvious and immediate. For example, these data are exclusively produced by people with location-enabled devices. These are not equally available within and across society, with, for example, the very old and very young being less likely to be captured by such data, and considerable variations exist in the willingness of individuals to share location-based information in different cultures and countries (Li et al. 2013; Krasnova and Veltri 2011). The first question we must ask therefore with respect to representativeness is who can, and is willing to, contribute data to platforms such as Twitter, Flickr, Instagram, and Google's location-based services. Some uses of space—for example, for play by small children—are likely to be underrepresented, especially if children are encouraged to spend time outside without their parents.

The second question that we can ask with respect to representativity relates to the ways in which these data are linked to space. Typically, we use other geographic datasets, for example, produced by national mapping agencies and/or volunteers (e.g., OpenStreetMap and GeoNames), and the content of these datasets will profoundly influence any analysis. For example, OpenStreetMap is known to have biases related to gender in terms of the categories of objects mapped (Gardner et al. 2020), while place-name density in GeoNames reflects geopolitical events (Acheson et al. 2017). A third question that we can ask with respect to these issues relates to culture and language. The underlying models used to capture, for example, emotions are often based on Western (and more specifically Anglo-Saxon) notions and assume universal emotions shared across cultures. Furthermore, many methods to analyze text focus on English as a starting point, despite clear evidence that this has limitations with respect to the ways in which we understand the world (Blasi et al. 2022).

12.4.2   *Methods for Comparative Analyses*

As stated earlier (Sect. 12.3), behavior is a complex, dynamic entity that can be studied and understood only with the help of simplifying abstractions and decomposition. Decomposition includes interchangeable selection of aspects and facets to put in focus and division of behaviors into episodes. Depending on the analysis goals, there can be two strategies for decomposing behaviors into episodes: partitioning of the entire behavior by dividing its time span into intervals, e.g., of a chosen fixed length and extraction of episodes with particular properties. The latter strategy is achieved by means of temporal queries, which select multiple disjoint time intervals such that the query conditions fulfil during these intervals. The pieces of behaviors contained within the selected time intervals are extracted as episodes for analysis. A set of primitives for making temporal queries has been proposed by Andrienko et al. (2021a). The query primitives enable selection of sets of time intervals containing situations with specified characteristics and, moreover, further selection of sets of intervals having certain temporal relationships to the previously selected intervals. This can be used, in particular, for considering selected episodes stepwise or for studying what happened before or after them.

Comparative visual analysis of selected aspects of behaviors can be enabled by juxtaposed representation of two or more behaviors along a time axis. This approach was taken by Chen et al. (2021) for comparison of streams of text messages published on social media by different politicians. The authors created an imaginative visual design where the flow of time is represented as a river and bridges across the river are built from significant keywords extracted from the texts. This technique has been called "co-bridges." The visualization supports both qualitative (common and distinct keywords) and quantitative (stream volume, keyword frequencies) comparisons. Moreover, it is possible to compare two or

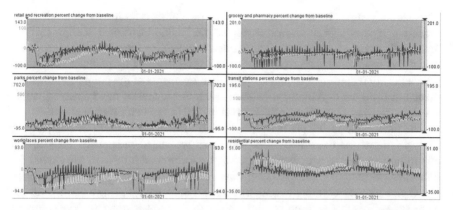

Fig. 12.5 Comparison of population mobility behaviors in three countries (Germany in blue, Italy in red, and the UK in yellow) using multiple time graphs showing dynamics of different attributes

more co-bridges by juxtaposing them. These may be, for example, co-bridges corresponding to different themes of the politicians' agendas.

While alignment along a common time axis is essential for comparing behaviors as dynamic entities, the representation of behaviors depends on the type of data under analysis. One of our recent work directions was comparative analysis of behavior characteristics expressed by multiple time-variant numeric attributes, i.e., by multivariate time series. An approach is being developed using, among others, the publicly available data of COVID-19 Google Mobility Trends.[1]

The company Google summarizes anonymized data provided by apps such as Google Maps into statistics showing how collective mobility behaviors of people in different countries were changing throughout the COVID-19 pandemic. The data consist of daily counts of visitors to specific categories of places (e.g., grocery stores, parks, train stations, etc.) relative to baseline days before the pandemic outbreak. Hence, the population mobility behaviors in different countries are expressed by time series of six attributes.

In comparing the behaviors in different countries, it is insufficient to compare the dynamics of each attribute separately from others, although this can be done relatively easily by means of usual time graphs or line charts, as shown in Fig. 12.5. This visualization does not support holistic perception of patterns of joint development of the attributes.

Consideration of temporal evolution of complex data, in particular, combinations of values of multiple attributes, is supported by the visualization technique called time curve (Bach et al. 2016); a similar method was simultaneously proposed by van den Elzen et al. (2016). The approach relies on embedding of data corresponding to different time units in a low-dimensional (typically 2D) space based on similarities between the data in terms of an appropriate similarity metric.

---

[1] https://ourworldindata.org/covid-google-mobility-trends.

The class of computational methods for data embedding is known as dimensionality reduction methods. Well-known examples are multidimensional scaling (MDS), Sammon's mapping, t-SNE, etc. The result of embedding (also called projection) is visualized as a scatterplot where time units are represented by points. Consecutive time units are represented by lines. Spatial arrangements of the points and lines are treated as different development patterns, such as gradual or rapid changes, stable states, oscillation, stagnation, etc. (Bach et al. 2016; van den Elzen et al. 2016).

The time curve technique is by itself poorly suited for the task of comparative analysis of two or more behaviors which, as we stated previously, calls for an opportunity to see the behaviors aligned along a common time axis. We extend the time curve technique in the following way. We paint the background of a projection plot (i.e., a plot representing a result of behavior embedding) using a continuous 2D color scale. Thereby, each point on the plot receives a specific color. The colors of the points can be transmitted to other displays, in particular, to a timeline display suitable for aligned representation of several behaviors. This generic approach is demonstrated in Fig. 12.6 using the COVID-19 mobility data for Germany, Italy, and the UK shown in Fig. 12.5.

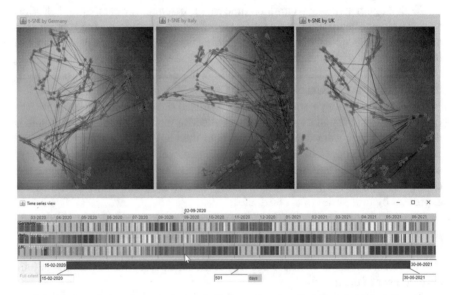

Fig. 12.6 Upper part: the population mobility behaviors in Germany, Italy, and the UK are represented by means of data embedding (projection) according to the time curve technique. To enable aligned representation of the behaviors along a common time axis, continuous coloring is applied to the backgrounds of the projection plots. Lower part: a joint representation of the three behaviors in a timeline display, where the horizontal dimension represents the time flow. The behaviors are represented by horizontal bars with segments painted in the colors of the corresponding points in the projection plots. In this example, the Saturdays and Sundays are filtered out (shown in the timeline view as gray segments), which enables disregarding the weekly fluctuations and focusing on long-term patterns of change

The character of the color variation in the timeline view is representative of the character of the joint development of the multiple attributes. Abrupt changes of the color tone along the timeline signify sudden or rapid changes in a behavior, bar pieces painted in very similar color shades correspond to stable states in the behavior development, and smooth changes of shades correspond to gradual transitions between states.

12.4.3 Emojis to Study Sentiment, Emotion, and Context of Events

Since geosocial media are used to state opinions, express emotions, or document experiences, they contain a lot of subjective information. The recognition of such subjective phenomena is usually done via natural language processing, which is by now quite sophisticated but can hardly recognize irony or sarcasm, for example, and is often applied limited to one or a few languages. Promising solutions have been achieved in this context with emojis, which have become extremely popular in geosocial media and are available in steadily growing numbers.

Hauthal et al. (2019) used a Twitter dataset to investigate reactions to the political event Brexit in terms of opinions and emotions, using emojis in two different approaches. In the first approach, emojis and hashtags were combined. Hashtags, established in political campaigns before the referendum, indicate which sub-topic of the overall Brexit debate is addressed in a tweet, i.e., leave or remain. A sentiment toward these topics in terms of positive or negative was detected by emojis appearing in the same tweet. For this, emojis showing a positive or negative facial expression were considered, based on the official categorization of emojis by Unicode. The combination of a hashtag and an emoji results in the rejection or support of the UK leaving the European Union. A spatial comparison of these analysis results with the actual referendum results on NUTS1 level (the highest level in the hierarchical classification used to clearly identify and classify the spatial reference units of official statistics in the Member States of the European Union) showed a higher consistency than a pure hashtag-based consideration without including emojis.

In the second approach, emojis showing faces or persons with a countenance or gesture were not only considered on a positive-negative scale but were assigned to emotional categories based on a classification according to Shaver et al. (1987), which includes love, joy, surprise, fear, sadness, and anger (see Fig. 12.7). Each category of this classification is allocated emotional terms that are most likely to be mentioned when people are asked to name those emotions. The official Unicode names of all previously used emojis were matched with these terms and assigned to the corresponding emotional categories. Emotions were then examined comparatively before and after the announcement of the Brexit referendum results, with only sadness showing a significant increase overall and fear decreasing slightly. Spatially, the increase in sadness is evident in two out of three NUTS1 regions,

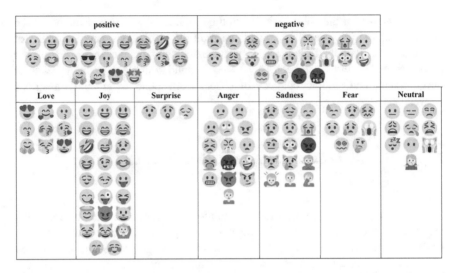

Fig. 12.7 Classification of Unicode Emoji (List v11.0) by authors into positive, negative, and according to emotional categories based on Shaver et al. (1987)

where votes did not match the overall referendum result, and an increase in joy in five out of nine NUTS1 regions where the hopes were fulfilled.

Another use of emojis to investigate subjectivity, in this case perception, was implemented in a study by Hauthal et al. (2021). In a global Instagram dataset about sunrise and sunset, the measure typicality was applied. Typicality is a relative measure specifically tailored for geosocial media that determines how typical a particular object of interest (e.g., emoji or hashtag) is within a sub-dataset compared to the total dataset. Sub-datasets may be formed spatially, temporally, thematically, etc. Typicality is calculated by the normalized difference of two relative frequencies and returns a positive (= typical) or negative (= atypical) value. Typicality was used to identify emojis in the previously mentioned global Instagram dataset that provide information about the context of the user while observing the event. On the one hand, these emojis deliver information about activities performed and, on the other hand, also about perceived landscape features in the immediate surroundings. It was found that emojis provide more detailed information in this regard than the hashtags contained in the same dataset. Moreover, location-specific emojis were identified, which are chosen depending on the location, and match the features of the physical environment, as shown by matching them with geographic attributes. This proves that emojis are not randomly chosen but provide insights not only into the user's situational context but also into their perception and thus appreciation of certain aspects of the environment.

These studies demonstrate the potential of emojis to provide insights into the subjectivity of geosocial media users in a relatively straightforward way. A further increase in the use of emojis as well as an increasing variety of them can still be expected, which will open up further possibilities and applications.

The conceptual framework presented in Sect. 12.3 strongly influenced both of the presented studies. In the study on Brexit, the framework led the way in allowing us to look at the reactions contained in the underlying dataset from numerous different perspectives and thus obtain in-depth results. Furthermore, the framework significantly influenced the development of the typicality measure in the second study described, as the sub-datasets required for the calculation can be formed following the four facets, which is the particularity and novelty of this measure.

12.5 Application-Oriented Workflows

12.5.1 Activity Analysis for Landscape and Urban Planning

VGI is also increasingly recognized as an important resource in the fields of landscape and urban planning, for example, to support the analysis of visitation patterns, assessing collective values, or improving human well-being through fair and equitable design of public green spaces (Ghermandi and Sinclair 2019). To this end, landscape and urban planners must first assess "what" is collectively valued, "where," by "whom," and "when" to understand the how and why of human behavior, as introduced in Sect. 12.3. However, the reproducibility of human behavior research is often impaired because samples, populations, and the phenomena being observed change between studies (Gruebner et al. 2017). This is particularly true for VGI and geosocial media, which are noisy, biased, difficult to fully sample, and often shared through incompletely documented and opaque application programming interfaces (APIs). In addition to these core challenges, protecting user privacy is becoming increasingly important when working with user-generated content (danah boyd and Crawford 2012). For this reason, Dunkel et al. (2023a) sought to develop a robust and transferable "workflow template," for assessing human activities and subjective landscape values through geosocial media worldwide—without compromising user privacy.

For demonstration purposes, an event type with a strong temporal and spatial consistency was chosen that allowed for a significant reduction in the number of "incidental variables" in the study while at the same time maintaining sample volume. Sunset and sunrise were among the few events that met these criteria. In addition, improving results reproducibility ideally requires an experiment with two maximally separated datasets and "finding relationships in the same direction and of similar strength" (Laraway et al. 2019, p. 38) in both. This was difficult to implement with more "newsworthy" topics. The consistent global and long-term footprint of the sunset and sunrise offered an opportunity to maximize the sample size while also providing a basis for reproducing the results using two datasets, albeit not universally representative but independent, collected from Instagram and Flickr. Despite the narrow topic of this study, the shared expected frequencies (e.g., Flickr 300 million post counts dataset for a 100×100 km grid, Dunkel et al. 2023b) can also be very useful for calculating chi in studies of other phenomena at global scales, for differently sampled data, e.g., based on a different set of search terms.

Importantly, sunset and sunrise events are entirely ephemeral yet have a profound, measurable impact on human perception and interaction with the environment. Unlike many other events, the ability to perceive sunset and sunrise is tightly bound to time, but almost completely decoupled from space. Photographs of these events function as evidence of presence in place and time. While the immediate reaction to taking a photograph of a sunset or sunrise is trivial, people will take into account all previous experiences, learned behaviors, expectations, etc. when reacting. Individual photographs therefore reflect different memorable experiences that function as artifacts of different preference contexts. The narrow thematic filter of sunset and sunrise allowed for a focused description and evaluation of these preference contexts. Based on the four-facet context model (Dunkel et al. 2019), reactions to sunset and sunrise were examined in terms of where, who, what, and when.

The study first asked whether it was possible to compare the relative importance of sunset and sunrise reactions, independent of overall visitation frequency, for different locations worldwide. Visualizing relative user frequency was critical because geosocial media tends to be skewed toward highly populated locations and cities. The goal was to provide a balanced evaluation assessment of sunset and sunrise reactions across different rural and urban regions. Several visualization methods were tested, such as based on a relative ranking method for individual locations (Fig. 12.8) and different metrics, such as user counts, post counts, or user days (Wood et al. 2013). The final workflow uses the signed chi equation, proposed by Clarke et al. (2007), to visualize over- and under-frequentation with respect to these two events and aggregated using HyperLogLog for a global grid with a resolution of 100×100 km.

Globally, sunrise events are often associated with east coasts (e.g., Italy, Sardinia in Fig. 12.8) or mountainous regions (e.g., the Alps), while sunset events are photographed on west coasts. The study also observed a strong ranking order between Flickr and Instagram reactions, despite the fact that both platforms have different user groups. In other words, we actually expected a much stronger effect of the platform on the results, and our work shows, at least for Instagram and Flickr and the selected events, that results can be reliable and reproducible across platforms. Still, for some locations, the incentives of the social media platforms themselves can have a significant impact on what gets shared and by whom. On Instagram, for instance, the Burning Man festival in Nevada ranks second worldwide for sunrise reactions. Out of a total of about 70,000 total visitors (Wikipedia, Burning Man,[2] 2022), 1295 ($\pm$30) users shared sunrise images on Instagram during the short period of the 2017 Burning Man festival, compared to only 54 ($\pm$2) Flickr users for the same location over a 10-year period—a pattern that can be explained by the different user groups of these platforms. Finally, the use of abstracted, estimated,

---

[2] https://en.wikipedia.org/wiki/Burning_Man#2013_to_2019.

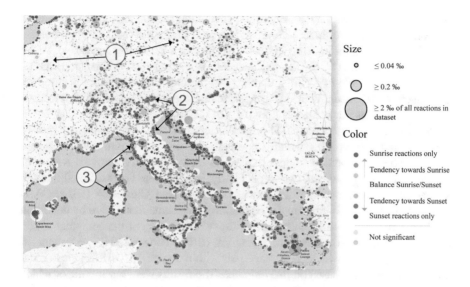

Fig. 12.8 Relative ranking of places based on reactions to the sunset and sunrise in Central Europe. This figure shows a preview visualization of the collected raw data that was later further aggregated to a 100 × 100 km grid in Dunkel et al. (2023a). Some repeated patterns are already observable here, such as the larger cities (1) commonly featuring a balanced representation of sunset and sunrise reactions, whereas sunrise reactions dominate along eastern-facing coasts and in mountainous regions (2) and sunset reactions being predominantly shared from along western-facing coasts (3)

non-personal data based on HyperLogLog, as suggested by a scoping study (Dunkel et al. 2020), was a practically feasible solution that supports a shift toward privacy-preserving and ethically aware data analysis in human preference research. The analysis process and anonymized data are made available in a repository, allowing transparent verification, replication, and transfer to other events or datasets (data repository; Dunkel et al. 2023b).

Even though individual people perceive landscapes and their attributed values differently, there are landscapes which the majority of people perceive as scenic and beautiful (Bell 2012). These prolific landscapes (e.g., Preikestolen in Norway or Wildkirchli in Switzerland) are often depicted by characteristic motif images, which are clusters of images all taken from a similar viewpoint and angle. Which landscapes become popular is driven by propagation of landscape or nature appreciation through travel guides or art from the romantic area, popularizing a selective subset of landscapes, thus not a new phenomenon. Today, tourism agencies and other influencers (e.g., celebrities, companies, movies, songs) can shape landscapes through social media promotion by planting seed images that people will try to recreate and, by doing so, form new motifs. By reaching millions of people and potentially influencing their future visiting plans, this social media-induced tourism can have drastic physical consequences on the local environment, infrastructure, and people (Simmonds et al. 2018). Being able to monitor the spatiotemporal emergence of motifs as a proxy for induced tourism is crucial to support local decision-makers to tackle the potential increase in visitation rates to a respective landscape. In the paper by Hartmann et al. (2022), we created an operationalizable conceptual model of motifs that is able to identify, extract, and monitor prone landscapes based on geotagged social media data. More specifically, the proposed pipeline leverages creative-commons Flickr images from the YFCC100M dataset (Thomee et al. 2016) within the European Nature 2000-protected areas, which represent a network of breeding and resting sites within important landscapes for rare and threatened species. The core methodological process to identify motifs within a corpus of 2.1 million images involved two steps. Firstly, images were downsampled through spatial clustering by using Hierarchical Density-Based Clustering (HDBSCAN) (McInnes et al. 2017) since images belonging to the same motif were by definition in close proximity to one another. Secondly, with the help of the computer vision algorithm Scale-Invariant Feature Transform (SIFT) (Lowe 2004), we calculated image similarities between each image pair within a spatial HDBSAN cluster and clustered them again based on that outcome. The results were our motifs, of which we found a total of 119 in our study sites across Europe. Analysis of the motifs revealed that 65% depict cultural elements such as castles and bridges, whereas the remaining 35% contain natural features that were biased toward coastal elements like cliffs. Ultimately, the early detection of emerging motifs and their monitoring allows the identification of locations subject to increased pressure, which enables managers to explore why sites are being visited and to take timely and appropriate actions (e.g., allocation of infrastructure such as toilets and rubbish disposals or visitor routing).

Not only descriptive textual information and emojis can be used for the analysis of geosocial media data, but it is also possible to use the image information directly. As an application for urban bicycle infrastructure planning, an object recognition algorithm based on convolutional neural networks was used to identify bicycles and potential parking spaces. The research and development work was carried out as a cooperation of a Young Research Group within the framework of the priority program VGIscience (Knura et al. 2021; Zahtila and Knura 2022). The research on object recognition was carried out in the COVMAP project (see Chap. 5); the processing of social media data and the development of methods for visual analysis were realized by the projects EVA-VGI (this chapter) and TOVIP (see Chap. 10).

12.5.2   Exploring People's Mobility Behavior

In search for a workable approach to analysis of multiple behavior episodes characterized by multivariate time series, Shirato et al. (2021) made an attempt to apply topic modelling. For this purpose, the patterns of variation of different attributes, or features, are represented by symbolic codes, which can be treated as words. The expected role of topic modelling is to reveal co-occurrences of such

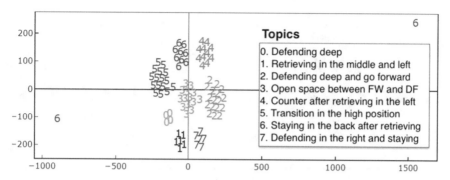

Fig. 12.9 Episodes from a football match are represented by dots in a 2D projection space obtained by applying t-SNE to the vectors of the topic weights. The dot colors represent the dominant topics

patterns. To transform a time series of numeric values into a symbolic code, the values are compared to the median of the time series and encoded by symbols '−', '=', or '+' depending on their relative position with respect to the median. Hence, each episode can be represented by a combination of the variation codes of the multiple variables. A method for probabilistic topic modelling, such as Latent Dirichlet Allocation (LDA), is applied to these combinations, which are treated as "texts" where the variation codes are "words." The resulting "topics" show which variation patterns of different variables tend to occur together in the same episodes (see Fig. 12.9). The topic modelling method also assigns vectors of topic probabilities to the episodes. Using these vectors, the episodes can be clustered and/or arranged in a projection space according to similarities of the variation patterns and further explored by means of various existing methods.

The approach was tested using football data as an example. The features that were involved in the analysis reflected widths of empty spaces between team players on different levels of separation from the goal that is under attack. The topics corresponded to combinations of dynamic patterns of the changes of the widths on the different layers. To support interpretation of the results of topic modelling, the representation of the topics in the form of a table was combined with a map of the football pitch, where the behavior patterns were summarized as the average positions and areas of movement of the team centers (see Fig. 12.10).

The experiment showed that application of topic modelling to episodes characterized by multivariate (and, possibly, multifaceted) data has potential for behavior analysis. The research in this direction is worth being continued.

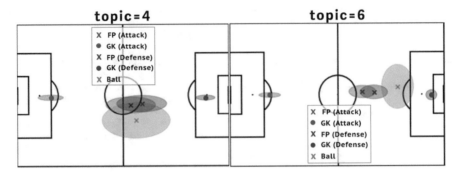

Fig. 12.10 Distributions of centroids after transitions for topic 4 and topic 6. Colors represent the objects (blue, defense; red, attack; green, ball). The cross (x) means the average position of centroids, and the ellipse means the standard deviation of centroids for each topic

12.6 Conclusion

The research approaches presented in the chapter were designed for the visual analysis of social media posts and trajectories of football players to study people's reactions and behavior. As a starting point, a conceptual behavioral model is introduced to describe an actor in a certain period of time with regard to its spatial, thematic-attributive changes under the influence of events and external context factors. For the analysis, behavior is broken down into episodes, which refer to short periods of time and support the description with simple patterns. In order to model the external influences on behavior, a context model is also proposed, which, in addition to spatial, temporal, and attributive influences, also takes into account events that have taken place over a period of time and the behavior of other people.

Based on the conceptual model, generic methods are presented, which allow analyzing the behavior. The temporal query approach enables the investigation within in individual episodes. Time curves were used for a holistic analysis of several attributes with regard to their development over time. An extension of the time curve is proposed by using continuous coloring derived from a dimensional reduced attribute space. A method developed in the project for the visual analysis of text messages in social media data uses a "co-bridge" metaphor. Significant keywords are connected over time to enable both a quantitative comparison with regard to common and different keywords and a qualitative analysis with regard to stream volume and keyword frequencies. Furthermore, the use of emojis as contextual indicators was examined. It has been found that emojis are suitable for describing the spatial context of people in terms of perceived objects and activities taking place at these locations. Various normalization methods such as Chi-Value, "typicality," and metrics such as "user count per day" were used to account for heterogeneity and bias in social media data.

Two main areas of application were considered, on the one hand, the perception of environmental phenomena as input for questions of urban and landscape planning and, on the other hand, the analysis of the movement behavior of people on the example of football games. The study of a global phenomenon such as sunrise and sunset based on Instagram and Flickr allows conclusions to be drawn about consistency and reproducibility, as well as motives why the events were documented. Based on the four-facet context model, sunset and sunrise responses are examined for the where, who, what, and when. In addition, responses from different groups are compared with the aim of quantifying differences in behavior and information spread. As a second field of application, the movement behavior of people was analyzed using visitor statistics for selected locations during the COVID-19 pandemic and movement data from football games. Therefore, time-dependent abstracted and aggregated representations are created to compare collective behavior patterns during the pandemic in different countries in the first case and different football teams or teams in different stages of the game in the second case.

Acknowledgments This research was funded by the Deutsche Forschungsgemeinschaft (DFG, German Research Foundation) within Priority Research Program 1894 *Volunteered Geographic Information: Interpretation, Visualization and Social Computing* (VGIscience, EVA-VGI, BU 2605/8-2) and the Swiss National Science Foundation (Project No 200021E-166788).

References

Acheson E, De Sabbata S, Purves RS (2017) A quantitative analysis of global gazetteers: patterns of coverage for common feature types. Comput Environ Urban Syst 64:309–320

American Psychological Association. Behavior, n.d. https://dictionary.apa.org/behavior. visited on 2012-03-20

Andrienko G, Andrienko N, Anzer G, Bauer P, Budziak G, Fuchs G, Hecker D, Weber H, Wrobel S (2021a) Constructing spaces and times for tactical analysis in football. IEEE Trans Vis Comput Graph 27(4):2280–2297. https://doi.org/10.1109/TVCG.2019.2952129

Andrienko N, Andrienko G, Miksch S, Schumann H, Wrobel S (2021b) A theoretical model for pattern discovery in visual analytics. Visual Inform 5(1):23–42. ISSN 2468-502X. https://doi.org/10.1016/j.visinf.2020.12.002

Bach B, Shi C, Heulot N, Madhyastha T, Grabowski T, Dragicevic P (2016) Time curves: folding time to visualize patterns of temporal evolution in data. IEEE Trans Vis Comput Graph 22(1):559–568. https://doi.org/10.1109/TVCG.2015.2467851

Bai Q, Dan Q, Mu Z, Yang M (2019) A systematic review of emoji: current research and future perspectives. Front Psychol 10:2221. ISSN 1664-1078. https://doi.org/10.3389/fpsyg.2019.02221

Bell S (2012) Landscape: pattern, perception and process. ISBN 9780203120088. https://doi.org/10.4324/9780203120088

Blasi DE, Henrich J, Adamou E, Kemmerer D, Majid A (2022) Over-reliance on English hinders cognitive science. Trends Cogn Sci 26(12):1153–1170

Burghardt D, Dunkel A (2022) Ethical analysis of geosocial data to balance social and individual interests. In: Proceedings, AutoCarto 2022

Chen S, Andrienko N, Andrienko G, Li J, Yuan X (2021) Co-bridges: pair-wise visual connection and comparison for multi-item data streams. IEEE Trans Vis Comput Graph 27(2):1612–1622. https://doi.org/10.1109/TVCG.2020.3030411

Clarke K, Wood J, Dykes J, Slingsby A (2007) Interactive visual exploration of a large spatio-temporal dataset: reflections on a geovisualization mashup. IEEE Trans Vis Comput Graph 13(6):1176–1183. https://doi.org/10.1109/TVCG.2007.70570

Collins C, Andrienko N, Schreck T, Yang J, Choo J, Engelke U, Jena A, Dwyer T (2018) Guidance in the human–machine analytics process. Visual Inform 2(3):166–180. ISSN 2468-502X. https://doi.org/10.1016/j.visinf.2018.09.003

Crawford K, Finn M (2015) The limits of crisis data: analytical and ethical challenges of using social and mobile data to understand disasters. GeoJournal 80(4):491–502

Boyd D, Crawford K (2012) Critical questions for big data. Inform Commun Soc 15(5):662–679. https://doi.org/10.1080/1369118X.2012.678878

Davis JD (2016) The four types of social media engagement. In: The Union Metrics Blog.

Duarte R, Araújo D, Correia V, Davids K (2012) Sports teams as superorganisms. Sports Med 42(8):633–642

Dunkel A, Andrienko G, Andrienko N, Burghardt D, Hauthal E, Purves R (2019) A conceptual framework for studying collective reactions to events in location-based social media. Int J Geograph Inform Sci 33(4):780–804. https://doi.org/10.1080/13658816.2018.1546390

Dunkel A, Löchner M, Burghardt D (2020) Privacy-aware visualization of volunteered geographic information (VGI) to analyze spatial activity: a benchmark implementation. ISPRS Int J Geo-Inform 9 (10). ISSN 2220-9964. https://doi.org/10.3390/ijgi9100607

Dunkel A, Hartmann MC, Hauthal E, Burghardt D, Purves RS (2023a) From sunrise to sunset: exploring landscape preference through global reactions to ephemeral events captured in georeferenced social media. PLoS ONE, 17(1). https://doi.org/10.1371/journal.pone.0280423

Dunkel A, Burghardt D, Hartmann M, Ross P, Eva H (2023b) Supplementary materials for the publication "From sunrise to sunset: exploring landscape preference through global reactions to ephemeral events captured in georeferenced social media"

Fonseca S, Milho J, Travassos B, Araújo D, Lopes A (2013) Measuring spatial interaction behavior in team sports using superimposed voronoi diagrams. Int J Perform Anal Sport 13(1):179–189. https://doi.org/10.1080/24748668.2013.11868640

Gardner Z, Mooney P, De Sabbata S, Dowthwaite L (2020) Quantifying gendered participation in openstreetmap: responding to theories of female (under) representation in crowdsourced mapping. GeoJournal 85(6):1603–1620

Ghermandi A, Sinclair M (2019) Passive crowdsourcing of social media in environmental research: a systematic map. Global Environ Change 55:36–47. ISSN 0959-3780. https://doi.org/10.1016/j.gloenvcha.2019.02.003

Goodchild MF (2007) Citizens as voluntary sensors: spatial data infrastructure in the world of web 2.0. Int J Spatial Data Infrastruct Res 2(2):24–32.

Gruebner O, Sykora M, Lowe SR, Shankardass K, Galea S, Subramanian S (2017) Big data opportunities for social behavioral and mental health research. Soc Sci Med 189:167–169. ISSN 0277-9536. https://doi.org/10.1016/j.socscimed.2017.07.018

Gudmundsson J, Horton M (2017) Spatio-temporal analysis of team sports. ACM Comput Surv 50(2). ISSN 0360-0300. https://doi.org/10.1145/3054132

Hartmann MC, Koblet O, Baer MF, Purves RS (2022) Automated motif identification: analysing flickr images to identify popular viewpoints in Europe's protected areas. J Outdoor Recreat Tourism 37:100479. ISSN 2213-0780. https://doi.org/10.1016/j.jort.2021.100479

Hauthal E, Burghardt D, Dunkel A (2019) Analyzing and visualizing emotional reactions expressed by emojis in location-based social media. ISPRS Int J Geo-Inform 8(3). ISSN 2220-9964. https://doi.org/10.3390/ijgi8030113

Hauthal E, Dunkel A, Burghardt D (2021) Emojis as contextual indicants in location-based social media posts. ISPRS Int J Geo-Inform 10(6). ISSN 2220-9964. https://doi.org/10.3390/ijgi10060407

Henriques G, Michalski J (2020) Defining behavior and its relationship to the science of psychology. Integr Psychol Behav Sci 54:328–353. https://doi.org/10.1007/s12124-019-09504-4

Highfield T, Leaver T (2016) Instagrammatics and digital methods: studying visual social media, from selfies and gifs to memes and emoji. Commun Res Practice 2(1):47–62. https://doi.org/10.1080/22041451.2016.1155332

Hu X, Tang J, Gao H, Liu H (2013) Unsupervised sentiment analysis with emotional signals. In Proceedings of the 22nd International Conference on World Wide Web, WWW '13, New York, NY, USA, 2013. Association for Computing Machinery, pp 607–618. ISBN 9781450320351. https://doi.org/10.1145/2488388.2488442

Kejriwal M, Wang Q, Li H, Wang L (2021) An empirical study of emoji usage on twitter in linguistic and national contexts. Online Soc Netw Media 24:100149. ISSN 2468-6964. https://doi.org/10.1016/j.osnem.2021.100149

Knura M, Kluger F, Zahtila M, Schiewe J, Rosenhahn B, Burghardt D (2021) Using object detection on social media images for urban bicycle infrastructure planning: a case study of dresden. ISPRS Int J Geo-Inform 10(11). ISSN 2220-9964. https://doi.org/10.3390/ijgi10110733

Kralj Novak P, Smailović J, Sluban B, Mozetič I (2015) Sentiment of emojis. PLOS ONE 10(12):1–22. https://doi.org/10.1371/journal.pone.0144296

Krasnova H, Veltri NF (2011) Behind the curtains of privacy calculus on social networking sites: the study of Germany and the USA

Laraway S, Snycerski S, Pradhan S, Huitema BE (2019) An overview of scientific reproducibility: consideration of relevant issues for behavior science/analysis. Perspect Behav Sci 42(1):33–57. ISSN 25208977. https://doi.org/10.1007/s40614-019-00193-3

Li J, Chen S, Chen W, Andrienko G, Andrienko N (2020) Semantics-space-time cube: a conceptual framework for systematic analysis of texts in space and time. IEEE Trans Vis Comput Graph 26(4):1789–1806. https://doi.org/10.1109/TVCG.2018.2882449

Li L, Goodchild MF, Xu B (2013) Spatial, temporal, and socioeconomic patterns in the use of twitter and flickr. Cartogr Geograph Inform Sci 40(2):61–77

Lowe DG (2004) Distinctive image features from scale-invariant keypoints. Int J Comput Vis 60(2):91–110

Luckmann T (2013) Theorie des sozialen Handelns. De Gruyter, Berlin, Boston. ISBN 9783110848922. https://doi.org/doi:10.1515/9783110848922

McInnes L, Healy J, Astels S (2017) hdbscan: hierarchical density based clustering. J Open Source Softw 2(11):205

Memmert D, Lemmink KA, Sampaio J (2017) Current approaches to tactical performance analyses in soccer using position data. Sports Med. https://doi.org/10.1007/s40279-016-0562-5

Olteanu A, Castillo C, Diaz F, Kıcıman E (2019) Social data: biases, methodological pitfalls, and ethical boundaries. Front Big Data 2:13

Shaver P, Schwartz J, Kirson D, O'connor C (1987) Emotion knowledge: further exploration of a prototype approach. J Pers Soc Psychol 52(6):1061

Shirato G, Andrienko N, Andrienko G (2021) What are the topics in football? Extracting time-series topics from game episodes. In: 2021 IEEE Visualization Conference. http://geoanalytics.net/and/papers/vis21poster.pdf

Simmonds C, McGivney A, Reilly P, Maffly B, Wilkinson T, Cannon G (2018) Crisis in our national parks: How tourists are loving nature to death. The Guardian. https://www.theguardian.com/environment/2018/nov/20/national-parks-america-overcrowding-crisis-tourism-visitation-solutions

Solove DJ (2013) Privacy self-management and the consent dilemma'(2013). 126 Harvard Law Review 1880, 2012–141

Teitler BE, Lieberman MD, Panozzo D, Sankaranarayanan J, Samet H, Sperling J (2008) Newsstand: a new view on news. In: Proceedings of the 16th ACM SIGSPATIAL International Conference on Advances in Geographic Information Systems, GIS '08, New York, NY, USA, 2008. ACM, pp 18:1–18:10–18:1–18:10. ISBN 978-1-60558-323-5. https://doi.org/10.1145/1463434.1463458

Thomee B, Shamma DA, Friedland G, Elizalde B, Ni K, Poland D, Borth D, Li, L-J (2016) YFCC100 m. Commun ACM 59(2):64–73. https://doi.org/10.1145/2812802

van den Elzen S, Holten D, Blaas J, van Wijk JJ (2016) Reducing snapshots to points: a visual analytics approach to dynamic network exploration. IEEE Trans Vis Comput Graph 22(1):1–10. https://doi.org/10.1109/TVCG.2015.2468078

Williams ML, Burnap P, Sloan L (2017) Towards an ethical framework for publishing twitter data in social research: taking into account users' views, online context and algorithmic estimation. Sociology 51(6):1149–1168. https://doi.org/10.1177/0038038517708140

Wood SA, Guerry AD, Silver JM, Lacayo M (2013) Using social media to quantify nature-based tourism and recreation. Sci Rep 3(1):1–7. https://doi.org/10.1038/srep02976

Zahtila M, Knura M (2022) Visualizing point density on geometry objects: application in an urban area using social media VGI. KN-J Cartogr Geograph Inform 72(3):187–200. https://doi.org/10.1007/s42489-022-00113-7

Chapter 13
Digital Volunteers in Disaster Management

Ramian Fathi and Frank Fiedrich

Abstract During disaster situations, social media is used extensively by the affected population for communication and collaboration, but there is also increased public sharing of important disaster-related information about the current situation. With the goal of utilizing this data and Volunteered Geographic Information (VGI) for disaster management, digital volunteers organized themselves into so-called Volunteer and Technical Communities (V&TC). In addition, professionalized digital volunteers have institutionalized Virtual Operations Support Teams (VOST) in established Emergency Management Agencies (EMA). While technical issues have dominated research in this area in recent years, questions about the motivation, organization, and impact of the analytical work of these volunteers have remained unanswered. In this chapter, we present five studies that address questions about the motivation of digital volunteers, organization, and collaboration requirements, the analytical impact of VOST, data biases in Crisis Information Management (CIM), and privacy-related topics. Overall, it could be shown that digital volunteers make a significant contribution during disaster management, in which they effectively process their analytical results and VGI for the management of disaster situations. However, human limitations and privacy-related methods need to receive greater attention in the future, both in research and in practice.

Keywords Digital volunteers · Disaster management · Virtual operations support teams · Social media · Social media analytics · Emergency operation center · Data bias · Information management

R. Fathi (✉) · F. Fiedrich
Chair for Public Safety and Emergency Management, School of Mechanical Engineering and Safety Engineering, University of Wuppertal, Wuppertal, Germany
e-mail: fathi@uni-wuppertal.de; fiedrich@uni-wuppertal.de

13.1 Introduction

The earthquake in Haiti 2010 can be seen as a starting point of digital volunteering in the context of disaster management and humanitarian assistance. The first digital volunteers were committed to helping affected populations by processing and providing Volunteered Geographic Information (VGI) for disaster response. Over time, they organized themselves and formed virtual groups. These emerging so-called Volunteer and Technical Communities (V&TC) opened the view for multiple new fields of research. Many research approaches paid attention to technical topics like data mining and creating better analytical tools for the volunteers. In contrast, less research has been conducted to determine which motivational factors drive people to volunteer digitally in disaster management and which organizational requirements exist to collaborate with established Emergency Management Agencies (EMA). Furthermore, questions about the impact of analytical results in this time-critical context arise. The underlying research focused on the volunteers themselves and organizational requirements for collaboration, more specifically their motivational, participative, and analytical factors. This chapter is based on research carried out in the project "Active Participation and Motivation of Professionalized Digital Volunteer Communities: Distributed Decision-Making and its Impact on Disaster Management Organizations."

After this brief introduction, the second section of this chapter presents a comprehensive study of the motivational factors of operationally active digital volunteers (Fathi and Fiedrich 2020). In a cross-organizational online survey, possible motives, individual organizational commitment, and potential incentive options were analyzed. In addition, two experienced digital team leaders of V&TC were interviewed in guided expert interviews about methods and measures for increasing motivational and organizational commitment. Based on the findings generated in this way, explanatory patterns for the motivation factors of digital volunteers can be derived on the one hand; on the other hand, beneficial and identification-generating measures can be identified.

Contrary to the V&TC, digital volunteers institutionalized Virtual Operations Support Teams (VOST), which are closely linked to established EMA. This development led to more in-depth research questions concerning the professionalized digital volunteers and their integration in decision-making processes using VGI in Emergency Operations Centers (EOC), which are discussed in Sects. 13.3 and 13.4. The organizational structure and technical requirements for succeeding and their decision-making processes in a time-critical environment were of interest. A research gap between the VGI created by VOST and decision-makers needs in the established EMA was acknowledged, as the digital volunteers started to collaborate with EMA. Therefore, the topic of voluntary digital participation for collaborative emergency management was explored in collaboration with the University of Stuttgart, whose research efforts mainly focused on visual analysis of VGI. In Sect. 13.2, we present a case study which was conducted with the project "VA4VGI-2" (Chap. 6), where structural, procedural, and technical requirements of integrating

VOST in EOC structures were investigated, applying a mixed-method approach (Fathi et al. 2020).

Overall, little attention was paid to VOST, who act as groups of data analysts with direct integration to EOC. Specific tasks of VOST include filtering, verifying, and analyzing social media data from various platforms and creating information products for decision-makers in EOC. These information products can contribute to the situational awareness of EOC members and to the decision-making processes by integrating actionable information. In a case study following the 2021 flooding in Germany, the aspects of analyzing social media by digital volunteers in VOST and the impact of the information products on situational awareness and decision-making were examined and are presented in Section 4 (Fathi and Fiedrich 2022).

Analytical and decision-making processes in the time-critical environment of EOC in disaster management are challenging due to the numerous disaster-related conditions like time-pressure and uncertainty. To examine the interplay of the conditions in disaster management, VGI, and biases in Crisis Information Management (CIM), a workshop experiment was conducted with digital volunteers and decision-makers (Paulus et al. 2022). A three-stage experiment on epidemic response was developed as the underlying scenario to analyze how biases can be mitigated by observing digital volunteers and decision-makers in the analytical and decision-making processes and is presented in Sect. 13.5. The findings of this case study suggest that debiasing efforts are strongly undervalued, and external analysts fail to debias data successfully in favor of rapid results. The biased data was then passed on to decision-makers in the form of information products, who make decisions based on biased data.

Section 13.6 addresses the challenge of privacy-aware data analytics in disaster management and discusses a collaborative work between this underlying project and "Privacy Aspects" (Chap. 14).

13.2   Motivational Factors of Digital Volunteers in Disaster Management

With the emergence of social media and VGI, various new research areas and new opportunities to use this open-access data increased, also in disaster management. Anyhow, limited resources characterize disaster, and EMA do not have enough staff to analyze big amounts of data to integrate VGI in their situational awareness and decision-making processes. Digital volunteers analyze social media data from, e.g., Twitter and proceed disaster-related VGI. The volunteers can work dislocated from the actual operational site and thus can be deployed almost instantly. Over time, the volunteers organized themselves and formed V&TC. These communities fostered research interest, especially in the fields of organization and technology. An existing research gap in the context of volunteering on a digital basis in disaster management are the motivational factors of V&TC members. Questions regarding the motivation of the digital volunteer, the barriers to participation, and the commitment to the V&TC have not yet been extensively answered. In the paper "Digital Volunteers in Disaster Management—Motivational Factors and Barriers of Participation" (Fathi and Fiedrich 2020), we aimed to understand what motivates the digital volunteers and which incentive options there are to further motivate them. Additionally, we looked into measures and methods that can be implemented by digital team leaders to motivate their team members. Lastly, the differences and correlations between the needs of the digital volunteers and the motives of the digital team leaders were examined.

Therefore, we used a mixed-methods approach of quantitative and qualitative social science methods. It was of special interest to capture and query the digital volunteers in their social contexts as well as their individuality. An online survey was conducted among different V&TC, to explore motivational factors and incentive options. The survey was designed under the use of the Volunteer Functions Inventory (VFI) introduced by Clary and Snyder (1999), which is a widely used questionnaire on volunteer motivation. In order to understand measures and methods to foster motivation of digital volunteers by leaders of V&TC, guideline-based expert interviews were carried out. The guidelines were designed based on the online survey, but the content was transferred to the perspective of team leaders.

It was found that digital volunteers are mostly motivated by their values, but also by the experience, they are gaining paired with fun-based intrinsic motivation. In contrast, having the prospect of a "career" within V&TC was less motivating. The participants strongly agreed to statements of organizational commitment and identification with their organization. More than 70% stated that they fully agreed to be proud to be part of their V&TC. The main barriers of digital volunteering were named as time, trust in one's own abilities, and Internet access. Especially during crises, time allocation becomes a challenge. Collaborating with other V&TC or a pool of digital volunteers, who can be acquired ad hoc, seems to be an appropriate method to allocate work of digital volunteers. The queried digital volunteers see potential for motivational enhancement rather in non-crisis times, for example, more feedback and additional online and offline community activities without a disaster context. Accordingly, the volunteers see incentive options in digital or analog exercises or events and appreciative measures. It became clear that feedback is very important to the digital volunteers, who, due to their dislocation, can only guess what impact and use the resulting information products and VGI have. Negative impact on the motivation of digital volunteers were identified as a lack of feedback and a lack of identification with the work or the V&TC. Feedback cannot only be given by the tasking EMA but also by the public or the digital team leaders (Fathi and Fiedrich 2020).

13.3   Virtual Operations Support Teams in Disaster Management

Virtual Operations Support Teams (VOST) are groups of professionalized digital volunteers, who are closely linked to EMA. During a VOST operation, the common tasks include monitoring and analyzing social media, verifying and geolocating information and developing crisis maps, recognizing and analyzing trends and sentiment in social media, and ad hoc tasks as assigned by the EOC. The actionable information identified and verified can then be provided to the EOC decision-maker to expand situational awareness and support decision-making processes. Furthermore, VOST integrate a liaison officer in the EOC structures. This enables to ensure effective communication and distribution of tasks between VOST and EOC. VOST pursue the goal of effectively integrating information products and VGI into decision-making processes through close organizational integration in the time-critical context of disaster management. This in turn led to multiple research questions concerning structural, procedural, and technical requirements for an effective collaboration between a virtual team of analysts and decision-maker in an EOC. These questions were to be explored in collaboration with the project "VA4VGI-2" (Chap. 6). As described in the paper "VOST: A case study in voluntary digital participation for collaborative emergency management" (Fathi and Fiedrich 2020), the main goal was to understand the decision-making processes, which emerged by integrating a VOST into the structures of an EOC. An exploratory case study was conducted as field research during the start (Grand Départ) of the Tour de France in Düsseldorf, Germany in 2017. Especially of interest were the requirements of structure and procedure for a successful collaboration, technical requirements and the evaluation of existing technical tools for social media analytics, the identification of the actual tasks that needed performing during the operation, and structural, organizational, and technical implications for future decision-making systems in EOC.

The VOST operation at the Grand Départ in Düsseldorf was in the scope of a pilot project with the German Federal Agency for Technical Relief ("Technisches Hilfswerk" – THW). The THW VOST consists of 20 digital volunteers which were appointed as THW members and thus act as team members in a governmental EMA rather than a loosely coupled group of digital volunteers in V&TC. To make use of the insights provided by the case study, multiple methods for data collection and analysis were conducted. These methods comprised participant observation during the two-day operation, focus group discussions and informal interviews with decision-makers and VOST members at different stages of the operation, analysis of the tasks performed by the VOST during the operation, analysis of the organizational setup, and technology use and decision-making processes of the VOST. For this particular operation, the following VOST working priorities were identified as a result of the focus group discussions: identification of critical crowd densities and flows; detection of unusual events; image analysis of social media;

developing of a crisis map for spatial analysis; identification of false information, rumors, and fake news; and scenario-dependent tasks.

The structural and procedural requirements for a successful collaboration were identified as a division in small VOST working groups to simplify the distribution of tasks. This allows specialized subgroups to be formed to respond to dynamic operational situations, e.g., verification groups, and to ensure information exchange on the level of team and group leaders, individual group briefings to implement adjustments quicker, and, most importantly, the implementation of a liaison officer. The technical requirements are especially a reliable user experience and custom-tailored tools for the use during an operation to alleviate the high mental workload. Putting new tools to the test in real-world operations seems to be a beneficial way to ensure advanced algorithmic tools. The most time-consuming and mentally challenging work at the same time poses collection, filtering, and documentation of user-generated content from social media platforms. Advanced mining tools are crucial to verify social media data in a time-critical environment. Situation monitoring is a highly repetitive and demanding task, which can only be carried out by a digital volunteer for a certain amount of time. Nonetheless, the biggest challenge seems to be the velocity and the volume of user-generated social media data. Additional disaster-related data becomes available all the time during the analysis, which presents a challenge for real-time social media analytics. To address the questions of what disaster-related information is processed by a VOST during a disaster management and what impact it has on members of an EOC, another case study was conducted.

13.4 Social Media Analytics by Virtual Operations Support Teams in Disaster Management

Climate change poses numerous challenges and risks, including a significant increase in extreme weather events such as flooding (IPCC 2021). With this, the need for crisis communication and social media analytics in times of disaster rises accordingly. VOST address this need, integrated into EOC structures in times of crises, and their goals are to increase the situational awareness of decision-makers and to provide actionable information to improve decision-making in a time-critical environment. To examine these efforts, a case study was carried out, using the data collected by 22 VOST analysts during the 2021 flood in Wuppertal, Germany (Fathi and Fiedrich 2022). The city was severely flooded in July 2021; parts of the city had to be evacuated, and warning sirens were set off (Zander 2021). The EOC operated in cooperation with the VOST of the German Federal Agency for Technical Relief (THW VOST), which was deployed virtually but was directly connected to the EOC through a liaison officer, who was physically present in the EOC. This operation thus raised the research question, how VOST can support situational awareness and generate actionable information for EOC decision-making processes by integrating social media analytics practices. The research question was explored by analyzing the data generated during the THW VOST operation and by a survey among EOC decision-makers on the impact of the information provided by VOST on their decisions and situational awareness.

Unlike other research, which focuses, e.g., mainly on social media big data analysis, decision-making processes, or developing machine learning approaches, this study aims for examining closely the real-world VOST integration. Case studies, such as the underlying, can provide valuable insights of virtual teams in the disaster management context. In order to analyze the data generated by the digital volunteers during the operation, the following tasks were performed: data cleaning, summarizing categories, visualization of the data, and subsequently a comparative quantitative analysis and contextualization of the data. Furthermore, three different parameters, the format, the source, and the mean value of the prioritization, were used for an in-depth analysis. The survey of decision-makers was conducted among the EOC members, which collaborated with the THW VOST and used their information products during the flood response in Wuppertal. The prerequisite was that the interviewee had worked with VOST information products during the operation. Nine decision-makers from the EOC met these criteria, and all of them participated in the survey.

To classify information categories in the VOST dataset, which was identified by VOST volunteers during the flood response, 536 social media posts from eight different social media platforms were analyzed. Additionally, 42 posts from websites (e.g., traditional media) were collected. The dataset was classified in 23 different categories. The largest category was found to have emerged after the flood, namely, spontaneous community engagement (see Fig. 13.1). The earlier phases of the flood were dominated by categories like level of the river, warning, or flooded roads.

Posts from categories that could have had a direct impact were forwarded by VOST during the operation as actionable information to the EOC decision-maker. These social media posts were prioritized as "highly-relevant." The information in the format of videos was found to have a higher priority than information in the form of texts or images. In a category analysis over time, it was found that real-time disaster events, such as the activation of the warning siren, are simultaneously apparent in social media data.

VOST impact on situational awareness was explored by querying EOC directors and executives who collaborated with the THW VOST during the flood. All statements in the survey were rated with a strong overall agreement. The statement with the highest degree of agreement was: "Information from VOST contributes to expanded situational awareness." The necessity of a liaison officer in the EOC was also strongly agreed to. The lowest level of agreement was given the statement that VOST information can forecast developments of future situations. The second category of statements concerned the VOST impact on decision-making, e.g., to ensure people-centered risk and crisis communications and contributed to confidence in decision-making.

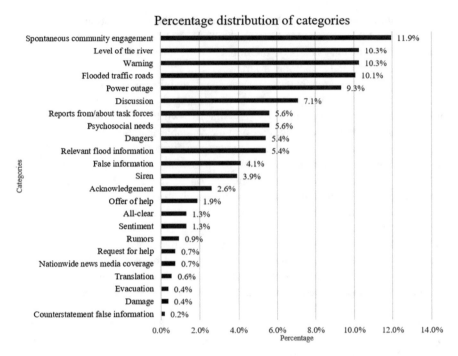

Fig. 13.1 Percentage distribution of social media information during the flood response 2021. (Source: Fathi and Fiedrich 2022)

The results underline that situational awareness and decision-making is supported by VOST information and VGI on three different levels: perception, comprehension, and projection. Based on this distinction of situational awareness from Endsley (1988), it can be concluded that statements that can be assigned to the first two levels (perception and comprehension) receive high agreement. However, VOST information during the dynamic flood situation supports less to project the future flood-situation in a more precise way. Statements in this category received less agreement. This condition can be explained by the observation that VOST information and VGI provide new and complementary information that must be processed, comprehended, and projected by EOC decision-makers onto future scenarios in addition to that provided by other sources (feedback from responders, emergency calls, etc.). Due to the accompanying conditions in disaster management (e.g., time pressure, uncertainty), the cognitive load on decision-makers is very high so that the use of VGI can create biases in data analysis and decision-making.

13.5   Data Bias in Crisis Information Management

Due to various conditions, humanitarian crises and disaster management especially challenge digital volunteers' data analysis. Crisis Information Management (CIM) is characterized, e.g., by time criticality and uncertainty. Additionally, resources are limited, and the cognitive load is high. This makes analysts and decision-makers prone to inducing biases in the data and cognitive processes. When undetected, biases remain untreated and lead to decisions based on biased information, which in turn can lead to an inefficient response. To find out more about the interplay of data and cognitive biases, an exploratory three-stage experiment on epidemic response was conducted (Paulus et al. 2022). A scenario-based workshop was held in The Hague in 2020 with experienced crisis decision-makers and digital volunteers from various V&TC and VOST, which entailed stage 1 and 2. For stage 3, the same participants were additionally addressed in an online survey. The experimental scenario was an epidemic outbreak in three countries. The first stage included an observation of digital volunteers, who were provided with different datasets with biased data, e.g., in the infection spread. The observation was set out to be in a fictional but realistic setting to avoid interference in a real epidemic response. In stage 2 of the experiment, the decision-makers were provided with the VGI and information products (e.g., maps) created in stage 1 and had to make decisions on treatment center placements. Stage 3 of the experiment was an online survey onsite of the workshop. It aimed to explore whether confirmation bias leads to path dependencies of former decisions based on biased information. In the survey, they were able to select the information they viewed as most important for future decision-making from a list of datasets.

The results of the experiment show that in the first stage, the participants failed to debias data, even though biases were detected. Debiasing efforts were undervalued in favor of immediate results. The information products created based on biased data in stage 1 were then forwarded to the decision-makers in stage 2, who made their decisions based on biased information. Even though the decision-makers in all three groups put enormous pressure on the digital volunteers to find out on which datasets the information products were generated, they did not succeed in ensuring that the decisions were not based on biased information. Confirmation bias was detected in stage 3; the reliance on conclusion reached with biased data was reinforced by it. Thus, biased assumptions remained undetected. The main causes for biased data remaining untreated are the described conditions of data analysis and decision-making in the context of disaster management. The realistic scenario design made it possible to recreate the general conditions, e.g., by simulating time pressure and uncertainty. Mindfulness debiasing efforts have been found to be effective to counteract these conditions and therefore pose a promising strategy to mitigate data and cognitive biases in future disaster management (Paulus et al. 2022).

13.6 Privacy-Aware Social Media Data Processing in Disaster Management

Social media analytics by Emergency Management Agencies for relief purposes has become a common practice in the last few years. In the time-critical environment of disaster and life-threatening situations, privacy is often perceived as a secondary problem. Nonetheless, avoiding unnecessary data retention is important to protect the social media users' privacy by, e.g., preventing subsequent abuse. In times of crisis, social media users are especially vulnerable, e.g., by sharing names, addresses, or other personal information, for example, searching for missing relatives or friends.

In a cross-project effort, expertise from the present project and "Privacy Aspects" (Chap. 14) were combined for a joint case study (Löchner et al. 2020). The study examined the extent to which VOST can integrate privacy-aware methods and algorithms, in particular HyperLogLog (HLL), in their operational work in disaster management. To investigate the practicality of privacy-aware methods and HLL, a case study with digital volunteers from two VOST has been conducted. For this, a focus group discussion addressing opportunities, challenges, and implementation barriers was held. The focus group discussion with participants from THW VOST and VOST Baden-Württemberg aimed to document the expertise of the participants, who are experienced in the field of social media analytics.

The most important finding was that the focus group discussion revealed no disadvantage against using privacy-friendly methods and HLL for VOST. The algorithm does not distract the data analysis process, since the VOST work starts after the data processing via HLL. Overall, HLL was found to be an appropriate technology to ensure privacy-aware social media data processing. Opposing to the initial assumptions that HLL use might be conflicting with gathering data for creating information products, several benefits of the algorithm were come up upon. These benefits include improved working with big datasets, which might lead to a more widespread use of HLL and thus improved privacy-awareness among digital volunteers.

13.7 Conclusion and Outlook

In this chapter, various facets of digital volunteering using and processing VGI in the disaster management sector were highlighted. In particular, social, organizational and analytical factors were examined and discussed in five different papers. It was shown that the motivation of V&TC is particularly value-based and that the commitment to the virtual team is pronounced. However, measures can be derived, especially for future developments, in order to sustainably establish digital volunteering. Professionalized teams in the structures of established EMA have been institutionalized worldwide, but detailed research on the requirements for collaboration between a virtual team and an EOC has been lacking. In Sect. 13.3, a joint study with the project VA4VGI-2 (Chap. 6) was presented. From the results, it can be concluded that an effective integration of digital volunteers and their analytical skills into established disaster management structures is achievable, although specific requirements have to be considered.

The analytical value for situational awareness and decision-making of EOC members during a specific disaster management situation could be highlighted in Sect. 13.4. In conclusion, the section shows that VOST analysts were able to identify, verify, and categorize a large amount of disaster-related information from eight different social media platforms in various formats. During the 2021 flood, this information contributed to an expanded situational awareness of the decision-makers in an Emergency Operation Center and made it possible, for example, to conduct crisis communication in a more people-centered manner. Thus, it could be shown that during dynamic disaster situations, the involvement of digital volunteers was of major relevance for the operation management. Furthermore, it could be shown that the virtual structures of a VOST are able to effectively support an EOC even in an acute disaster situation. Nevertheless, analysts and decision-makers in such situations are accompanied by conditions that can bias results and decisions. In order to investigate data and cognitive biases in the work of digital volunteers and decision-makers with VGI, a paper was presented in Chap. 5 in which this aspect was investigated in a three level experiment. It could be shown that debiasing efforts were not pronounced enough, and thus biased information was considered in decision-making. It can be deduced that in future exercises and trainings, debiasing efforts need to receive more attention in order to ensure the integration of digital volunteers and VGI in the future. This is also accompanied by privacy-aware analytics of social media in disaster situations, where the affected population is particularly vulnerable. To investigate possible methods and an algorithm developed in the project "Privacy Aspects" (Chap. 14) in the use of VOST, a case study was conducted with two VOST (Sect. 13.6). Thereby, it could be examined that the use of privacy-aware methods and algorithms in operational use is reasonable and possible.

Overall, it could be shown that digital volunteers make a significant contribution during disaster management, in which they effectively process their analytical results and VGI for the management of disaster situations. However, human limitations and privacy-aware methods need to receive greater attention in the future, both in research and in practice. In addition, questions remain about how situational pictures will need to be designed by digital volunteers in the future. In a project funded by the "German Federal Office of Civil Protection and Disaster Assistance" called "#sosmap" (Fiedrich 2022), it will be investigated from August 2022 to what extent psychosocial situation pictures can be created for EOC by analyzing social media.

Acknowledgments This research was supported by the German Research Foundation DFG within Priority Research Program 1894 Volunteered Geographic Information: Interpretation, Visualization and Social Computing (VGIscience, projectnumber: 314672086).

References

Clary EG, Snyder M (1999) The motivations to volunteer: theoretical and practical considerations the motivations to volunteer. Curr Directions Psychol Sci 8(5):156–159. ISSN 0963-7214. https://doi.org/10.1111/1467-8721.00037

Endsley MR (1988) Design and evaluation for situation awareness enhancement. Proc Hum Fact Soc Annual Meeting 32(2):97–101. ISSN 0163-5182. https://doi.org/10.1177/154193128803200221

Fathi R, Fiedrich F (2020) Digital freiwillige in der katastrophenhilfe - motivationsfaktoren und herausforderungen der partizipation. In: Hansen C, Nürnberger A, Preim B (eds) Mensch und Computer 2020. https://doi.org/10.18420/MUC2020-WS117-406

Fathi R, Fiedrich F (2022) Social media analytics by virtual operations support teams in disaster management: situational awareness and actionable information for decision-makers. Front Earth Sci. https://doi.org/10.3389/feart.2022.941803

Fathi R, Thom D, Koch S, Ertl T, Fiedrich F (2020) Vost: a case study in voluntary digital participation for collaborative emergency management. Inform Process Manag 57(4). ISSN 03064573. https://doi.org/10.1016/j.ipm.2019.102174

Fiedrich F (2022) https://www.buk.uni-wuppertal.de/de/forschung/laufende-projekte/sosmap/

IPCC (2021) Climate change 2021: the physical science basis. contribution of working group I to the sixth assessment report of the intergovernmental panel on climate change [Masson-Delmotte V, Zhai P, Pirani A, Connors SL, Péan C, Berger S, Caud S, Chen Y, Goldfarb L, Gomis MI, Huang M, Leitzell K, Lonnoy E, Matthews JBR, Maycock TK, Waterfield T, Yelekçi O, Yu R, Zhou B (eds)]

Löchner M, Fathi R, Schmid D, Dunkel A, Burghardt D, Fiedrich F, Koch S (2020) Case study on privacy-aware social media data processing in disaster management. ISPRS Int J Geo-Inform 9(12):709. https://doi.org/10.3390/ijgi9120709

Paulus D, Fathi R, Fiedrich F, van de Walle B, Comes T (2022) On the interplay of data and cognitive bias in crisis information management. Inform Syst Front. ISSN 1387-3326. https://doi.org/10.1007/s10796-022-10241-0

Zander U (2021) Starkregenereignis in Wuppertal. https://lernplattform-babz-bund.de

Chapter 14
Protecting Privacy in Volunteered Geographic Information Processing

Marc Löchner, Alexander Dunkel, and Dirk Burghardt

Abstract Social media data is used for analytics, e.g., in science, authorities, or the industry. Privacy is often considered a secondary problem. However, protecting the privacy of social media users is demanded by laws and ethics. In order to prevent subsequent abuse, theft, or public exposure of collected datasets, privacy-aware data processing is crucial. In this chapter, we show a set of concepts to process social media data with social media user's privacy in mind. We present a data storage concept based on the cardinality estimator HyperLogLog to store social media data, so that it is not possible to extract individual items from it, but only to estimate the cardinality of items within a certain set, plus running set operations over multiple sets to extend analytical ranges. Applying this method requires to define the scope of the result before even gathering the data. This prevents the data from being misused for other purposes at a later point in time and thus follows the privacy by design principles. We further show methods to increase privacy through the implementation of abstraction layers. As another additional instrument, we introduce a method to implement filter lists on the incoming data stream. A conclusive case study demonstrates our methods to be protected against adversarial actors.

Keywords Privacy · Social media · Data retention · HyperLogLog

14.1 Introduction

Social media services like Twitter or Instagram are used to communicate and share information worldwide, which generates a rich set of data. Since a large part of this data is publicly available, it can be used beyond the features of the social media services itself, especially by third parties.

M. Löchner (✉) · A. Dunkel · D. Burghardt
Technische Universität Dresden, Dresden, Germany
e-mail: marc.loechner@tu-dresden.de; alexander.dunkel@tu-dresden.de;
dirk.burghardt@tu-dresden.de

© The Author(s) 2024
D. Burghardt et al. (eds.), *Volunteered Geographic Information*,
https://doi.org/10.1007/978-3-031-35374-1_14

The main problem with social media data utilization for applications other than their dedicated use case is that explicit consent from the social media user is usually missing. While most users are aware that their content is publicly available on the Internet, they do not assume that data is frequently recycled for other purposes such as scientific, commercial, or administrative use (Boyd and Crawford 2012). Accordingly, this demands an exceptional strong focus on their privacy.

In contrast to other environments, the data to be protected is already public (Williams et al. 2017). In the view of third parties, that data can be utilized for any purpose, including those that oppose the user's interest (Zhou et al. 2008). But data can also be used with good intentions (Daly et al. 2019), whereas "good" could be defined by "in the user's Interest." For example, social media has shown a valuable source of information in crisis mapping, emergency response, or public planning (Fiedrich and Fathi 2021; Dunkel 2021).

In order to support ongoing development of positive use cases, scientists need to respect and actively protect social media users' privacy. Scientists need to take explicit control over data that they expose and prevent accidental disclosures.

An approach to support the adoption of accidental disclosure prevention techniques is to *prevent* the gathering of privacy-relevant data in the first place. We specifically aim at providing methods for the use of social media data following the *privacy-by-design* principles (Cavoukian et al. 2009).

In this chapter, we show a set of concepts that enable to process social media data with social media user's privacy in mind. We present a data storage concept that implements an algorithm called HyperLogLog (HLL) (Flajolet et al. 2007) to not store raw social media data but only statistics about their occurrence. We further show that while losing precision of the data, privacy can even be increased by applying multiple layers of abstraction on the data. For a context-dependent treatment of privacy and to cover edge cases, we further introduce a model to implement filter lists on the incoming data stream. A conclusive case study demonstrates our methods to be protected against adversarial actors.

14.2   Fundamentals

14.2.1   Related Work

Issues and challenges related to privacy arise everywhere, where social media data is involved. Following up, we link to research projects within this book, which are primarily based on processing social media data and therefore our research is relevant for.

The EVA-VGI project (see Chap. 12) studies the heterogeneity, quality, subjectivity, spatial resolution, and temporal relevance of geo-referenced social media data. Focusing on the integration of spatial, temporal, topical, and social dimensions combined with an explicit link between events and reactions, they present conceptual approaches and methods that enable a privacy-aware visual analysis of VGI in general and geo-social media data in particular. The project has taken advantage of the results of our research by implementing HLL on datasets related to their publications (Dunkel et al. 2020).

Similarly, the VA4VGI project (see Chap. 6) describes how geo-aware filtering and anomaly detection on geo-referenced social media data can be a significant information source for stakeholders in journalism, urban planning, or disaster management. They present tag maps that provide overview-first, details-on-demand, visual summaries of large amounts of social media data over time and thus visualize their temporal evolution.

Closely related to the former is the DVCHA project (see Chap. 13). The overall objective of their research is to study the implications of social media data for the efficiency of disaster management. Focusing on so-called Virtual Operations Support Teams (VOST), their research addresses motivation, success factors, and improvement of distributed decision making processes based on disaster-related real-time social media data.

In a collaboration with the DVCHA project, we carried out a case study, in which we explored the deployment of HLL into disaster management processes (see Sect. 13.6). We developed and conducted a focus group discussion with VOST members, where we identified challenges and opportunities of working with HLL and compared the process with conventional techniques (Löchner et al. 2020). Findings showed that deploying HLL in the data acquisition process of VOST operations will not distract their data analysis process. Instead, several benefits, such as improved working with huge datasets, may contribute to a more widespread use and adoption of the presented technique, which provides a basis for a better integration of privacy considerations in disaster management.

14.2.2   On Privacy Aspects

From a generic point of view, *privacy* is the freedom to fully or partially retreat oneself in a self-controlled manner. There are always multiple forms of definitions of the term *privacy*, stretching from personal to a cultural point of views (Solove 2008). It is important to distinguish between the *right to privacy* and the *concept of privacy* (Hildebrandt 2006). The *right* is clearly formed by laws, whereas the *concept* is rather vaguely determined based on subjectively perceived personal values. Privacy is often sacrificed voluntarily in exchange for perceived benefits and sometimes violated by others, either intentionally or accidentally (Reyman 2013).

Privacy by design as a set of principles is a relevant objective in the conception of applications in general. As Cavoukian et al. (2009) state, privacy must be approached from a design-thinking perspective. It must be incorporated in technologies not as an optional on-top feature but as a fundamental characteristic of organizational priorities, project objectives, design processes, and planning operations. Concepts built upon these principles are hard to break in terms of privacy violations.

A contemporary method to protect data has been presented as *differential privacy* (DP) (Dwork 2008) and adopted frequently (Desfontaines and Pejó 2020). DP adds certain amounts of random data to a set of real data set, in order to make real data indistinguishable from the random data and thus protect it from being identified as such. However, DP still requires the original data to be available to process. Furthermore, DP requires developing new concepts and models for each data set, which is very inefficient when dealing with really large sets of data.

In the geo-community, there is a wide range of concepts known to protect privacy in terms of location data. Some techniques are based on anonymity, e.g., *mix zones* (Beresford and Stajano 2003) or *k-anonymity* (Ciriani et al. 2007). Others are based on obfuscation, e.g., *imprecision* (Duckham and Kulik 2005), or policy like *restriction* (Hauser and Kabatnik 2001). All of these approaches require the possession of original raw data. Processed data sets are unable to be updated with subsequent data, which requires reprocessing of the entire data set upon updates. This is very inefficient when dealing with large amounts of social media data.

In the context of social media data, the consideration of privacy, ethics, and legal issues should play an important role. The statement "Privacy of user data and information should be considered in the initial design of VGI systems" (Mooney et al. 2017) can be extended to platforms and methods for the analysis and further processing of social media data in general.

Kounadi et al. (2018) discuss privacy threats related to inference attacks on *geosocial network data*. They provide protection recommendations for sharing these sorts of data and publishing resulting visualizations. Keßler and McKenzie (2018) proposed in a total of 21 theses to reflect on the current state of *geoprivacy* from a technological, ethical, legal, and educational perspective. They provide various examples of how common it has become to share location and how it can be used and misused.

14.2.3 Data Retention

Processing social media data is to a relevant extent based on operating analytics software, which provides automatic analysis on gathered social media data stored in local databases. Their user interfaces take input to be crawled for in the stored data and return, for example, statistics of post occurrences in any context. Depending on the situation, only parts of that information may be relevant (see Sect. 14.3.1). Still, the entirety of every post has been and remains stored in local databases.

This means that if a data item is being deleted on the site of the corresponding social media service, it still resides at the place where it has been downloaded to. Technically, that practice meets the requirements to be termed *data retention*. We define this term as such: preserving data for an indefinite time period with no specific purpose for any individual data item but with the assumption to make use of the information in entirety at a later point in time.

The term is being discussed in the public mostly in conjunction with telecommunication analysis and surveillance. European Digital Rights public interest group states that "data retention practices interfere with the right to privacy at two levels: at the level of retention of data, and at the level of subsequent access to that data by law enforcement" (Rucz and Kloosterboer 2020).

We introduce the term in a broader and more technical environment to emphasize the explosive nature of recklessly dealing with personal data, which social media data is (European Commission 2018). According to the above definition, the term is valid for any case of storing and retending personal data in stocks. Wright et al. (2020) use it even to describe any storage of data underlying scientific studies.

Owning a set of data requires great responsibility in terms of data security. It opens up risks of possible abuse, theft, or accidental public exposure (Miller 2020). Breaking it down to a simple rule, it can be stated that "the more data you have, the more data you can lose" (Guillou and Portner 2020).

Beyond governmental agencies and law enforcement, also commercial players, journalists, researchers, or nonprofit organizations face challenges when storing individual-related data like those from social media. Stieglitz et al. (2018) discovered that the volume of data was most often cited as a challenge by researchers. Wang and Ye (2018) summarize common techniques for social media analytics in natural disaster management and coin the term *mining* for that matter.

Furthermore, the social impact of misusing large sets of data is well-known. The Cambridge Analytica scandal is one of the examples that show how massive data sets can be alienated (Berghel 2018). The company used personal information from millions of Facebook users without their consent to derive information about their political points of view and then microtarget personally tailored political advertisements to them. They claimed to have a major impact on the 2016 US presidential election, which can be regarded as a threat to democratic legitimacy (Dowling 2022).

Users of social media services start to realize that all of their data is not only publicly available but made use of by third parties. Data retention drives forgetfulness as a social concept at risk (Blanchette and Johnson 2002). The *chilling effect*, people slowly increasing self-discipline and restriction of their communication behavior due to becoming aware of digital surveillance, and panopticism (Manokha 2018; Büchi et al. 2022) are described consequences.

Nevertheless, the huge amount of data raised by social media services being a tremendous privacy thread is only one side of the coin. Large sets of social media data can also be beneficial for the public. The work of humanitarian organizations depends on publicly available data that is authentic and relevant. Especially, VOSTs rely on the availability of public social media data (Kuner and Marelli 2020); therefore, its prosperity must be preserved. A gradual retreat of users from social media services in favor of closed, "antisocial" messaging groups (Leetaru 2019; Wilson 2020) must be prevented.

14.2.4   HyperLogLog

One of our contributions to this issue presented in this chapter is based on storing data using an algorithm called *HyperLogLog* (HLL). This algorithm is a cardinality estimator first introduced by Flajolet et al. (2007).

Its fundamental strength is the ability to *estimate* the distinct count of a multiset (cardinality) and store it in a data structure, which does not allow the extraction of individual elements. This is done by storing only hashes of data items instead of the original raw data and identifying them by counting leading zeros of the binary representation of their hashes. The algorithm is able to predict how many distinct items have been added to the HLL set, based on the maximum number of leading zeros observed. This makes processing data using HLL very efficient in terms of processing time and storage space. It is not possible to search for prior unknown information in an HLL set, for example, the usernames of all the posts that have been gathered. This makes implementing HLL follow the *privacy by design* principle.

14.3   Concepts

14.3.1   Privacy-Aware Storage

The key aspect for our approach is to make it impossible to relate to the original social media data from a given processed data set (*privacy by design*). Therefore, we propose to utilize the cardinality estimation algorithm HyperLogLog (HLL) described in Sect. 14.2.4 to gathered store social media data.

To provide a minimal example of the process, we introduce a scenario, in which the difference in spatial occurrences of social media posts including a certain hashtag should be visualized. The result should be a choropleth map of areas according to the amount of post occurrences within that area (see Fig. 14.1). Areas are defined by a *GeoHash*, a hierarchical grid-like geocode identification concept (Niemeyer 2008; Morton 1966).

To store the occurrence of posts in an area, it is only necessary to *count* the number of distinct occurring posts, their *cardinality*. Reflecting, this unveils that storing the entirety of a social media post is unnecessary. It is sufficient to memorize its unique identifier (ID), which has been assigned by the social media service it originates from.

However, storing the ID in clear text in the database will allow identifying the post and thus the author of a post later on. The characteristics of HLL in turn enable to store data like the ID of a post in a set without the ability to regain it without prior knowledge about its existence in the set. Storing post IDs in an HLL set related to their geohash will only reveal their cardinality. Posts that occur later in the stream and match the same geohash will be added to this HLL set, which increases its cardinality by one for each new post. The geohash itself representing the post's

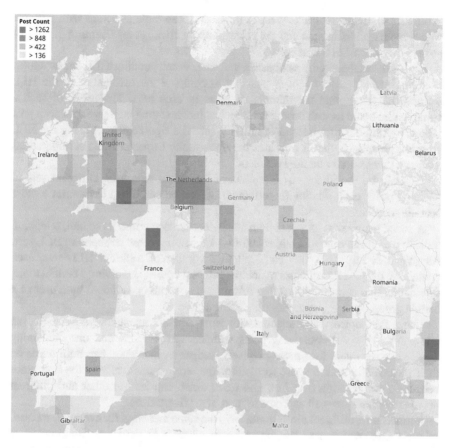

Fig. 14.1 Example of a map showing areas with different occurrences of posts containing #omicron hashtag on Twitter from January through March 2022. Map data: OpenStreetMap contributors Color distribution: Head/Tail Breaks (Jiang 2013)

Table 14.1 Exemplary database table structure showing four records (each stands for one area represented by the geohash) and the corresponding HLL set containing the post IDs

geohash	id
w41s	\x128b7fdf939b45ec2ef0ca
6yws	\x128b7fbfd17eca803517d2
c29s	\x128b7fe00ef312fcf023c9
75cs	\x128b7fcc47a6c00361c5e7

originating area is stored as the index of the database record (see Table 14.1). The resulting HLL data structure represents all posts matching a certain term from a certain area, while it is impossible to derive the post IDs back from it.

Using HLL, we do *not* store the post IDs itself but calculate hashes from them and store them in an array of counters that represent the set of post IDs (see

Sect. 14.2.4). Table 14.1 shows an example database table structure with geohash values representing an area and the corresponding HLL set representing the IDs of posts that occurred in that area.

Having a database with geohashes and their corresponding HLL set as shown exemplarily in Table 14.1, it is possible to compute the cardinality of the HLL set and thus determine the number of posts in each area. The result of such a computation could as well be achieved by just incrementing an integer per seen post ID and storing the sum instead of an HLL set. The significance of using the HLL algorithm instead is that it provides the opportunity to perform the set operations *union* and *intersection* on the HLL sets.

This can be useful for combinations of individual data sets. Different sets of gathered posts, each relating to certain terms, can be combined to monitor a more specific scenario.

A social media post as a data item can be broken down into its spatial, temporal, topical, and social components, each of which can be stored as separate HLL sets. As shown in Fig. 14.2, this can lead to a number of different HLL sets, each containing the post IDs of posts matching different criteria: involving a certain topic, originating in a certain area or in a certain time period, or authored by a user of a certain group.

Using the topical facet exemplarily in a disaster management scenario, an intersection of a set containing posts with the terms `fire` and one containing `forest` posts could lead more precisely to disaster incidents than both terms on their own. It still makes sense to monitor the terms individually in the first place because a combination of `fire` and `accident` can lead to other and different disaster incidents, as well as `forest` and `accident` does.

Furthermore, different terms could have the same meaning, for example, `flood`, `high tide`, `wave`, and `tsunami` could all refer to the same situation. So, a union of HLL sets on posts over these terms can provide more comprehensive information about disasters. Likewise, terms in different languages could also be

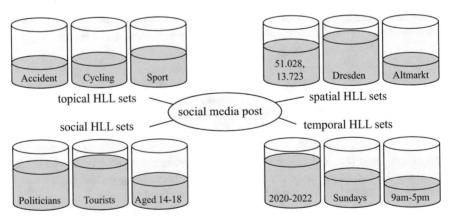

Fig. 14.2 Examples of HLL sets derived from the four facets of a social media post

monitored in combination. This, for example, enables VOSTs (see Sect. 14.2.1 and Chap. 13) to monitor larger, multiple languages involving areas like border triangles or including smaller countries like Benelux or the Baltics.

This concept provides privacy by design because it does not store the post IDs in a readable way. It only stores a statistical derivative resulting from the characteristics of the HLL algorithm (see Sect. 14.2.4) and complies to the *privacy by design* principles. The following subsections cover how it can be extended even further by applying extended concepts to adjust the level of privacy protection.

14.3.2 Abstraction Layers

The concept of *abstraction* has been widely used in the geo-community to visualize spatial information scale dependent on different degrees of detail (Burghardt et al. 2016). We re-dedicate these generalization methods from geovisualization to privacy protection.

Herein, we present a model to improve privacy for social media users, in particular in the context of data collection. It aims at withdrawing precision from the data by deriving multiple abstraction layers of it. Applying these layers, we are able to quantitatively describe different levels of privacy. By deploying methods of *generalization* and thus decreasing precision of the data, we can increase privacy, and vice versa.

Figure 14.3 shows a visual representation of this model, following the four-facet representation to characterize a social media post, introduced by Dunkel et al. (2019). The bottom layer in each facet is formed by the original data. Each following layer represents an increase in privacy protection for the user. This way, we have the ability to adjust the level of detail of the data in a fine-grained and context-dependent way. Each of the layers is described in detail in the following subsections.

14.3.2.1 Spatial Facet

In the spatial facet, the original data is usually represented by a *coordinate* in latitude and longitude or a tiny area surrounding that coordinate. A first abstraction from it

Fig. 14.3 Abstraction layers for each facet

can be an arbitrarily named *place* that includes the coordinate, e.g., a market square or a park. The next abstraction could be an administrative or functional *region* or other territories enclosing the place, e.g., a city or metropolitan area, county, or state. Cities can also be regarded as intermediate layers below regions. The next abstraction layer could be a *country* or another even broader defined, e.g., natural, political, administrative, or religious region.

When applying this model to the HLL-based storage concept presented in Sect. 14.3.1, it is crucial to note that even the lowest layer needs to be an area to be able to count multiple posts within it. If using a point coordinate instead, chances tend toward zero for multiple posts hitting that exact coordinate. An alternative approach with clustering techniques would be necessary there.

In an application implementing this model, the database index would be the geohash, place, city, or country, depending on the layer of abstraction. The corresponding HLL sets include the hashed post IDs of posts that originate in that respective area. An appropriate visualization of that data would be a map showing visually differentiable areas (see, e.g., Fig. 14.1).

14.3.2.2   Temporal Facet

Abstractions in the temporal facet are clearly defined by common time units. The basic layer is the timestamp of the publication of a post, abstracted as the day of the publication or a month or even a year.

In analogy to the spatial facet, when implementing this model, the database index must be a time range rather than a point in time, to be able to associate multiple matching posts with it. To visualize just the temporal facet, a timeline is the preferred graphical representation. Usually, this facet requires to also filter for a certain topic first, to prevent just visualizing *all* social media posts occurring within a certain time period.

14.3.2.3   Topical Facet

The topical facet is characterized by applying topic modeling techniques (Kherwa and Bansal 2019) to find abstractions of terms. The basic layer can be defined by the *terms*, the original content in the post, e.g., The River Elbe has burst its banks in Dresden today. A more general layer can be rendered in the overall *subject* of the post, e.g., Dresden flood. Another abstraction layer can be the *domain* of the Natural Disaster.

In an implementation, the database index will represent these terms, subjects, or domains, and the corresponding HLL sets hold the associated post IDs. It is not trivial to generate more generic terms for specific posts, but topic modeling techniques can help with that task. A word cloud can visualize these terms in different sizes (Hearst et al. 2019) depending on the cardinality of posts associated with them.

14.3.2.4 Social Facet

The social facet relates to users of social media. When running analyses on this object, it is crucial to note that we are switching the focus. While we are trying to avoid storing personal data around an object in the other facets, here we want to achieve the opposite: counting appearances of data that relates to a single person or a group. An exemplary analysis would be to count the number of posts per user or group. In this scenario, it is especially useful to apply abstraction layers in order to gain privacy for a single user. In analogy to the temporal facet, it is also useful to filter posts for a certain topic in beforehand.

In the basic layer, the creator of a social media post, the *user*, is targeted. The database index would be the username or id, and the corresponding HLL set consists of the post IDs. Combining, e.g., all of the user account's *followers* to a group and regarding posts of all of them could apply as the first abstracted layer. Through network analysis (Maireder et al. 2014), we can define *clusters* to be objects in a second step of abstraction. Another layer of abstraction could be the consideration of different social network *platforms* (Cosenza 2022), which have a distinct user base, which might originate from different cultural backgrounds, e.g., Twitter, Instagram, WeChat, and VKontakte.

Implementing groups of users is more challenging than in other layers. The group of followers in the second layers consists of a list of user IDs eventually, which could again be stored in an HLL set and get an ID assigned to. IDs of multiple groups are then stored in HLL sets and can be combined or contrasted with other group IDs, defining clusters accordingly.

All the described layers are only examples and can be replaced by other structures. Also, the number of abstraction layers can be chosen arbitrarily, as the granularity of the data can change.

It should be noted that abstraction layers do not only gain privacy for the social media users, but they also diminish the precision of the data. This makes applying abstraction to social media data be a compromise between privacy and precision.

14.3.3 Filter Lists

Storing social media data using HLL to be processed in analytics software forms the basement of privacy protection. Applying generalization methods as described in Sect. 14.3.2 provides further opportunities to adjust data precision. However, there are edge cases that require special handling. For instance, even the existence of a single specific term, a specific time, location, etc. may provide hints that can be repurposed or combined with other (e.g., external) information to compromise user privacy in certain situations. Following the principle that different data must be treated differently (Almås et al. 2018), we seek to contribute to a systematic approach to fine-tuning privacy preservation and analytical flexibility.

There are two main approaches to adjusting privacy—utility trade-offs with HLL and abstraction layers. First, *stop and allow lists* can be used during the generation of the HLL set to enable context-dependent data protection through filtering. Second, *threshold values* can be defined flexible to influence the granularity of the HLL set indexes and, based on that, the degree of anonymity. Table 14.2 lists examples for each context in the framework, where accuracy (utility) may be traded in favor of a higher degree of privacy, similar to the broader data sensitivity spectrum proposed by Rumbold and Pierscionek (2018).

Whether stop lists or allow lists are preferable depends on the context of application. Allow lists are more restrictive and require less effort from the analysts, by automatically excluding all terms, times, locations, etc. that are not explicitly considered beforehand. For the spatial context, for instance, unless worldwide data is required, allow lists are frequently used, to limit data collection to a specific area, region, place, etc. Conversely, stop lists can be added selectively on top, to exclude places that are known to be related to vulnerable groups or sensitive contexts (e.g., hospitals, party locations). Similarly, filter lists for specific terms, hashtags, or emoji can be defined for the topical context.

For topical contexts, the openness of possible references complicates defining holistic stop lists ahead of time. As an example, Fig. 14.4 shows a map generated from terms, hashtags, and emoji used on the social media services Twitter, Flickr, and Instagram at a public vantage point and park. The syringe emoji could indicate drug use, which may lead to further onsite investigation by, e.g., authorities, with potential unexpected consequences of the user perspective. Obviously, this is an edge case for social-individual privacy because both positive (society) and negative (user) consequences are imaginable. One solution would be to assign the specific emoji to a thematic broader *emoji class*, e.g., the umbrella group of "medical emoji"[1] (see Sect. 14.3.2). As another solution, the syringe emoji could be classified ahead of time, for increased sensitivity, leading to, e.g., a greater spatial granularity reduction on data ingestion, or exclusion, preventing having to deal with this ambiguous ethical edge case in advance.

Lastly, as the second approach to enable systematic user privacy with HLL, threshold values may be defined, similar to what is known from other disciplines, such as the HIPAA Privacy Rules for health data publications (Malin et al. 2011) or census statistics (Szibalski 2007, p.142). Allshouse et al. (2010), for instance, use geomasking in combination with k-anonymity, to define a lower threshold of $k = 5$ (people), which is a rule of thumb size in geoprivacy (Kamp et al. 2013). Comparable best-practice threshold values could be defined for HLL sets of different sizes, e.g., suggestions by Desfontaines et al. (2019), with smaller sets indicating lesser privacy protection due to a scarce context collapse. In the spatial context, this could be implemented by using quadtrees, for example, to split and aggregated social data into sub-sections (quads), based on pre-defined thresholds, where the resolution is automatically decreased for areas of lesser data density.

---

[1] Unicode Consortium, unicode.org/emoji/charts-13.0/full-emoji-list.html#medical.

Table 14.2 Example of sensitive context factors for which no data analysis might be carried out

Type of context	Example of sensitive context factor	Reference
Spatial context	– Home location	(Georgiadou et al. 2019), (Kim et al. 2021)
	– Hospitals	(Ağır et al. 2016), (Kim and Kwan 2021)
	– Related to specific events (concert grounds, party locations)	(Such et al. 2017)
Temporal context	– Nighttimes	(Nikas et al. 2018)
	– Past and archived content, time collapse	(Brandtzaeg and Lüders 2018)
	– During specific events (e.g., new year, Thanksgiving, 4th of July)	(Such et al. 2017)
Topical context	– Activists, protesters, dissidents	(Uldam 2018)
	– Health issues (e.g., related to diabetes or corona)	(Matković et al. 2021)
Social context	– Children	(Steinberg 2016), (Marwick and Boyd 2014)
	– LGBTQ+[a]	(Birnholtz et al. 2020)
	– Personal, social relationships	(Houghton and Joinson 2010)
	– Minorities (race and religion)	(Mashhadi et al. 2021)

[a] Lesbian, gay, bisexual, transgender, queer, and others

Fig. 14.4 A thematically sensitive emoji on drug use at selected locations

14.4 Case Study

Even though different implementations of HLL exist, all share a number of basic steps. At the core, the binary representation of any given character string is divided into *buckets*, for which the number of leading zeroes is counted (see Sect. 14.2.4). Because any given character string is first randomized, it is possible to predict how many distinct items must have been added to a given HLL set, based on the maximum number of leading zeroes observed. In other words, if multiple items are added to an HLL set, only the highest number of leading zeroes per bucket needs to be memorized. As a result, the cardinality estimation will only approximate counts.

As a side effect, there is a limited ability to check whether a specific user or ID has been added to a HLL set. In an adversarial situation, Desfontaines et al. (2019) refer to such a check as an *intersection attack*. Intersection attacks first require obtaining the hash of a targeted person or ID and then adding this hash to an HLL set. If the HLL set changes, an adversarial may be able to increase their initial suspicion by a certain degree. To better illustrate intersection attacks and how and under which circumstances the privacy of a user could become compromised in the presented two-component research setup, we briefly introduce two examples.

Alex is included in the YFCC100M dataset (Thomee et al. 2016) because he published 289 photos under Creative Commons Licenses between 2013 and 2014 on Flickr; 120 of these photos are geotagged. Given this information, it will be relatively easy to re-identify Alex. Sandy is an internal adversary. She could be someone working at an analytics service with full access to the database. Robert, on the other hand, is someone representing an external adversary, with access only to the published dataset. In the first example, the privacy of Alex is compromised if Sandy could increase or confirm her suspicion that Alex was not at his workplace in Berlin on 9 May 2012. In the second example, the privacy of Alex is compromised if Robert could increase or confirm his suspicion that Alex was indeed at least once at a specific location, e.g., contrary to what Alex claims. Finally, Alex could be someone who voluntarily contributed his pictures to the conceived analytics service or altruistically published Creative Commons photos on Flickr.

Consider that, at the moment of contribution, Alex may not have thought of the consequences for his privacy but later realized his mistake. With the use of raw data, even removing any compromising data from Flickr, this change would need to be reflected in any subsequent data collection, such as in the analytics service or the YFCC100M dataset. This is either impractical or impossible. The question is, therefore, whether it is possible to replace raw data workflows with a privacy-aware visualization pipeline, without significantly reducing utility.

Several factors must coincide for intersection attacks to be successful. Firstly, an adversarial must have access to HLL sets. In our system model, this can either be an internal adversary (Sandy), having direct access to the database, or an external adversary (Robert), having access only to published data. Furthermore, an adversary must be able to either compute hashes for a given target user or somehow gain access to a computed HLL set for the given user. The former is only possible if the secret key is compromised. The latter appears conceivable, in our example, if the adversary has some prior knowledge about other locations visited by a target user, and if the HLL sets of these locations ideally contain only the target user or a few other users. In the following, we explore this worst-case scenario, where both Sandy and Robert somehow got hold of an HLL set that only contains Alex's computed hashes.

For Sandy, this means in order to test whether Alex was not in Berlin on 9 May 2012, she either needs Alex's original user ID and the secret key to construct the hash or find another location that has only been visited by Alex on this date. In this unlikely scenario, the result of an intersection attack for all grid cells is shown in Fig. 14.5. Visible in the figure is that a large number of other grid cells show false positives for the intersection test, that is, these HLL sets did not change, even when updated with the particular user day-hash for Alex.

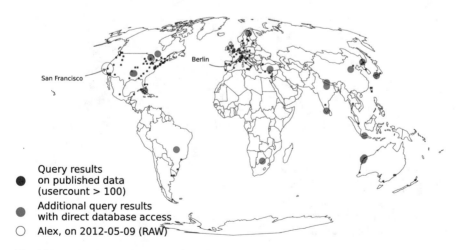

Fig. 14.5 Evaluation of scenario "Sandy" (Dunkel et al. 2020, CC-BY 4.0)

Since HLL prevents the occurrence of false negatives, and San Francisco is indeed among these locations, the result does include Alex's actual location on 9 May 2012. Depending on the size of the targeted HLL set, Sandy may then increase her suspicion by some degree. In the case of the grid cell for San Francisco, with 209,581 user days, this increase in posterior knowledge may be found to be negligibly small. In other words, even if there was no post from Alex on 9 May 2012, the intersection attack may have produced the same result. In conclusion, even in the worst scenario, having direct access to the database and a compromised secret key, Sandy could not gain any further affirmation.

Similarly, and rather incidentally, the positive grid cell for Berlin does indeed falsely suggest that Alex was in Berlin. This is not surprising given that larger HLL sets have a higher likeliness of showing false positives and Berlin is a highly frequented location. In other words, Alex benefits from the privacy-preserving effect of HLL.

In the second scenario, consider a situation in which Robert may have an a priori suspicion that Alex went to Cabo Verde. Alex, on the other hand, does not want Robert to know that he went surfing without him. Robert knows that Alex is participating in the conceived analytics service and, somehow, gains access to an HLL set containing only one hashed user ID from Alex. The results of the intersection attack for all grid cells are shown in Fig. 14.6. Since only 56 users have been to Cabo Verde in the YFCC100M dataset, the particular bin is not included in the published benchmark data, which is limited by a minimum threshold of 100 users. However, with direct access to the database, Robert could observe that Cabo Verde is among the locations revealed. In this case, Robert may gain some affirmation for his suspicion that Alex was in Cabo Verde. At the same time, a definite answer will not be possible, given the irreversible approximation of the HLL structure. For example, for the same intersection attack, for set sizes below 56

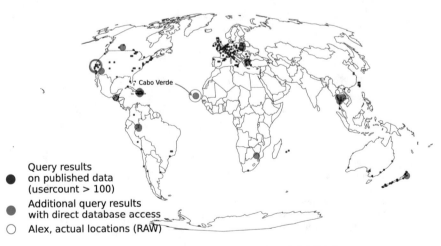

Fig. 14.6 Evaluation of scenario "Robert" (Dunkel et al. 2020, CC-BY 4.0)

users, there are 14 other grid cells that show false positives, down to 8 users. In other words, even though these HLL sets do not change when tested, Alex has never been to these locations.

While these two scenarios provide a base to understand how intersection attacks may be executed in a spatial setting, a valid question is how likely successful intersection attacks are overall. To some degree, this depends on questions of security, such as protecting the secret key or managing database access.

Another part is directly related to the distribution of collected data and the number of outliers that are present at each stage of data processing. If data is more clustered, users will generally receive more benefits from the privacy-preserving effects of HLL. This can be quantitatively substantiated with the given dataset (Dunkel et al. 2020).

14.5 Conclusion

The research presented in this chapter introduced a number of approaches to deal with privacy aspects in the process of social media data processing. Social media data is being used as a source of data for wide-ranging projects within and beyond the scope of this book (see Sect. 14.2.1). The relevance of privacy aspects in processing this kind of data and the range of related work are pointed out in Sect. 14.2.2. Furthermore, in Sect. 14.2.3, we discussed our focus on data retention as a potential threat for analysts. We defined the term and explained that we make use of that specific term to emphasize the explosiveness of dealing with personal data.

We showed that it is possible to preserve the privacy of social media users with the major concepts. As a basis for our first concept, we first introduced the cardinality estimation algorithm HyperLogLog in Sect. 14.2.4. In Sect. 14.3.1, the main part of this chapter, we introduced a concept to store social media data in a way that it is not possible to extract individual items from it but only to estimate the cardinality of social media data items within a certain set, plus running set operations over multiple sets to extend analytical ranges. Applying this method requires defining the scope of the result before even gathering the data and thus prevents the data from being misused for other purposes at a later point in time. This follows the *privacy-by-design* principle.

As an extension to the first concept, we proceeded by introducing a concept that is well known in the geographic community, generalization, in Sect. 14.3.2. By defining a number of abstraction layers, it is possible to even more reduce the data to be stored, depending on the required precision. The less precise data is needed, the fewer data needs to be stored. Finally, in Sect. 14.3.3, we explain the conceptual exclusion of edge cases by applying filter lists to the data set.

A closing case study in Sect. 14.4 explains the concept of intersection attacks and shows that under rare circumstances the HyperLogLog technology is vulnerable against them. The case study unveils that the larger the dataset, the less likely are intersection attacks. Since social media data is usually very large, implementing the HyperLogLog technology is an excellent approach to protect the data from being abused, thieving, or publicly exposed and thus preserves the privacy of social media users.

Acknowledgments This research was supported by the German Research Foundation DFG within Priority Research Program 1894 *Volunteered Geographic Information: Interpretation, Visualization and Social Computing* (VGIscience, EVA-VGI, BU 2605/8-2).

References

Ağır B, Huguenin K, Hengartner U, Hubaux J-P (2016) On the privacy implications of location semantics. In: Proceedings on Privacy Enhancing Technologies. https://doi.org/10.1515/popets-2016-0034

Allshouse WB, Fitch MK, Hampton KH, Gesink DC, Doherty IA, Leone PA, Serre ML, Miller WC (2010) Geomasking sensitive health data and privacy protection: an evaluation using an e911 database. Geocarto Int, 443–452. https://doi.org/10.1080/10106049.2010.496496

Almås I, Attanasio O, Jalan J, Oteiza F, Vigneri M (2018) Using data differently and using different data. J Dev Eff, 462–481. https://doi.org/10.1080/19439342.2018.1530279

Büchi M, Festic N, Latzer M (2022) The chilling effects of digital dataveillance: A theoretical model and an empirical research agenda. Big Data Soc. https://doi.org/10.1177/20539517211065368

Beresford AR, Stajano F (2003) Location privacy in pervasive computing. IEEE Pervasive Comput, 46–55. https://doi.org/10.1109/mprv.2003.1186725

Berghel H (2018) Malice domestic: The cambridge analytica dystopia. Computer, 84–89. https://doi.org/10.1109/mc.2018.2381135

Birnholtz J, Kraus A, Zheng W, Moskowitz DA, Macapagal K, Gergle D (2020) Sensitive sharing on social media: Exploring willingness to disclose prep usage among adolescent males who have sex with males. Soc Media Soc. https://doi.org/10.1177/2056305120955176

Blanchette J-F, Johnson DG (2002) Data retention and the panoptic society: The social benefits of forgetfulness. Inf Soc, 33–45. https://doi.org/10.1080/01972240252818216

Boyd D, Crawford K (2012) Critical questions for big data: Provocations for a cultural, technological, and scholarly phenomenon. Inf Commun Soc, 662–679. https://doi.org/10.1080/1369118x.2012.678878

Brandtzaeg PB, Lüders M (2018) Time collapse in social media: extending the context collapse. Soc Media Soc. https://doi.org/10.1177/2056305118763349

Burghardt D, Duchêne C, Mackaness W (2016) Abstracting geographic information in a data rich world. Springer, New York. https://doi.org/10.1007/978-3-319-00203-3

Cavoukian A et al. (2009) Privacy by design: The 7 foundational principles. Information and Privacy Commissioner of Ontario, Canada

Ciriani V, Di Vimercati SDC, Foresti S, Samarati P (2007) κ-anonymity. In: Secure data management in decentralized systems. Springer, New York, pp 323–353. https://doi.org/10.1007/978-0-387-27696-0_10

Cosenza V (2022) World map of social networks. https://vincos.it/world-map-of-social-networks. Accessed 19-Jul-2022

Daly, A, Devitt, SK, Mann, M (2019) Good Data, Theory on Demand, 29. Institute of Network Cultures, Amsterdam. https://networkcultures.org/blog/publication/tod-29-good-data/

Desfontaines D, Pejó B (2020) Sok: differential privacies. In: Proceedings on Privacy Enhancing Technologies, pp 288–313. https://doi.org/10.2478/popets-2020-0028

Desfontaines D, Lochbihler A, Basin D (2019) Cardinality estimators do not preserve privacy. In: Proceedings on Privacy Enhancing Technologies, pp 26–46. https://doi.org/10.2478/popets-2019-0018

Dowling M-E (2022) Cyber information operations: Cambridge analytica's challenge to democratic legitimacy. J Cyber Policy, 1–19. https://doi.org/10.1080/23738871.2022.2081089

Duckham M, Kulik L (2005) A formal model of obfuscation and negotiation for location privacy. In: International Conference on Pervasive Computing. Springer, pp 152–170. https://doi.org/10.1007/11428572_10

Dunkel A (2021) Tag maps in der Landschaftsplanung. Springer Fachmedien Wiesbaden, Wiesbaden, pp 137–166. https://doi.org/10.1007/978-3-658-29862-3_8

Dunkel A, Andrienko G, Andrienko N, Burghardt D, Hauthal E, Purves R (2019) A conceptual framework for studying collective reactions to events in location-based social media. Int J Geogr Inf Sci, 780–804. https://doi.org/10.1080/13658816.2018.1546390

Dunkel A, Löchner M, Burghardt D (2020) Privacy-aware visualization of volunteered geo-graphic information (vgi) to analyze spatial activity: A benchmark implementation. ISPRS Int J Geo-Inf. https://doi.org/10.3390/ijgi9100607

Dwork C (2008) Differential privacy: A survey of results. In: International Conference on Theory and Applications of Models of Computation. Springer, pp 1–19. https://doi.org/10.1007/978-3-540-79228-4_1

European Commission (2018) What is personal data? https://ec.europa.eu/info/law/law-topic/data-protection/reform/what-personal-data. Accessed 21-Nov-2022

Fiedrich F, Fathi R (2021) Humanitäre hilfe und konzepte der digitalen hilfeleistung. In: Sicherheitskritische Mensch-Computer-Interaktion. Springer, pp 539–558. https://doi.org/10.1007/978-3-658-32795-8_25

Flajolet P, Fusy E, Gandouet O, Meunier F (2007) Hyperloglog: the analysis of a near-optimal cardinality estimation algorithm. Discrete Math Theor Comput Sci. https://doi.org/10.46298/dmtcs.3545. https://dmtcs.episciences.org/3545

Georgiadou Y, de By RA, Kounadi O (2019) Location privacy in the wake of the gdpr. ISPRS Int J Geo-Inf. ISSN 2220-9964. https://doi.org/10.3390/ijgi8030157

Guillou C, Portner C (2020) Data retention - more than meets the eye. https://www.theprivacyhacker.com/2020/12/data-retention/

Hauser C, Kabatnik M (2001) Towards privacy support in a global location service. In: Proceedings of the IFIP Workshop on IP and ATM Traffic Management, pp 81–89

Hearst MA, Pedersen E, Patil L, Lee E, Laskowski P, Franconeri S (2019) An evaluation of semantically grouped word cloud designs. IEEE Trans Vis Comput Graph, 2748–2761. https://doi.org/10.31219/osf.io/3eutf

Hildebrandt M (2006) Privacy and identity. Privacy and the criminal law. Intersentia, Antwerp/Oxford

Houghton DJ, Joinson AN (2010) Privacy, social network sites, and social relations. J Technol Hum Serv, 74–94. https://doi.org/10.1080/15228831003770775

Jiang B (2013) Head/tail breaks: A new classification scheme for data with a heavy-tailed distribution. Prof Geogr, 482–494. https://doi.org/10.1080/00330124.2012.700499

Kamp M, Kopp C, Mock M, Boley M, May M (2013) Privacy-preserving mobility monitoring using sketches of stationary sensor readings. In: Joint European Conference on Machine Learning and Knowledge Discovery in Databases. Springer, pp 370–386. https://doi.org/10.1007/978-3-642-40994-3_24

Keßler C, McKenzie G (2018) A geoprivacy manifesto. Trans GIS. https://doi.org/10.1111/tgis.12305

Kherwa P, Bansal P (2019) Topic modeling: a comprehensive review. EAI Endors Trans Scal Inf Syst. https://doi.org/10.4108/eai.13-7-2018.159623

Kim J, Kwan M-P (2021) An examination of people's privacy concerns, perceptions of social benefits, and acceptance of covid-19 mitigation measures that harness location information: A comparative study of the us and south korea. ISPRS Int J Geo-Inf, 25. https://doi.org/10.3390/ijgi10010025

Kim J, Kwan M-P, Levenstein MC, Richardson DB (2021) How do people perceive the disclosure risk of maps? Examining the perceived disclosure risk of maps and its implications for geoprivacy protection. Cartogr Geogr Inf Sci, 2–20. https://doi.org/10.1080/15230406.2020.1794976

Kounadi O, Resch B, Petutschnig A (2018) Privacy threats and protection recommendations for the use of geosocial network data in research. Soc Sci, 191. https://doi.org/10.3390/socsci7100191

Kuner C, Marelli M (2020) Data analytics and big data. International Committee of the Red Cross, Geneva, Switzerland, pp 92–111

Löchner M, Fathi R, Schmid D, Dunkel A, Burghardt D, Fiedrich F, Koch S (2020) Case study on privacy-aware social media data processing in disaster management. ISPRS Int J Geo-Inf, 709. ISSN 2220-9964. https://doi.org/10.3390/ijgi9120709

Leetaru K (2019) The era of precision mapping of social media is coming to an end. https://www.forbes.com/sites/kalevleetaru/2019/03/06/the-era-of-precision-mapping-of-social-media-is-coming-to-an-end/

Maireder A, Schlögl S, Schütz F, Karwautz M, Waldheim C (2014) The european political twittersphere: Network of top users discussing the 2014 european elections. University of Vienna, Viena

Malin B, Benitez K, Masys D (2011) Never too old for anonymity: a statistical standard for demographic data sharing via the hipaa privacy rule. J Am Med Inf Assoc, 3–10. https://doi.org/10.1136/jamia.2010.004622

Manokha I (2018) Surveillance, panopticism, and self-discipline in the digital age. Surveillance Soc, 219–237. https://doi.org/10.24908/ss.v16i2.8346

Marwick AE, Boyd D (2014) Networked privacy: How teenagers negotiate context in social media. New Media Soc, 1051–1067. https://doi.org/10.1177/1461444814543995

Mashhadi A, Winder SG, Lia EH, Wood SA (2021) No walk in the park: The viability and fairness of social media analysis for parks and recreational policy making. In: ICWSM, pp 409–420. https://doi.org/10.1609/icwsm.v15i1.18071

Matković R, Vejmelka L, Ključević Ž (2021) Impact of covid 19 on the use of social networks security settings of elementary and high school students in the split-dalmatia county. In: 2021 44th International Convention on Information, Communication and Electronic Technology (MIPRO). IEEE, pp 1476–1482. https://doi.org/10.23919/mipro52101.2021.9597179

Miller V (2020) Understanding digital culture. SAGE Publications Limited, London, UK

Mooney P, Olteanu-Raimond A-M, Touya G, Juul N, Alvanides S, Kerle N (2017) Considerations of privacy, ethics and legal issues in volunteered geographic information. Map Citizen Sensor, 119–135. https://doi.org/10.5334/bbf.f

Morton GM (1966) A computer oriented geodetic data base and a new technique in file sequencing. International Business Machines Company, New York

Niemeyer G (2008) geohash.org is public! https://blog.labix.org/2008/02/26/geohashorg-is-public. Accessed 06-Sep-2022

Nikas A, Alepis E, Patsakis C (2018) I know what you streamed last night: On the security and privacy of streaming. Digit Investig, 78–89. https://doi.org/10.1016/j.diin.2018.03.004

Reyman J (2013) User data on the social web: Authorship, agency, and appropriation. Coll Engl, 513–533

Rucz M, Kloosterboer S (2020) Data retention revisited. https://edri.org/our-work/launch-of-data-retention-revisited-booklet/

Rumbold JM, Pierscionek BK (2018) What are data? A categorization of the data sensitivity spectrum. Big Data Res, 49–59. https://doi.org/10.1016/j.bdr.2017.11.001

Solove DJ (2008) Understanding privacy. Harvard University Press, Cambridge, MA

Steinberg SB (2016) Sharenting: Children's privacy in the age of social media. Emory LJ, 839

Stieglitz S, Mirbabaie M, Ross B, Neuberger C (2018) Social media analytics–challenges in topic discovery, data collection, and data preparation. Int J Inf Manag, 156–168. https://doi.org/10.1016/j.ijinfomgt.2017.12.002

Such JM, Porter J, Preibusch S, Joinson A (2017) Photo privacy conflicts in social media: A large-scale empirical study. In: Proceedings of the 2017 CHI Conference on Human Factors in Computing Systems, CHI '17. Association for Computing Machinery, New York, NY, USA, pp 3821–3832. ISBN 9781450346559. https://doi.org/10.1145/3025453.3025668

Szibalski M (2007) Textteil - Kleinräumige Bevölkerungs- und Wirtschaftsdaten in der amtlichen Statistik Europas. Wirtschaft und Statistik, 137–143

Thomee B, Shamma DA, Friedland G, Elizalde B, Ni K, Poland D, Borth D, Li L-J (2016) Yfcc100m: The new data in multimedia research. Commun ACM, 64–73. https://doi.org/10.1145/2812802

Uldam J (2018) Social media visibility: challenges to activism. Media Cult Soc, 41–58. https://doi.org/10.1177/0163443717704997

Wang Z, Ye X (2018) Social media analytics for natural disaster management. Int J Geogr Inf Sci, 49–72. https://doi.org/10.1080/13658816.2017.1367003

Williams ML, Burnap P, Sloan L (2017) Towards an ethical framework for publishing twitter data in social research: Taking into account users' views, online context and algorithmic estimation. Sociology, 1149–1168. https://doi.org/10.1177/0038038517708140

Wilson S (2020) The era of antisocial social media. https://hbr.org/2020/02/the-era-of-antisocial-social-media

Wright, DN, Demetres, MR, Mages, KC, DeRosa, AP, Jedlicka C, Stribling JC, Baltich Nelson B, Delgado, D (2020) How long should we keep data? An evidence-based recommendation for data retention using institutional meta-analyses. Samuel J. Wood Medical Library: Faculty Publications

Zhou B, Pei J, Luk W (2008) A brief survey on anonymization techniques for privacy preserving publishing of social network data. In: ACM Sigkdd Explorations Newsletter, pp 12–22. https://doi.org/10.1145/1540276.1540279

Printed in the United States
by Baker & Taylor Publisher Services